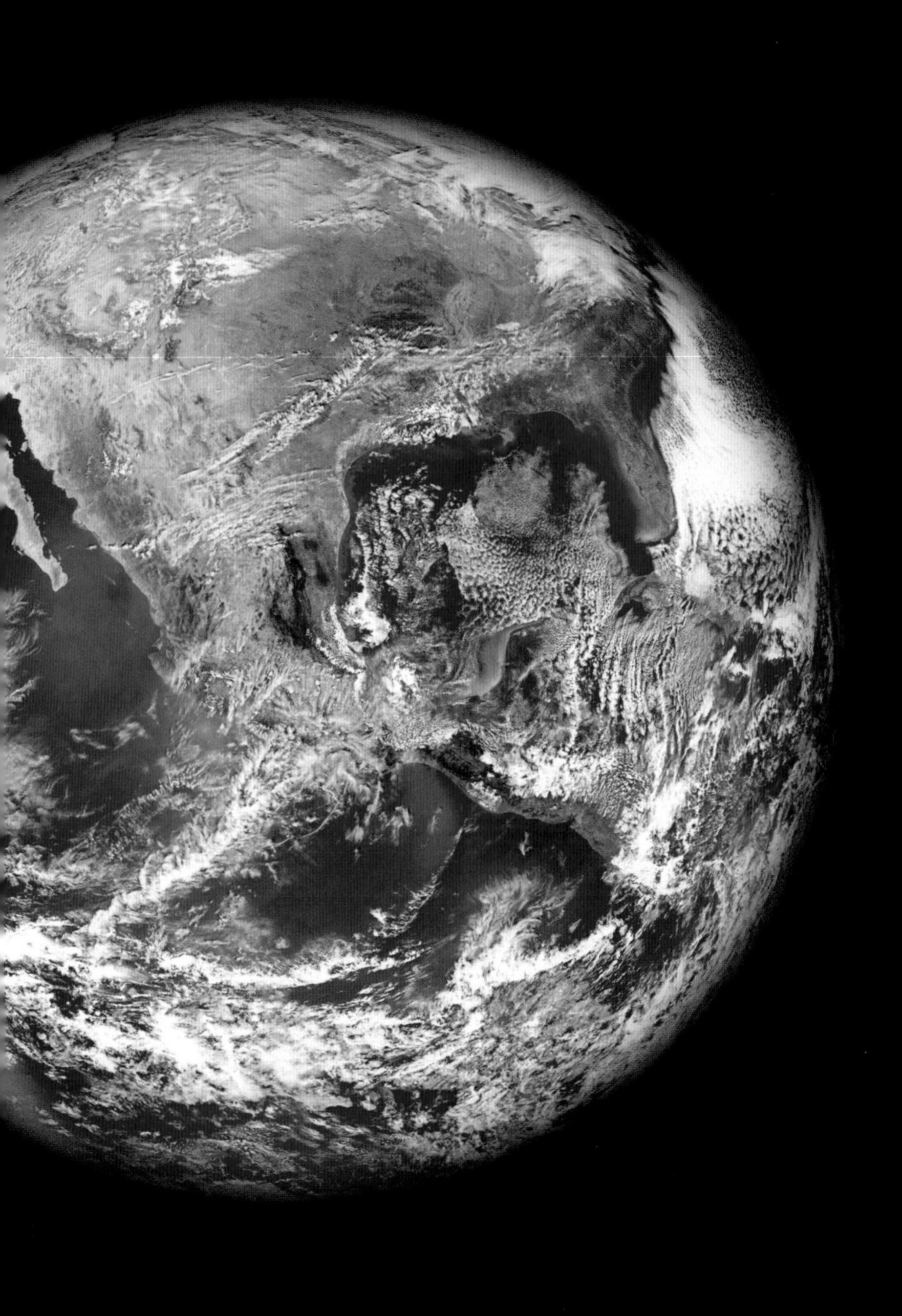

SPACE ATLAS

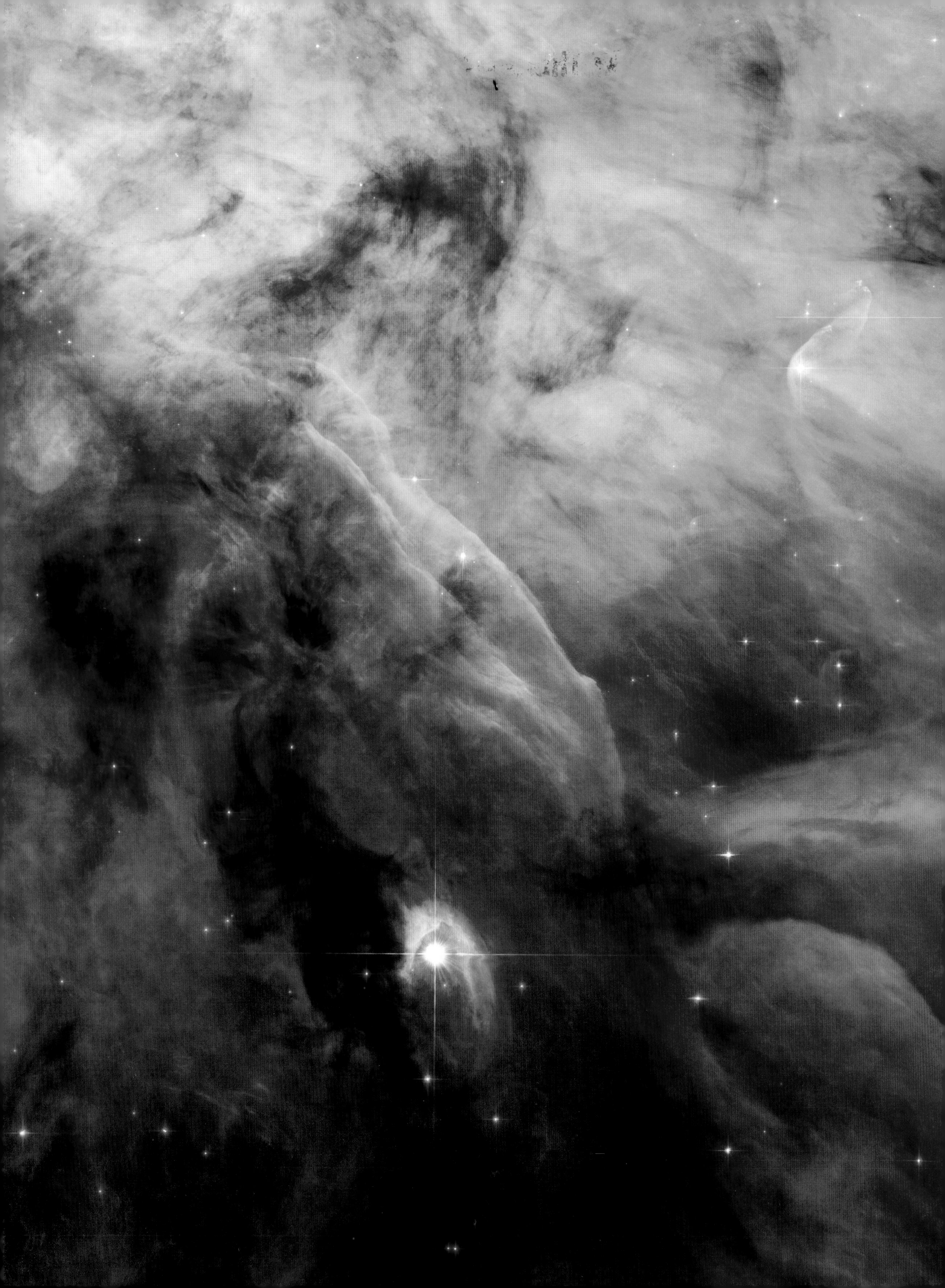

SECOND EDITION

SPACE ATLAS

MAPPING THE UNIVERSE AND BEYOND

JAMES TREFIL

WITH A FOREWORD
BY BUZZ ALDRIN

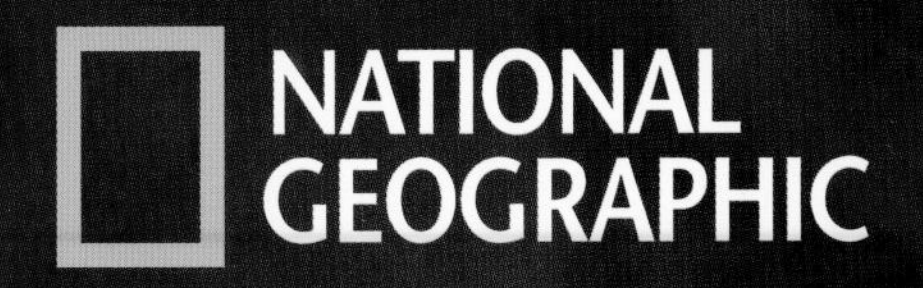

WASHINGTON, D.C.

CONTENTS

PREVIOUS PAGES: (Page 1) Western hemisphere of planet Earth. (Pages 2–3) Orion Nebula, with star LP Orionis at lower left.

An image compiled from both visible-light and x-ray sources reveals the bubble-shaped remnants of supernova SNR 0509-67.5, located in the Large Magellanic Cloud galaxy. Heated material is shown in green and blue; the pink shell represents the shock wave of the supernova moving through the interstellar medium.

THE SOLAR SYSTEM

Queen of the solar system, Saturn is famed for its spectacular ring system. This image, compiled in natural light from the Cassini orbiter, shows the complex structure of the rings and gaps and the sharp shadows they throw across the gas giant's atmosphere.

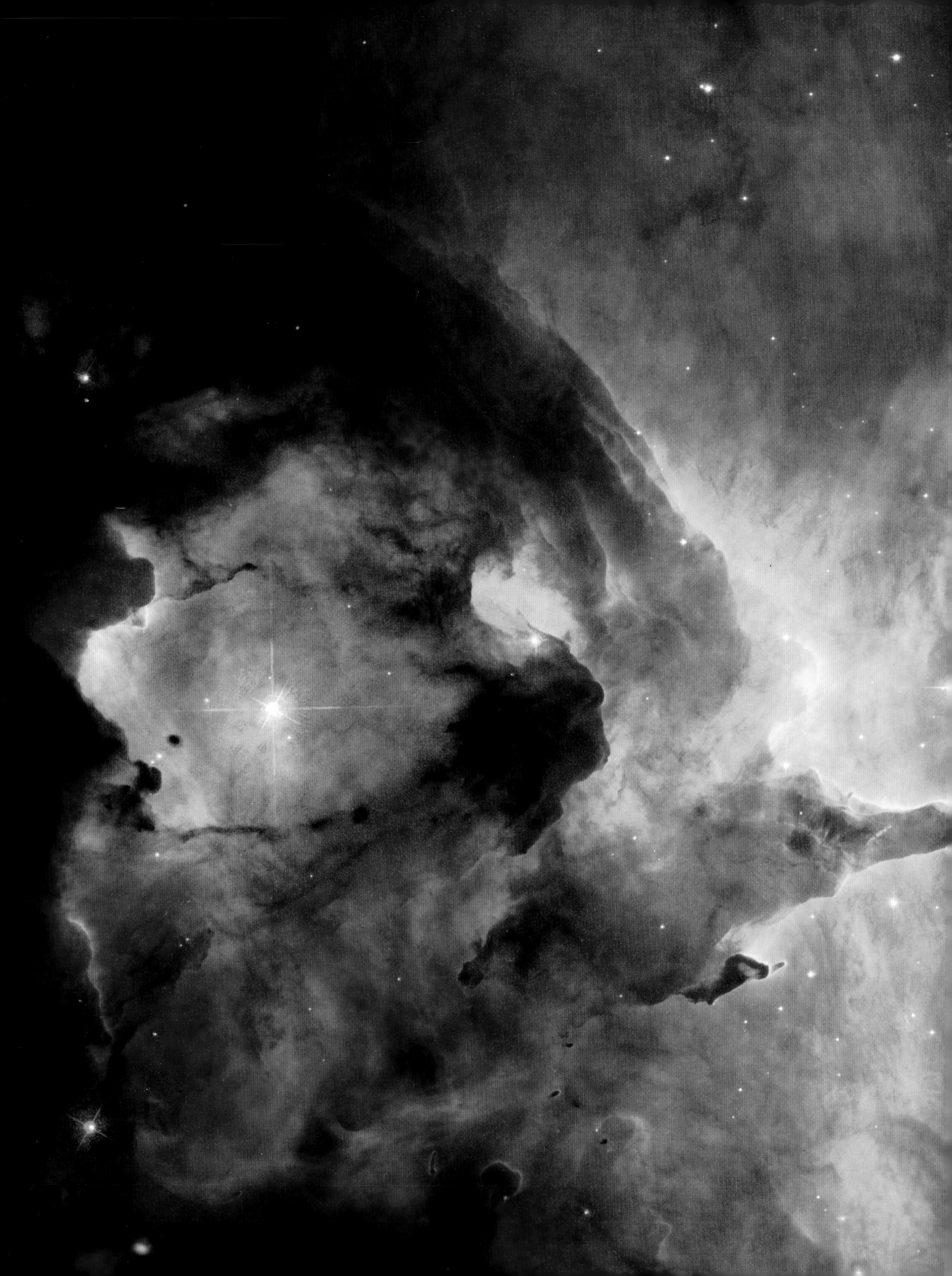

THE GALAXY

Brilliant stars and gaseous, star-forming regions make up the nebula NGC 6357. The brightest star in the image above is actually a double star, Pismis 24-1, with an enormous combined mass about 200 times that of our sun.

THE UNIVERSE

Violence illuminates nearby galaxy Centaurus A. With a massive black hole at its heart, the galaxy is colliding with, and destroying, a spiral galaxy. The collision creates areas of star formation at the dusty edges of the galaxy, seen here.

LANDING ON THE MOON
50 YEARS LATER

FOREWORD

BY BUZZ ALDRIN

You are holding in your hands the universe—a universe of knowledge that's captured in this unprecedented volume. There is power displayed in this book's pages of stunning images, maps, and mind-opening text. Presented here is a grand tour of time and space and the information gleaned by horizon hunting—a restlessness and hunger for new awareness and appreciation about the cosmos.

This book is testimony to what we know as well as acknowledgment of our ignorance. Consider the many tools of space exploration at our disposal today, from Earth-based and space-situated observatories to robotic probes flung outward across our solar system. These capabilities add new meaning to what Native American people have called sky-watching astronomers: "The men with the long eyes."

I consider myself a Global Space Statesman. In my travels, I've had the good fortune to convene with kings, queens, presidents, and prime ministers, march in countless parades, and be honored with medals. From these engagements, it has become evident to me that there's a calling for a strong, vibrant, and cutting-edge space agenda for the 21st century.

And as I travel around the world, I am often asked, What's the value of space exploration, and do we even need a space program? As my good friend, astrophysicist Stephen Hawking notes, we have made extraordinary progress in the last hundred years, "but if we want to continue, our future is in space." I agree.

Astronaut Buzz Aldrin walks on the surface of the moon near the leg of the lunar module *Eagle* lander during the milestone-making Apollo 11 mission in July 1969. Astronaut Neil Armstrong took this photograph during their joint exploration of the Sea of Tranquility region of the moon. The *Eagle* spacecraft and Armstrong can be seen clearly reflected in the visor of Aldrin's helmet.

Admittedly, I won't be around to see much of it. Still, I want to do everything in my power to lay the groundwork for our future space exploration.

There is a powerful collective curiosity about our neighborhood of planets, moons, asteroids—and beyond to exoplanets circling their parent stars. Which distant worlds may have habitable conditions, perhaps supporting extraterrestrial civilizations? That's just one query in a universe of questions that await answers.

There is no doubt that the scientific advancements that come from space research lead to products and technology that we use in our daily lives. One need look no further than cell phones, Global Positioning System (GPS) satellites, and myriad medical advances. These technologies wouldn't have been possible without investments in the space program. Likewise, a forward-looking space program fuels the minds of the next generations of space engineers, scientists, and others to take on the future. That means increasing support for *S*cience, *T*echnology, *E*ngineering, the *A*rts and *M*athematics—simply put, applying *STEAM* power!

I am proud of the fact that the United States has led the world in human space exploration for the past 50 years. That said, I am also deeply concerned about the recent erosion of America's space prowess. NASA's budget has decreased from approximately 4 percent of the U.S. federal budget in the 1960s to 0.4 percent today of government discretionary funds. I cannot stress enough that the nation's leadership role in human spaceflight within the coming decades must be restored and boosted to build upon the legacy we have achieved.

BACK IN TIME

My own off-Earth stomping ground—the moon—must not be ignored as a return destination. The moon is potentially an invaluable wellspring of oxygen, water, and propellant that could be developed by commercial companies. Working with international partners, we can make use of the moon to widen our reach and sharpen essential robotic skills, as well as train crews for early voyages beyond the moon—expressly to the planet Mars.

Before we get to human Mars missions, let's take a walk back in time.

The Apollo program was a team effort, one that embraced the collective talents of 400,000 people working together on a common vision. It was a unified enterprise, blending government and industry, innovation and teamwork to make a long-held dream a reality.

It was just eight years after President Kennedy's mandate to strive for the impossible that Neil Armstrong and I walked across the Sea of Tranquility's talcum-like lunar dust in that unforgettable summer of 1969. The first words that came to my mind were "magnificent desolation." It was a "magnificent" triumph for humans to set foot on another world. And yet there was the "desolation" of the lunar landscape with no sign of life—no atmosphere —and total blackness beyond the bright sunlit terrain. Standing there, looking at Earth, I was struck by the fact that everything I knew and loved was suspended overhead, on a far-away and ever-so-small, fragile blue sphere that was surrounded by the stark blackness of space.

Upwards of a billion people all over the world watched and listened as Neil and I walked across the lunar landscape from our Eagle lunar lander. We three, including our Apollo 11 colleague and command module pilot, Michael Collins, who was circling the moon, were farther from Earth than any human had ever been. Despite the distance, we felt connected to everyone on Earth, comforted by their participation as our incredible journey unfolded.

As we walked around our Eagle lander, Neil had the camera most of the time. It was my job to set up experiments. There are two images from our mission that have turned out to be classics. One is known as the "visor" photo, an image in which you can see a reflection of the *Eagle* spacecraft and Neil in the visor of my helmet. People have asked me why this photo is so powerful. I have three words: location, location, location. I did take a couple of photos, one that remains iconic to this day: my boot print on the moon.

Let me share an amusing story. Upon returning to Earth, because I was a government employee, I obediently filed a travel voucher for the trip to the moon and back. I had traveled from Texas to Florida, then to the moon, and was picked up in the Pacific Ocean for delivery to Hawaii, and returned to Texas. My travel voucher reimbursement claim added up to $33.31. The government, after all, had provided meals and accommodations from July 7 through July 27, 1969. As an interesting side note, the Apollo 11 crew had to sign customs forms for bringing back lunar rocks and moon dust samples.

Here's another tidbit that surfaced a few years ago. A document was written around the time of Apollo 11's liftoff to the moon by William Safire, who at the time was President Nixon's speechwriter. This internal White House essay, titled "In Event of Moon Disaster," said fate had ordained that the men who went to the moon to explore in peace "will stay on the moon to rest in peace." The Safire essay called us brave men, adding, "In their exploration, they stirred the people of the world to feel as one; in their sacrifice, they bind more tightly the brotherhood of man."

This just-in-case memo went on to suggest that prior to the President's statement, he should telephone each of the widows-to-be. Following the President's statement, at the point when NASA would have ended communications with us, it was recommended that a clergyman should adopt the same procedure as a burial at sea, commending our souls to "the deepest of the deep," concluding with the Lord's Prayer.

Fortunately, Nixon's words never had to be spoken, and the world welcomed us back from the moon as heroes. Amid all the excitement, of honors, parades, interviews, and debriefings, however, *we* knew that people were not cheering for three individuals but for what we *represented*—by the world coming together, humankind had achieved the impossible.

"Forever footprints" on the moon made on July 20, 1969, are likely to last in the lunar soil for millions of years. Aldrin took this photo to document the nature of the moon's dust and its compactness when walked upon.

UNITED STATES
USA
USA

ACCOMPLISHING THE IMPOSSIBLE

Apollo could never have happened without the cooperative efforts of many people working toward a shared goal. When you work together, you can sometimes accomplish the impossible.

Humankind has dreamed for centuries of reaching space—other planets, and even the stars. But it wasn't until the early 20th century that man first realized controlled mechanical flight when in 1903 the Wright brothers flew their plane at Kitty Hawk, North Carolina. Incidentally, my mother was born that same year. I guess you could say pioneering was my destiny. Her maiden name was Marion *Moon*.

It was less than 55 years later, in October 1957, that the Soviet Union achieved an impressive and unexpected technological feat. They boosted the first artificial moon into Earth's orbit. That Sputnik did far more than beep itself into the history books. As a reaction the following year, America formed the National Aeronautics and Space Administration (NASA) with the goal of opening up the frontier of space. Thus, the space age was born, and a space race was about to be set in motion.

In 1961, NASA launched America's first Mercury astronaut, Alan Shepard, on a 15-minute suborbital flight that touched the edge of space. Remember, the Soviet Union had already chalked up a bold triumph by sending the first human into space, cosmonaut Yuri Gagarin, for one full orbit around the Earth.

That powerful one-two success by the Soviets spurred President John F. Kennedy to ask the newly formed NASA what was possible in response. The answer was that it would take at least 15 years before we could put a man on the moon. As I learned recently, Kennedy had originally wanted us to go to Mars. But hearing that, jaws dropped at NASA. Space agency officials and top engineers at the time, after a busy weekend of intense calculation, informed the President that Mars was a little too hard. Instead, we should shoot for the moon as a far more realistic goal.

The Saturn V launch vehicle carrying the Apollo 11 crew roars skyward on July 16, 1969, from the Kennedy Space Center in Florida.

On May 25, 1961, a mere three weeks after Alan Shepard's suborbital excursion, President Kennedy boldly challenged America to commit to the goal of landing a man on the moon before the end of the decade! At that time the aggregate time in space for the United States' human spaceflight program was all of 15 minutes, with no manned orbital flight experience. The rockets and spacecraft needed to go beyond Earth's orbit didn't exist! Many thought fulfilling the Kennedy challenge was impossible. We didn't have the know-how.

What we did have, however, was a leader with vision, determination, courage, and confidence that the moon goal could be realized. We could get there. By publicly stating the objective and assigning a specific timetable, President Kennedy gave us no way out.

As if to double-down on his moon goal declaration, Kennedy later spoke at Rice University's stadium, emphasizing the challenges ahead:

But if I were to say, my fellow citizens, that we shall send to the moon, 240,000 miles away from the control station in Houston, a giant rocket more than 300 feet tall, the length of this football field, made of new metal alloys, some of which have not yet been invented, capable of standing heat and stresses several times more than have ever been experienced, fitted together with a precision better than the finest watch, carrying all the equipment needed for propulsion, guidance, control, communications, food and survival, on an untried mission, to an unknown celestial body, and then return it safely to Earth, re-entering the atmosphere at speeds of over 25,000 miles per hour, causing heat about half that of the temperature of the Sun...and do all this, and do it right, and do it first before this decade is out—then we must be bold.

FOLLOW-ON STEPS

The goals behind Project Apollo went beyond landing Americans on the moon and returning them safely to Earth. They included, as Kennedy had noted, harnessing the technological wherewithal to meet his moon goal and other national interests in space; achieving preeminence in space for the United States; developing the human capacity to productively explore and work in the lunar environment; and carrying out a program of scientific exploration of the moon.

Bear in mind, President Kennedy's original goal was to send *a* man to the moon and bring him back safely to Earth. That objective could have been met by a person landing on the moon, peering out the window, saying hello to Earth, perhaps deploying a science-gathering robot, and never setting foot on the lunar surface. That choice was a road not taken. Rather, a bigger goal was achieved, buddy-system-style; the result was two moon-walking astronauts.

There were six triumphant Apollo lunar landing missions between 1969 and the close of 1972: Apollo missions 11, 12, 14, 15, 16, and 17. The April 1970 mission of Apollo 13 did not land on the moon because of a malfunction en route, but the dramatic return to Earth of the crew—James A. Lovell, Jr., mission commander; John L. Swigert, Jr., command module pilot; and Fred W. Haise, Jr., lunar module pilot—drew global attention to the bravery and resolve of the astronauts and the perseverance of mission controllers.

In all, just 12 of us were able to kick up dust on the moon. The accumulated experience of myself and my fellow moonwalkers was limited: from Apollo 11's modest 2.5 hours to Apollo 17's campaign of lunar surface forays that added up to roughly 22 hours. But those six missions that landed on the moon returned a wealth of scientific data and brought back more than 2,200 lunar samples, weighing a total of nearly 840 pounds. Experiments performed on the lunar surface involved such things as soil mechanics, meteoroids, moonquakes, magnetic fields, and solar wind.

Reflecting on the Apollo 11 mission 50 years later, it's important to recall the early flights, the follow-on steps and giant leaps of my colleagues, and their respective adventures.

Sadly, my memories include the Apollo 1 tragedy of January 27, 1967. That mission was to be the first crewed flight of Apollo, but a fire swept through their Apollo command module during a launch-pad test. We lost astronauts Virgil Grissom, my close friend Edward White, and Roger Chaffee. It was a devastating day—for me, NASA, and the nation. The business of pioneering space was always a dangerous enterprise. Years later, that reality would be driven home again and again with the loss of 14 valiant explorers during the U.S. space shuttle program in 1986 and 2003, as well as by two spaceflight accidents in 1967 and 1971 that claimed the lives of four Russian cosmonauts.

Following the Apollo 1 fire, the program got back on course for the moon with the Apollo 7 engineering test in Earth orbit with Walter Schirra, Jr., as commander, R. Walter Cunningham as lunar module pilot, and Donn F. Eisele as command module pilot. That October 1968 11-day flight completed crucial tests of the Apollo spacecraft systems.

Aldrin unpacks experiments from the scientific equipment bay of the *Eagle* lander. Scientific equipment deployed on the moon included a laser-ranging retroreflector and a lunar dust detector.

Apollo 8 put us on a true trajectory for the moon. The crew consisted of Frank Borman as commander, William A. Anders as lunar module pilot, and James A. Lovell, Jr., as command module pilot. It was the first flight to take humans in the vicinity of the moon and is rightly heralded as a bold step forward in the development of a lunar landing capability. Lifting off on December 21, 1968, the Apollo 8 mission took six days and included 10 orbits around the moon. The test of the navigation and propulsion systems for departing Earth and launching toward the moon proved to be a green light, thumbs up affair.

Apollo 9 was the first manned flight of the lunar module. The 10-day flight circling Earth in March 1969 was led by James A. McDivitt as mission commander, David R. Scott served as command module pilot, and Russell L. Schweickart took the seat of lunar module pilot. That mission demonstrated various important functions, including a complete rendezvous and docking profile and extravehicular crew operations. All systems performed satisfactorily. For the first time, the lunar landing module was tested as a self-sufficient spacecraft, performing active rendezvous and docking maneuvers that paralleled those scheduled for the following Apollo 10 lunar-orbit mission.

The Apollo 10 mission of May 1969 encompassed all aspects of an actual crewed lunar landing—minus the landing. It was the first flight of a complete, crewed Apollo spacecraft to orbit the moon. Thomas Stafford

esa

was commander, Eugene Cernan served as lunar module pilot, and John Young filled the command module pilot post. That daring mission included an eight-hour lunar orbit of the separated lunar module followed by a descent to about nine miles off the moon's surface before ascending for rendezvous and docking with the command and service module in about a 70-mile circular lunar orbit. All mission objectives were achieved.

UNKNOWN UNKNOWNS

The July 1969 Apollo 11 mission applied the lessons learned from previous missions, thus setting the stage for the first attempted landing of humans on another celestial body. There remained, however, many unknown unknowns before landing *Eagle* within the moon's Mare Tranquillitatis. Thanks to the timing of the crew rotation, Neil Armstrong, Michael Collins and I were chosen for the historic mission to be the first to attempt a landing on the moon.

When I learned of being on that first moon landing crew, I came home and told my wife about my mixed emotions. Truth be told, I'd just as soon have been on a later mission. We'd get more interesting things to do. On the other hand, I also could envision all the laudatory speeches around the world upon our return. I did have conflicted feelings, but ultimately, my fighter pilot background kicked in, and I loyally saluted my assignment.

There was no way I could have turned it down. No two ways about it, I was a very lucky person. NASA had estimated a 60 percent chance of landing on the moon without having to abort—and a 95 percent likelihood that we three would return home safely. We sure liked those odds!

Even today, President Kennedy's pledge to go to the moon remains entrancing—not because it was easy, but because it was hard. Moreover, he gave us a decade . . . and we beat that. The Apollo 11 mission showed the ability of a nation to envision an enormously ambitious goal, make it a priority, and create the necessary technology to make it a reality.

Our success with Apollo 11 in turn bolstered subsequent moon landings:

Apollo 12: Charles Conrad, Jr., mission commander; Richard F. Gordon, command module pilot; Alan L. Bean, lunar module pilot. The Apollo 12 lunar module *Intrepid* made a precision landing on the lunar surface on November 19, 1969, in Oceanus Procellarum. This precision landing was highly significant because it showed that we could target landing points of great scientific interest, even in rough terrain.

Apollo 14: Alan B. Shepard, Jr., mission commander; Stuart A. Roosa, command module pilot; Edgar D. Mitchell, lunar module pilot. Apollo 14's *Antares* lunar module landed on February 5, 1971. The mission had several objectives: investigate the lunar surface near a preselected point in the moon's Fra Mauro formation, deploy and activate an Apollo lunar surface experiments package, further develop the ability to work in the lunar environment, and obtain photographs of candidate exploration sites.

Apollo 15: David R. Scott, mission commander; James B. Irwin, lunar module pilot; Alfred M. Worden, command module pilot. The Apollo 15's *Falcon* lunar module touched down on July 30, 1971, in the Hadley-Apennine region of the moon. This mission was the first of the three "J" missions designed to explore the moon over longer periods, over greater ranges, and with more

There is increasing interest in returning to the moon among both governments and commercial firms. The goal is to establish longer-term experimental stations such as this one envisioned by an artist.

instruments for scientific data acquisition than on previous Apollo missions. For the first time, a lunar rover would greatly extend the distance a crew could travel over the lunar surface.

Apollo 16: John W. Young, mission commander; Thomas K. Mattingly II, command module pilot; Charles M. Duke, Jr., lunar module pilot. On April 21, 1972, the lunar module *Orion* landed at the western edge of the Descartes Mountains, accomplishing the first landing in the central lunar highlands. Once again, a lunar rover was utilized to traverse the lunar surface. Over the course of nearly three days, samples were collected from 11 sites, including a deep-drill core from seven feet below the moon's surface.

Apollo 17: Eugene A. Cernan, mission commander; Ronald E. Evans, command module pilot; Harrison H. Schmitt, lunar module pilot. The lunar module *Challenger* reached the moon on December 11, 1972, stirring up dust as it reached the Taurus-Littrow Valley on the eastern rim of Mare Serenitatis. This mission set several records, including the longest total hours spent moonwalking and rover driving. Among the primary objectives of this mission were obtaining samples of highland material that were older than the Imbrium impact and investigating the possibility of young, explosive volcanism in this region. The flight of Apollo 17 concluded the first phase of human exploration of the moon.

CYCLING PATHWAYS TO MARS

With the 20th century Apollo moon landings behind us, and at my age, I find myself contemplating my legacy. I want to be remembered for more than just stirring up moon dust. Specifically, I want my legacy to include the establishment of a permanent human settlement on Mars in the 21st century.

For many years now, I have been blueprinting an approach toward establishing a permanent base on Mars by 2040. In cooperation with universities and space veterans around the country, this expeditionary approach is called "Cycling Pathways to Occupy Mars"—or CPOM for short.

CPOM, as a map to Mars habitation, leverages the best ideas of the emerging commercial space sector, the prowess of other spacefaring nations, academia, and nonprofit institutes. I have been planning this space transportation system of CPOM since 1985. With the help of Purdue University and my Buzz Aldrin Space Institute at the Florida Institute of Technology in Melbourne, I have a team helping me to fulfill and develop these concepts that I have worked on for more than three decades.

Even though we landed on the moon 50 years ago, we must not ignore the moon now in our quest for Mars settlement. In fact, building a base on the moon will provide an essential training ground for a future Mars base. I am a strong advocate for international cooperation in space. The United States should help other advanced nations—notably those of Europe, Russia, Japan, and China— reach the moon and build a base there. Simply put, the moon will enable us to go to Mars.

Here are the essentials behind my Cycling Pathways to Mars.

The first critical component to the plan is Bigelow Aerospace's BA330, an expandable habitat that is leaner than the International Space Station and less expensive. The BA330 habitat will help establish a foothold in low-Earth orbit and can co-orbit with China's space station, which is due to be operational in the 2020s.

The next step is to build a cycler, a rigid-core module docked at the center of two BA330s. Multipurpose crew vehicles, such as the U.S.'s *Orion* and similar spacecraft, could dock at one of four ports on the cycler.

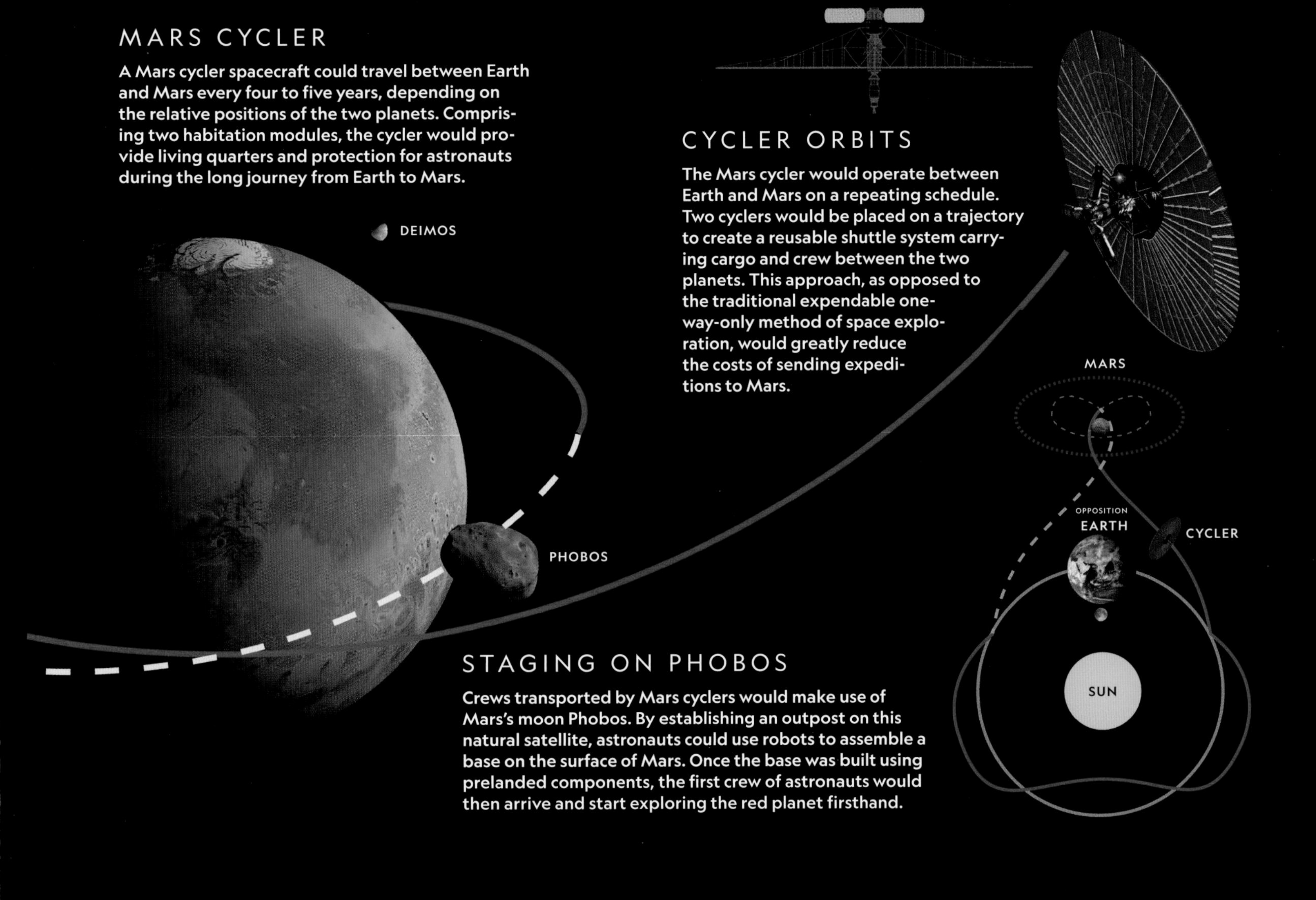

It would be modular in design and could be upgraded with different habitats.

This Earth–moon lunar cycler would allow regular travel between the Earth and moon by using their respective gravitational fields. Each cycle would take about a month. Spacecraft, such as *Orion*, could rendezvous with and depart from the cycler, like a train leaving a station. When the cycler approached the moon, spacecraft could undock and either enter lunar orbit or land on the lunar surface.

Aldrin's plan for human exploration and colonization of Mars envisions cycling spacecraft to establish and support a sustained and growing civilization on the red planet.

In the next phase, building on the design of the low-Earth orbit cycler, two rigid habitation modules would replace the inflatable Bigelow BA330 expandable modules. These habitats would protect pioneering travelers embarking on deep space missions.

Exploiting cis-lunar space—the space between the Earth and the moon—provides several big boosts to Mars exploration. First, spacecraft propellant fuel could be processed from lunar ice. By mining fuel from the moon, we avoid having to launch propellant from Earth into space, and we gain the advantage of being able to refuel in space, thus lowering the cost of space travel altogether.

Cis-lunar space is also invaluable for testing exploration modules and other new applications ultimately needed for Mars exploration. From cis-lunar space, we would be able to construct an international lunar base using remotely controlled robots. Once perfected, the same methods could be used to remotely construct, perhaps from Phobos, a moon of Mars, the first base on the red planet.

THE RECYCLING APPROACH

Developing a successful cycler system between Earth and the moon sets us on a path to the more formidable task of a cycler system between Earth and Mars. As I see it, we should cycle two spacecraft in the space between the two planets, a kind of interplanetary taxi service made of two large habitation modules in which space travelers would spend the long journey.

Crews from cyclers would undock and rendezvous with the Martian moon Phobos, from where they would remotely construct the Mars base and demonstrate safe entry, descent, and touchdown of unpiloted landers on the Martian terrain. This would all be done prior to dispatching crews to the landscape of Mars. Thus before any astronaut sets foot on Mars, the base will have been prepared and the risks greatly diminished.

A new class of Mars exploration equipment will greatly expand the breadth and depth of remote surveying. With just a second or less of connectivity between human controllers and hardware on the red planet, high-tech robotic probes will be able to quickly traverse zones of exploration, dive down into lava tubes, even rappel off cliffs. What's more, by making use of virtual reality, immersive technology, and artificial intelligence, off-Mars astronauts will be able to control with a joystick advanced rovers and other craft on the planet in real-time. I have experienced the use of avatars—electronic Marswalkers that are manipulated by a user in a virtual space. Those avatars can serve as field exploration parties that meander virtually across Mars. These techniques will allow us to determine where best to establish and then robotically assemble our first home-away-from-home—and all before the arrival of a crew.

So how do we accomplish the journey to and settlement of Mars? By working together—mirroring the tenacity of those that made Project Apollo a success—we can make this vision a reality. We can establish a permanent and growing settlement on Mars.

Reaching for the red planet—and establishing long-term working conditions, such as these imagined by a NASA artist—demands the best of our scientific and technological prowess.

OCCUPYING MARS

I believe our ultimate goal should be to settle the planet Mars, and I sense that that future could be closer than we think. Celebrating the 50th anniversary of humanity's first moon landing would be an opportune time for the U.S. President to say, "I believe that this nation should commit itself to lead international crews to occupy Mars in two decades." I'm not talking here about a Mars visit along the lines of the flag-and-footprints approach as was done with Apollo. Crews that land on Mars should not depart until the next people arrive; that's how we will keep the red planet occupied.

That should be our objective. I strongly object to making a humans-to-Mars program a look-alike Apollo moon project: put people on the surface of the planet, proclaim success, have the crew set up experiments, plant a flag, and quickly bring the explorers back to Earth. This type of scenario for Mars would be too vulnerable to cancellation and abandonment.

Humankind's road to Mars will not be easy. How best to make a Mars settlement self-sufficient will be among the major challenges that will need to be addressed early on. Shipping life-sustaining supplies from Earth would be prohibitively expensive. If settlements are to survive, we will need to figure out how to live off the land, using local Martian resources—water, soil, and other assets, some of which, I am sure, have yet to be identified.

Living off the un-earthly real estate that is Mars is no easy assignment. It means adopting an approach dubbed In-Situ Resource Utilization, or ISRU. That is space-speak for humans using Mars's natural resources not only to endure but also flourish. Water and atmospheric carbon dioxide on Mars are the most valuable resources available to us in the pioneering of the planet. These two basic resources can yield propellants, life-sustaining support, help in growing crops, and protection from lethal doses of radiation.

On any to-do list are excavating, extracting, refining, and putting to use Martian resources for construction needs. A settlement on Mars can be safe, affordable, and self-sufficient. Achieving self-sufficiency will require a technological toolkit of frontier technologies that includes robotics, machine intelligence, nanotechnology, synthetic biology, and 3D printing. For instance, plastics can be fashioned from carbon, hydrogen, and oxygen processed from Martian water and the planet's atmosphere. Extensive use of local materials on Mars is conceivably a game changer to pioneer, and in due course, settle Mars.

The red planet may well become the proving ground for many new technologies that not only advance Earth's own independence but promote Mars to become the supply source for fuels, oxidizers, life support, spare parts, replacement vehicles, habitats, and other products for enhancing spacefaring beyond low-Earth orbit and beyond Mars itself.

HUMANS: A MULTIPLANET SPECIES

We mustn't kid ourselves. A sustainable civilization on Mars would be extremely complex. It will be necessary to think through almost every aspect of human society and redesign it for another planet.

Developing a lifestyle worth living on Mars demands space transportation, power production, and food supplies, as well as construction materials to assemble a sprawling Mars settlement. That surely means tapping both robotic and human labor. I foresee that the settlement on Mars begins as the most international endeavor ever conceived. But I do not think it will be long before the inhabitants lose their national identities and become Martians.

My vision is that within the 21st century, the sandy face of the red planet is going to be dotted by the first footfalls of humans. The adventure ahead is historic. The upshot of creating a sustained presence there is that humanity becomes a multiplanet species. What better way to herald the 50th anniversary of Apollo than to renew our passion for space, to build the bridges and pathways that tie the Earth, our moon, and Mars together?

Humanity needs to explore, to push beyond current limits, just like we did in 1969. Apollo was the story of people at their very best. We started with a dream, and we did the impossible. I believe the impossible can be done again, but we need to roll up our sleeves and go for it.

There is no doubt that Mars is a world of surprises. Exploring and settling that planet assuredly promises us many teachable moments, about the red planet—and also about ourselves.

How often have you heard, "Let's shoot for the moon?" Well, I did that once, and now I'm shooting for Mars. But I've still got the moon in my travel plans as I look toward the future. I am shooting for the moon to lead to Mars. And now is the time for the world to go for it again.

Aldrin on July 20, 1969, inside the Apollo 11 lunar module (above) and in 2016 (opposite).

ABOUT BUZZ ALDRIN

BY LEONARD DAVID

If space was going to be our next new frontier, then Buzz Aldrin wanted to be a part of getting there.

Buzz Aldrin grew up in Montclair, New Jersey. His mother, Marion Moon, was the daughter of an Army chaplain, and his father, Edwin Eugene Aldrin, was an aviation pioneer. Buzz graduated one year early from Montclair High School, and he attended the U.S. Military Academy at West Point, graduating third in his class with a bachelor's degree in mechanical engineering.

Aldrin took his first flight at age two with his father. Surrounded by the influence of aviation, he entered the Air Force after graduating from the Military Academy. He served as a jet fighter pilot in the Korean War, where he flew 66 combat missions and shot down MIG-15s, later to be decorated with the Distinguished Flying Cross.

Following the Korean War, Aldrin was stationed in Germany in the late 1950s. The Cold War was escalating, and tensions were high. After his tour of duty in Germany flying F100 jets, he earned his doctorate of science in Astronautics at the Massachusetts Institute of Technology (MIT) and wrote his thesis on manned orbital rendezvous.

Buzz Aldrin's MIT thesis devised a technique for two piloted spacecraft to meet in space. Little did anyone know—including Aldrin—how critical that work would later be on the successful landing of humans, including him years later, on the moon.

Aldrin's first application to become a NASA astronaut was turned down. He was not a test pilot. Determined, he applied again. This time, his jet fighter experience and NASA's interest in Aldrin's concept for space rendezvous led them to accept him in 1963 to become part of the third group of astronauts. He was the first person in that elite corps with a doctorate. The docking and rendezvous techniques he devised for spacecraft in Earth and lunar orbit became critical to the success of the Gemini and Apollo programs and are still used today. His astronaut peers dubbed him "Dr. Rendezvous" for his expertise.

It was Aldrin who pioneered underwater training techniques to simulate spacewalking. In 1966 on the Gemini 12 Earth-orbiting mission, Aldrin carried out a path-breaking spacewalk—extravehicular activity (EVA)—setting a new EVA record at the time of 5½ hours of spacewalking. During that mission he also took the first "selfie" in space!

On July 20, 1969, Neil Armstrong and Buzz Aldrin made their historic Apollo 11 landing, becoming the first two humans to set foot on another world. An estimated 600 million people—at that time, the world's largest television audience in history—witnessed this unprecedented heroic endeavor.

Upon returning from the moon, Aldrin was decorated with the Presidential Medal of Freedom and numerous awards all over the world. In 2011, along with his Apollo 11 crew mates Neil Armstrong and Michael Collins, he received the Congressional Gold Medal.

Buzz Aldrin is the author of nine books, most recently his *New York Times* and *Washington Post* bestseller, *No Dream Is Too High: Life Lessons From a Man Who Walked on the Moon,* with Ken Abraham, and his children's book, *To the Moon and Back: My Apollo 11 Experience,* with Marianne Dyson, as well as *Mission to Mars: My Vision for Space Exploration,* published in 2013 and coauthored with veteran space reporter Leonard David—all published by National Geographic.

INTRODUCTION

The universe used to be such a simple place. • After all, for most of human history, the people who thought about such things pictured Earth as sitting still in the center of creation, while the heavenly bodies such as stars and planets moved around it. In ancient mythologies, Earth was usually flat, and the motion of the sun across the sky was due to the actions of a god or goddess. Starting in about the fifth century B.C., though, a new way of thinking about the universe began to develop around the eastern Mediterranean—a way that didn't depend on the whims of the gods, a way that began to take humanity out of what Carl Sagan called the "demon-haunted world." Greek philosophers began to construct models of the universe that may seem primitive to us but had the strange new feature of operating solely according to natural laws, without supernatural intervention. Many scholars consider this development to be the beginning of science.

All these models shared two basic, unquestioned assumptions. One was that Earth sat motionless at the center of the universe, while everything else—sun, moon, and planets—swung around it. The second assumption was that in the heavens, which were pure and eternal, everything moved in a circular path. (This assumption is based on the notion that a circle is the most perfect geometrical shape and therefore the appropriate figure for the realm of the pure.) In these models, the stars and planets were embedded in solid crystalline spheres whose rotations governed their motion across the sky. (This, incidentally, explains why comets were such a problem for early astronomers, since their orbits would shatter those spheres. It's one reason why Aristotle argued that comets had to be burning vapors in Earth's atmosphere.) Eventually, these models became quite complicated, with the planets embedded in small spheres that rolled within larger spheres.

Given the cozy nature of that model, it's a little surprising that there was a spirited debate in the ancient world centered on the question of whether the universe was finite or infinite in extent. The philosopher Archytas of Tarentum (428–327 B.C.) made an interesting argument that the universe must be boundless. Suppose, he argued, that the universe really has a boundary, an edge. Then a spearman could walk up to that edge and throw a spear outward. The spear would have to land somewhere, and that landing place would be outside the boundary. No matter how far out you make the boundary, the spearman can always find something outside of it. Therefore, he argued, the universe must have no boundary, but must be infinite.

Following Archytas's argument, we can identify events that expanded the human view of the universe, each event like another flight of a spear. In what follows, in fact, we will encounter three such spearmen, each of whom widened the universe we live in.

THE FIRST SPEARMAN

The first was the Polish cleric Nicolaus Copernicus (1473–1543). He produced the first serious model of the solar system in which the sun was at the center and Earth moved in orbit, like the other planets. In his book *On the Revolutions of the Celestial Spheres*, he wrote, "[I]t will be realized that the sun occupies the middle of the universe. All these facts are disclosed to us by the principle governing the order in which the planets follow one another, and by the harmony of the entire universe, if only we look at the matter, as the saying goes, with both

An illustration of an armillary sphere—a model of the heavens, with Earth at the center and the constellations of the zodiac around the edge—shows the standard depiction of the universe until the age of Copernicus. Flanking the sphere are diagrams of the theories of astronomers Ptolemy and Tycho Brahe.

Polish astronomer Nicolaus Copernicus overturned the astronomical world with a new, sun-centered vision of the universe. His landmark book, *On the Revolutions of the Celestial Spheres,* was published in 1543.

CL·PTOLEMAEI
TERRA
LUNA
VENERIS
SOLIS
MARTIS
JOVIS

eyes." With Copernicus, the human universe suddenly changed. No longer were Earth and humanity at the very center of existence. Human beings were now inhabitants of just one of many bodies circling the sun. Throughout this book the so-called Copernican principle will come into play—the notion that there is nothing particularly special about humanity and the planet that gave it birth.

With Copernicus, the universe also became much larger. No longer did human beings live in a cosmos bounded by the sky hanging a few miles over their heads and Earth under their feet. Astronomers after Copernicus came to realize that the solar system is huge compared with Earth. Imagine Earth as a sphere the size of a city block located in New York City. The sun would then be a large sphere somewhere around the Mississippi River, and the outer planets would be in Asia—quite a shift in perspective for people who had spent their entire existence on that city block.

THE SECOND SPEARMAN

The second spear was thrown by a German astronomer named Friedrich Bessel in the early years of the 19th century. Using state-of-the-art telescopes, he was the first to determine the distance to nearby stars. Once again, the universe expanded enormously. If we imagined the solar system crammed into a space the size of a football field, then a typical star would be in a city several hundred miles away.

As the century wore on, astronomers realized that our own sun was just one star—a pretty ordinary one at that—in a mighty city of stars they called the Milky Way. Our own sun and solar system, which had loomed so large in our minds, were now relegated to just one system among billions. Astronomers began to realize that not all stars are the same, and they started cataloging what they saw. They also noticed faint patches of light in the sky that they called nebulae, although their telescopes lacked the ability see what those clouds were made of. The world was getting ready for the third spearman.

THE THIRD SPEARMAN

His name was Edwin Hubble, and he was an American working at a brand-new telescope on California's Mount Wilson in the 1920s. With that telescope Hubble was able to examine nebulae in detail, picking out individual stars embedded in them. From this he was able to determine how far away those nebulae are. Once again the spear flew outward. He found that many nebulae were, in fact, their own gigantic cities of stars, like the Milky Way. Hubble established that the universe was composed of what we now call galaxies. In a sense, this was simply an extension of what Copernicus taught us so long ago. Earth is just one planet among many in the solar system, the sun just one star among many in the Milky Way, the Milky Way just one galaxy among billions in the universe.

But that wasn't the only thing that Hubble discovered. He found that those other galaxies are moving away from us, that the universe is expanding. This discovery led, in turn, to our current best picture of the origins of the universe, the scenario we call the big bang. In this account the universe began in an unimaginably hot, dense state about 14 billion years ago and has been expanding and cooling ever since. The amazing thing is that scientists have produced models that can reliably trace backward through this process, back to a fraction of a second after the initial event.

The five galaxies known as Stephan's Quintet are actually four interacting galaxies and a fifth (the bluish spiral at upper left) that is younger and much closer to our own. The central two galaxies are colliding and in the process giving birth to new stars, seen in the blue star clusters around their edges. The gravitational influence of the nearby galaxies has also distorted the arms of the barred spiral galaxy at top right.

Today it may be that another spearman is approaching the boundaries of our universe—his or her identity is still unclear. If, however, some modern theories prove to be successful, it may turn out that our own universe is just one among a multitude of universes in what scientists are starting to call the multiverse. Such a development would, of course, be the ultimate vindication of Nicolaus Copernicus although he could scarcely have imagined such an outcome.

FOUR UNIVERSES

This book is organized to follow the progress of Archytas's spearmen. Think of our world as being composed of a series of nested "universes," each seeming to encompass all of creation until a spearman comes along and takes us into the next one.

The first, most intimate universe is, of course, our own solar system. Before the invention of the telescope, it consisted of the six innermost planets, and most

scientific attention was focused on understanding how the planets moved—essentially, astronomers wanted to know where the planets were. In the 19th century this began to change, until today we focus on how the planets are built—on what (as opposed to where) they are. Furthermore, we have discovered that the solar system is a lot more complex than those early scientists had imagined. Starting with Galileo's discovery of Jupiter's moons, we began to realize that there was a lot more to the solar system than just a few planets. Each moon was, in effect, a new world with its own history, its own characteristics, its own mysteries. Even the cold, dark region beyond Pluto revealed structures and complexities of which we had never dreamed. This new solar system is the subject of the first section of this book.

THE GALAXY

The second universe is our own Milky Way galaxy. As was the case with the solar system, scientists began with the simple tasks—Where are stars located? How bright are they? and so forth. And as with the solar system, the 19th century slowly brought a new set of questions: What are stars made of and how do they work? It wasn't until the 1930s, though, that the new science of nuclear physics revealed the source of a star's energy in nuclear reactions, and it began to sink in that the stars are not eternal: They have a life cycle—a beginning, a middle, and an end—like everything else. In fact, we have come to see the Milky Way (and other galaxies as well) as a gigantic factory that converts the primordial hydrogen of the universe into the heavier elements from which planets and human beings (among other things) are made. In the process, we have discovered all sorts of new and interesting objects out there, from black holes to planetary systems around other stars. We also discovered that most of the matter in the Milky Way and other galaxies is not the familiar stuff from which we are made, but something new called dark matter. The exploration of this particular universe is the subject of the second section of this book.

THE UNIVERSE

The third universe is the massive collection of galaxies to which we usually apply that term. The study of the beginning and end of our universe has engaged scientists over the last few decades. We have traced our way back to the beginning using the tools of elementary particle physics and to the end using the tools of observational astronomy. Against all expectations, astronomers in the 1990s discovered that the expansion of the universe is not slowing down but speeding up. This knowledge, in turn, has led to the realization that most of the matter in the universe is in an unknown form that has been named dark energy. The fate of the universe depends on what dark energy is and what its properties are. In the third section of this book, we explore the universe as best we know it today.

Finally, for our fourth and last universe, we leave the realm of hard data and enter the speculative world of the theoretical physicist. Some modern theories suggest that our universe is just one among a huge multitude of universes—what some theorists call a multiverse. With this excursion, we will have completed our tour of the universes that make up the environment in which we live.

Studies of the collection of galaxies known as Pandora's Cluster yield clues to the presence of mysterious dark matter. The colliding galaxies in the cluster make up only about 5 percent of its mass, gases (colored here in red) another 20 percent. The rest of the cluster's mass (colored in blue) is dark matter. Dark matter is not visible, but it can be detected by its gravitational influence.

COSMIC JOURNEYS

Since the first moon landing, a veritable armada of spacecraft have left Earth to explore the solar system. Each line in these drawings represents one of those missions.

MISSIONS TO INNER SOLAR SYSTEM

- NASA — FAILED MISSION
- U.S.S.R./RUSSIA — FAILED
- EUROPEAN SPACE AGENCY OR OTHER EUROPEAN COUNTRY — FAILED
- JAPAN — FAILED
- CHINA — FAILED
- INDIA

DEEP SPACE MISSIONS

- PIONEER – NASA
- VOYAGER – NASA
- GALILEO – NASA & EUROPEAN SPACE AGENCY
- CASSINI-HUYGENS – NASA & EUROPEAN SPACE AGENCY
- NEW HORIZONS – NASA
- JUNO – NASA

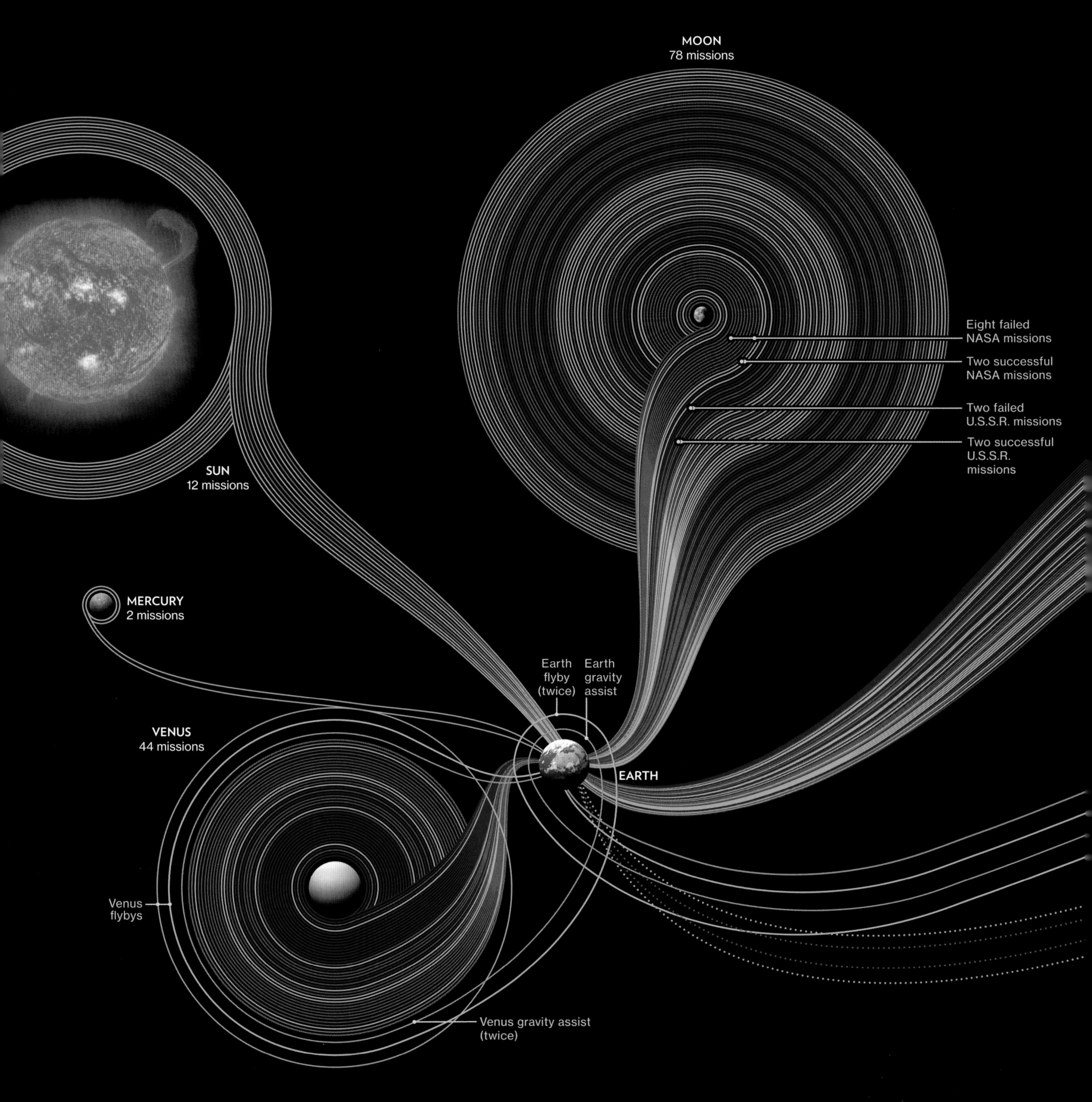

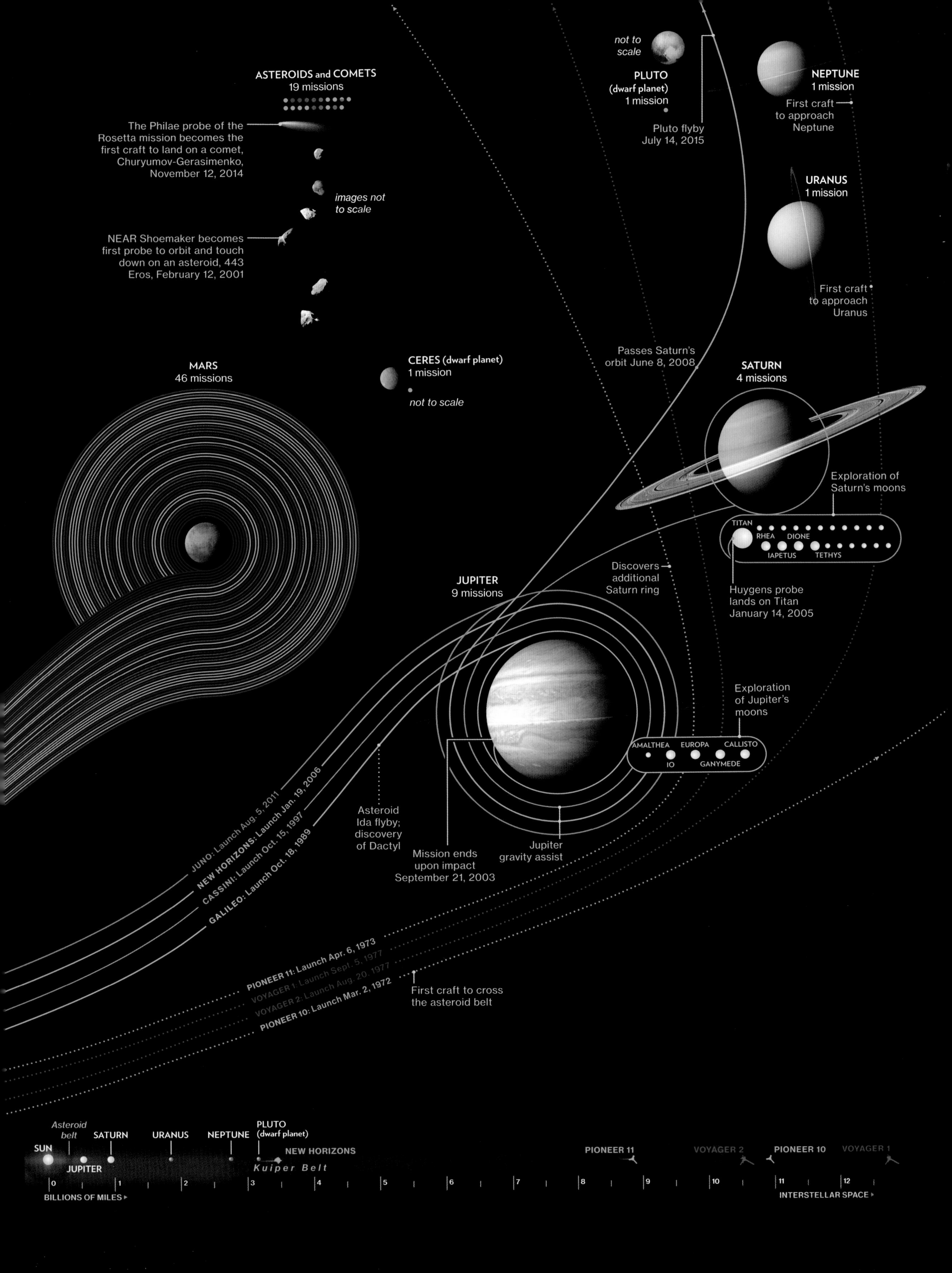

ASTEROIDS and COMETS
19 missions
The Philae probe of the Rosetta mission becomes the first craft to land on a comet, Churyumov-Gerasimenko, November 12, 2014
images not to scale
NEAR Shoemaker becomes first probe to orbit and touch down on an asteroid, 443 Eros, February 12, 2001
not to scale
PLUTO (dwarf planet) 1 mission
Pluto flyby July 14, 2015
NEPTUNE 1 mission
First craft to approach Neptune
URANUS 1 mission
First craft to approach Uranus
MARS 46 missions
CERES (dwarf planet) 1 mission
not to scale
Passes Saturn's orbit June 8, 2008
SATURN 4 missions
Exploration of Saturn's moons
TITAN
RHEA
DIONE
IAPETUS
TETHYS
Huygens probe lands on Titan January 14, 2005
Discovers additional Saturn ring
JUPITER 9 missions
Exploration of Jupiter's moons
AMALTHEA
IO
EUROPA
GANYMEDE
CALLISTO
Asteroid Ida flyby; discovery of Dactyl
Mission ends upon impact September 21, 2003
Jupiter gravity assist
JUNO: Launch Aug. 5, 2011
NEW HORIZONS: Launch Jan. 19, 2006
CASSINI: Launch Oct. 15, 1997
GALILEO: Launch Oct. 18, 1989
PIONEER 11: Launch Apr. 6, 1973
VOYAGER 1: Launch Sept. 5, 1977
VOYAGER 2: Launch Aug. 20, 1977
PIONEER 10: Launch Mar. 2, 1972
First craft to cross the asteroid belt
SUN
Asteroid belt
JUPITER
SATURN
URANUS
NEPTUNE
PLUTO (dwarf planet)
NEW HORIZONS
Kuiper Belt
PIONEER 11
VOYAGER 2
PIONEER 10
VOYAGER 1
0
1
2
3
4
5
6
7
8
9
10
11
12
BILLIONS OF MILES
INTERSTELLAR SPACE

THE NIGHT SKY

Have you ever had the experience of seeing the night sky in its full glory, far away from city lights? If so, you remember the brilliance of the stars against the velvet blackness of the sky.

For most of the history of *Homo sapiens,* this is what the sky looked like every night. No wonder early people grouped the stars together into constellations and incorporated them into their mythology. No wonder that the earliest astronomers were motivated by the (false) idea that the stars and planets had a profound effect on the affairs of humans.

This interest grew into an endeavor that modern astronomers would recognize. On bamboo books in China and clay tablets in Babylon, we find records of naked-eye observations of the night sky. The Greeks took these observations, added their own, and produced a picture of the night sky in which stars and planets circled Earth on crystal spheres.

Fifteen hundred years later, the Polish cleric Nicolaus Copernicus changed all that, introducing the notion that Earth circled the sun, rather than vice versa. Once Galileo turned his telescope on the heavens, our vision of the night sky changed forever. No longer was Earth the center of the universe. Our planet was just one among many. Our moon was a world of craters and mountains, and other planets had their moons as well, circling them like miniature solar systems.

Today we see the few thousand stars in the night sky as a small sample of the billions of stars in our own Milky Way galaxy and our own galaxy as one of billions in our universe. Even so, today's sky-watchers still use sky maps that resemble those of long ago. Stars are plotted out against a dome-like sky and contained in 88 official constellation regions. The next eight pages should allow you to find and appreciate the brightest stars and objects in the sky just as our ancestors did, thousands of years ago.

An 18th-century illustration of the constellations of the northern hemisphere shows the figures that have been assigned to the night skies since antiquity. Although it's hard actually to detect these shapes in the stars, they form a useful guide to the sky, and today's astronomers still use constellations to map the stars.

HOW TO READ THE SKY MAPS

Unlike terrestrial maps, which show land from above, star charts represent the sky from below. These four maps depict the skies over Earth's northern and southern hemispheres, with the north or south celestial poles at top center. Objects around the edge of the maps might be seen from either hemisphere.

To pinpoint locations, astronomers use celestial coordinates: Right ascension is similar to longitude and measured in hours, minutes, and seconds denoted by Roman numerals around the rims of the star charts. Declination corresponds to latitude, measuring position north or south of the celestial equator in degrees and minutes. Parallels of declination are concentric blue circles on the charts.

Astronomers organize the sky into 88 constellation regions, marked here by yellow boundary lines. Main stars in a constellation have Greek letter designations: alpha for the brightest, beta for second brightest, proceeding more or less in order of magnitude, or apparent brightness (see keys)

HÆMISPHÆ
LATUM BO
ANTI
RIUM STEL
REALE
QVUM.
Apud GERARDUM VALK, et PETRUM SCHENK, Amstelædami.

Zodiac
Declination
Ecliptic (path of the sun across the sky)
Constellation boundaries
New General Catalogue object
Constellation figure
Right ascension
Messier object

I NORTHERN SKY SUMMER–FALL

Stars and constellations on this map are highest in the northern hemisphere sky from July to December.

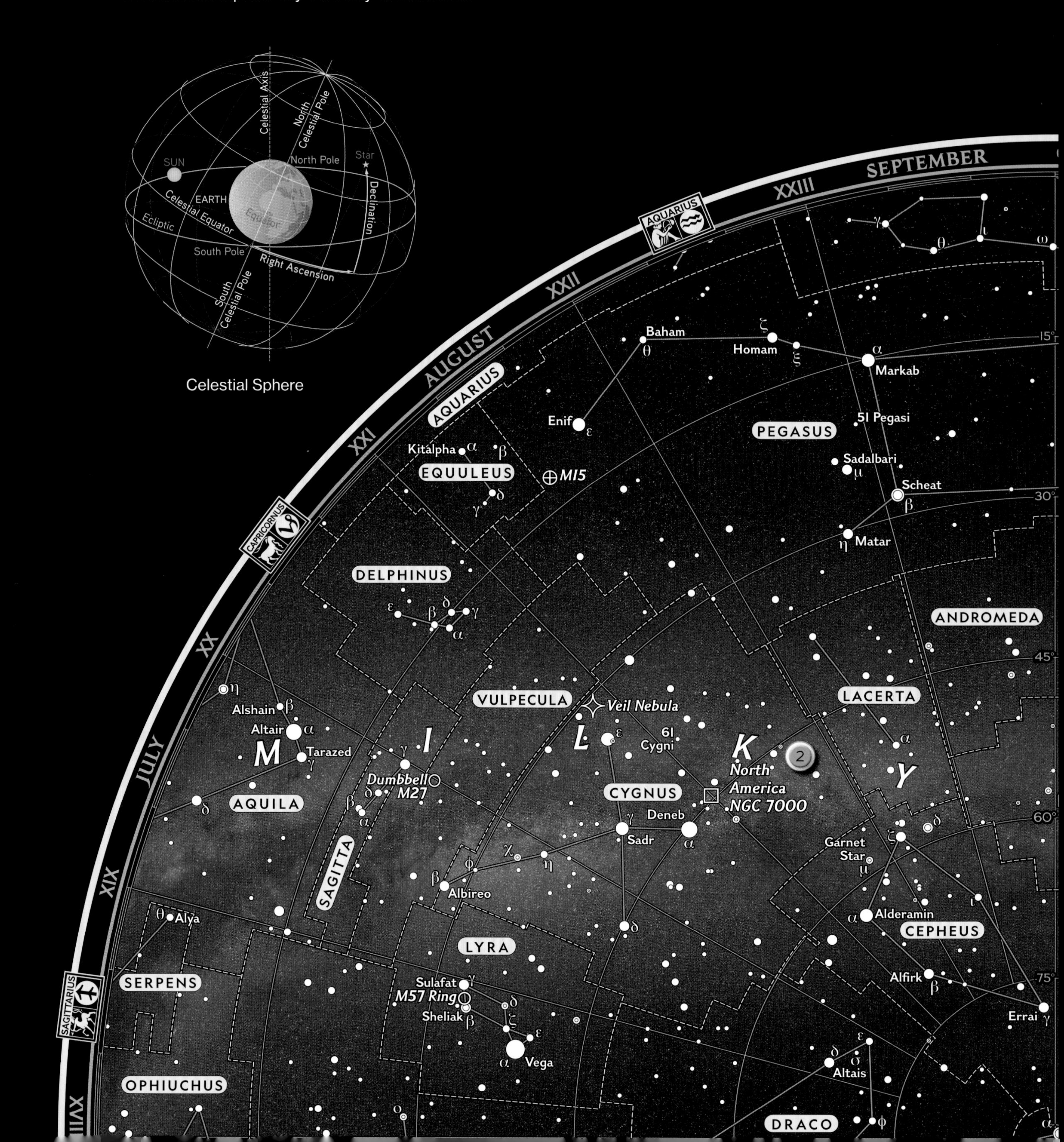

Celestial Sphere

KEY FEATURES

1. **POLARIS: The North Star**
2. **MILKY WAY**
3. **ANDROMEDA: The only northern galaxy visible to the naked eye**

Star magnitude: -1, 0, 1, 2, 3, 4, 5

Variable star

- Open star cluster
- Globular star cluster
- Galaxy
- Diffuse nebula
- Planetary nebula
- Supernova remnant
- Constellation boundary
- Ecliptic

CARTOGRAPHER'S NOTE: To find an object on these charts, such as the bright star Vega, move clockwise around the rim of the chart to right ascension 18 (XVIII) hours, 36 minutes. Look toward the pole to declination +38 degrees to find the star.

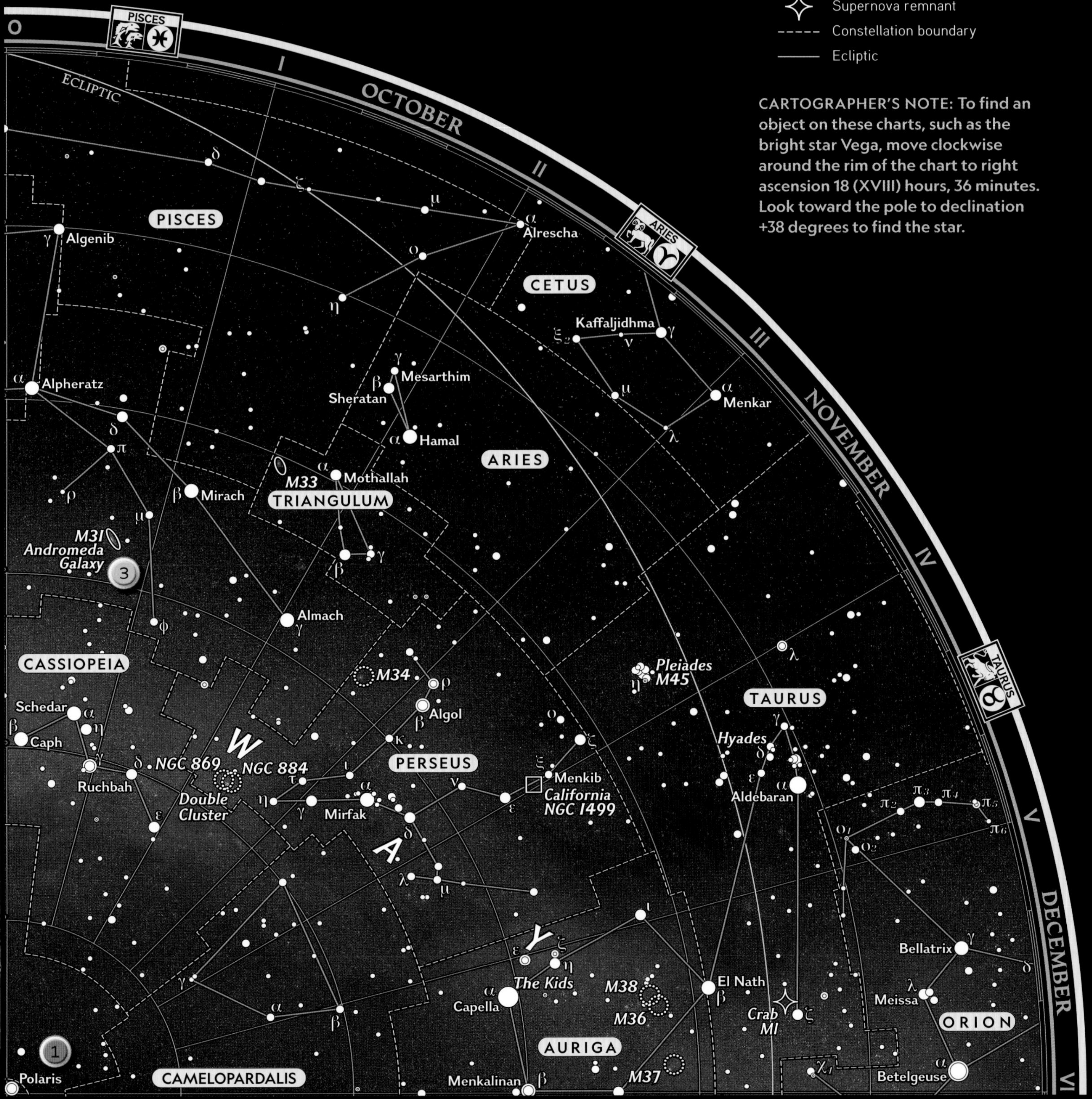

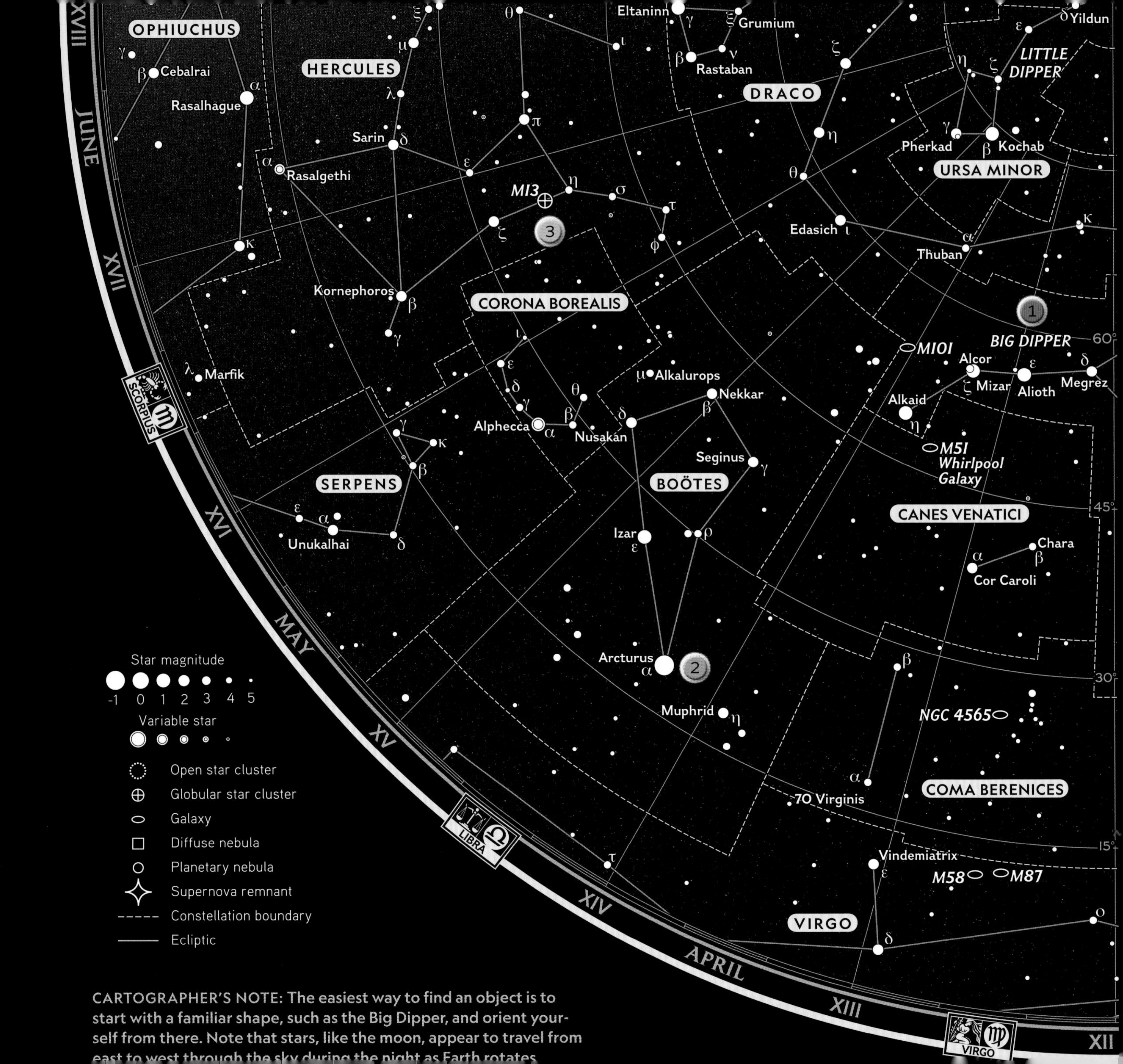

CARTOGRAPHER'S NOTE: The easiest way to find an object is to start with a familiar shape, such as the Big Dipper, and orient yourself from there. Note that stars, like the moon, appear to travel from east to west through the sky during the night as Earth rotates.

KEY FEATURES

1. **BIG DIPPER:** This star pattern, called an asterism, is part of Ursa Major.
2. **ARCTURUS:** At the tip of Boötes, this is the fourth brightest star in the sky.
3. **M13:** This globular cluster of stars is the brightest in the northern skies.

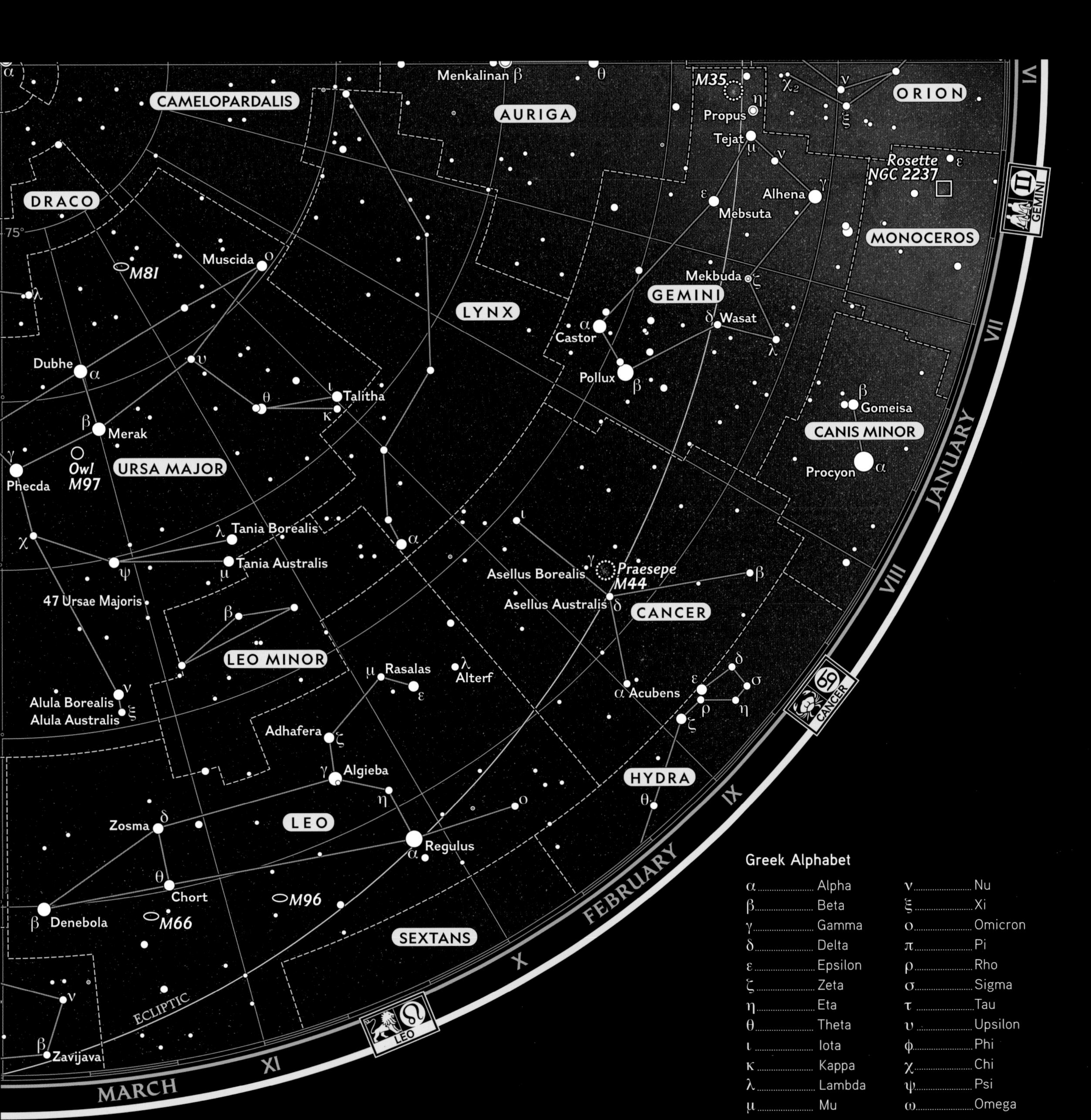

Greek Alphabet

α	Alpha	ν	Nu
β	Beta	ξ	Xi
γ	Gamma	ο	Omicron
δ	Delta	π	Pi
ε	Epsilon	ρ	Rho
ζ	Zeta	σ	Sigma
η	Eta	τ	Tau
θ	Theta	υ	Upsilon
ι	Iota	φ	Phi
κ	Kappa	χ	Chi
λ	Lambda	ψ	Psi
μ	Mu	ω	Omega

1 SOUTHERN SKY SUMMER–FALL

Stars and constellations on this map are highest in the southern hemisphere sky from July to December.

CARTOGRAPHER'S NOTE: Stars and constellations around the edges of this map will be visible from much of the northern hemisphere. There is no prominent southern polestar, but the southern sky features the brightest stars and some of the most dramatic celestial bodies.

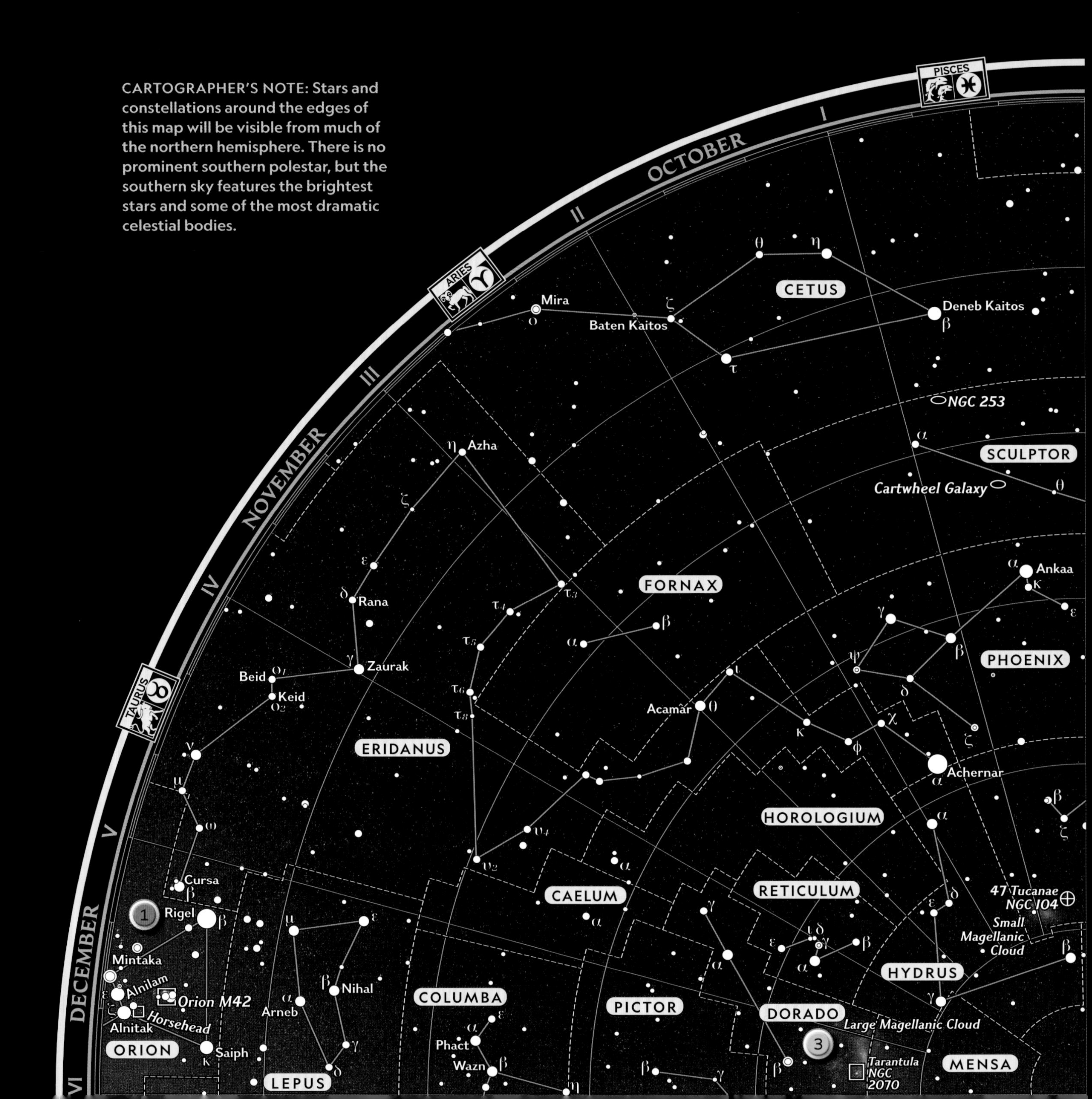

KEY FEATURES

1. **ORION:** One of the most prominent constellations, visible north and south
2. **HELIX NEBULA:** A large and spectacular planetary nebula
3. **LARGE MAGELLANIC CLOUD:** One of two galaxies visible to the naked eye from the southern hemisphere

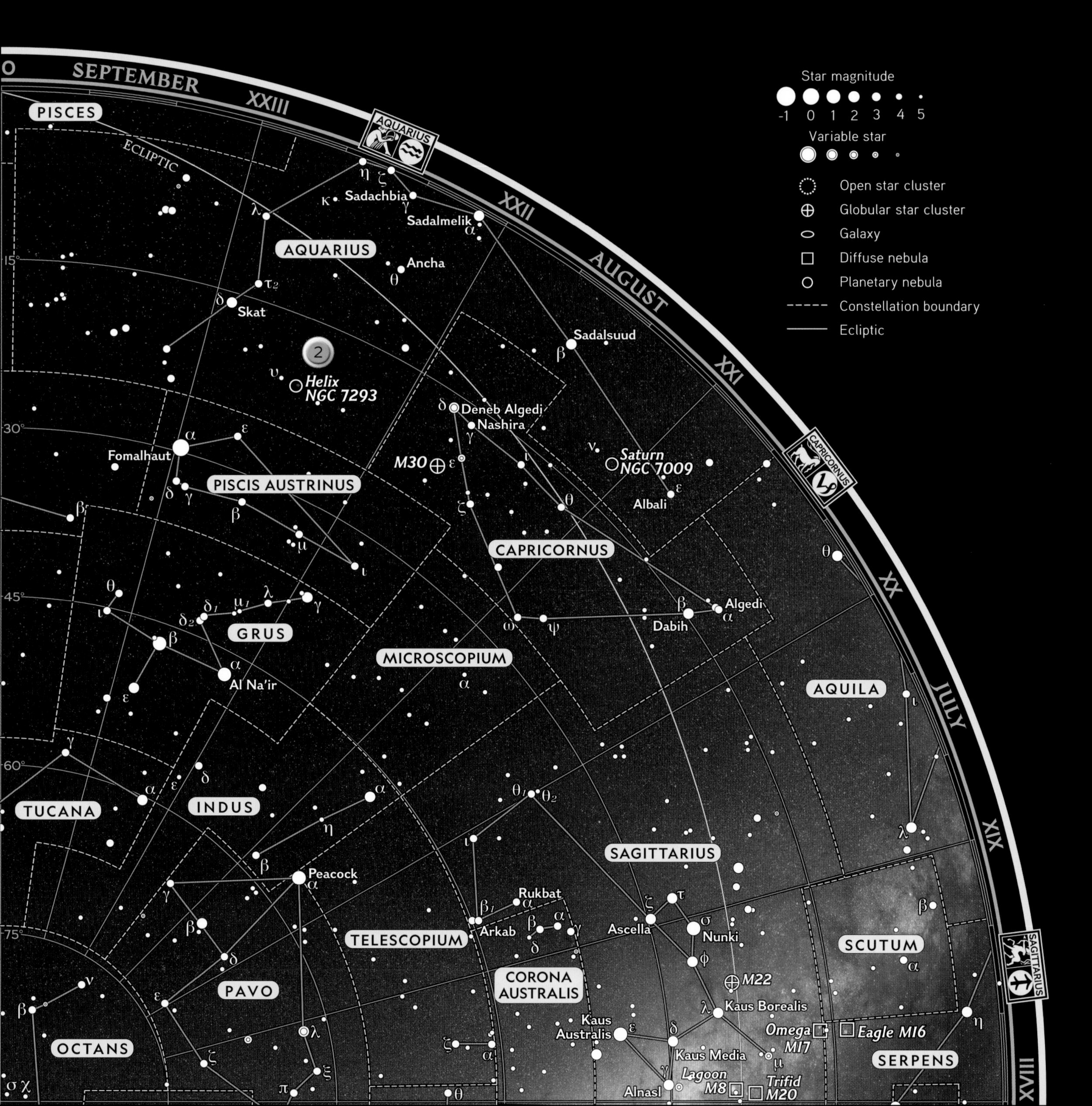

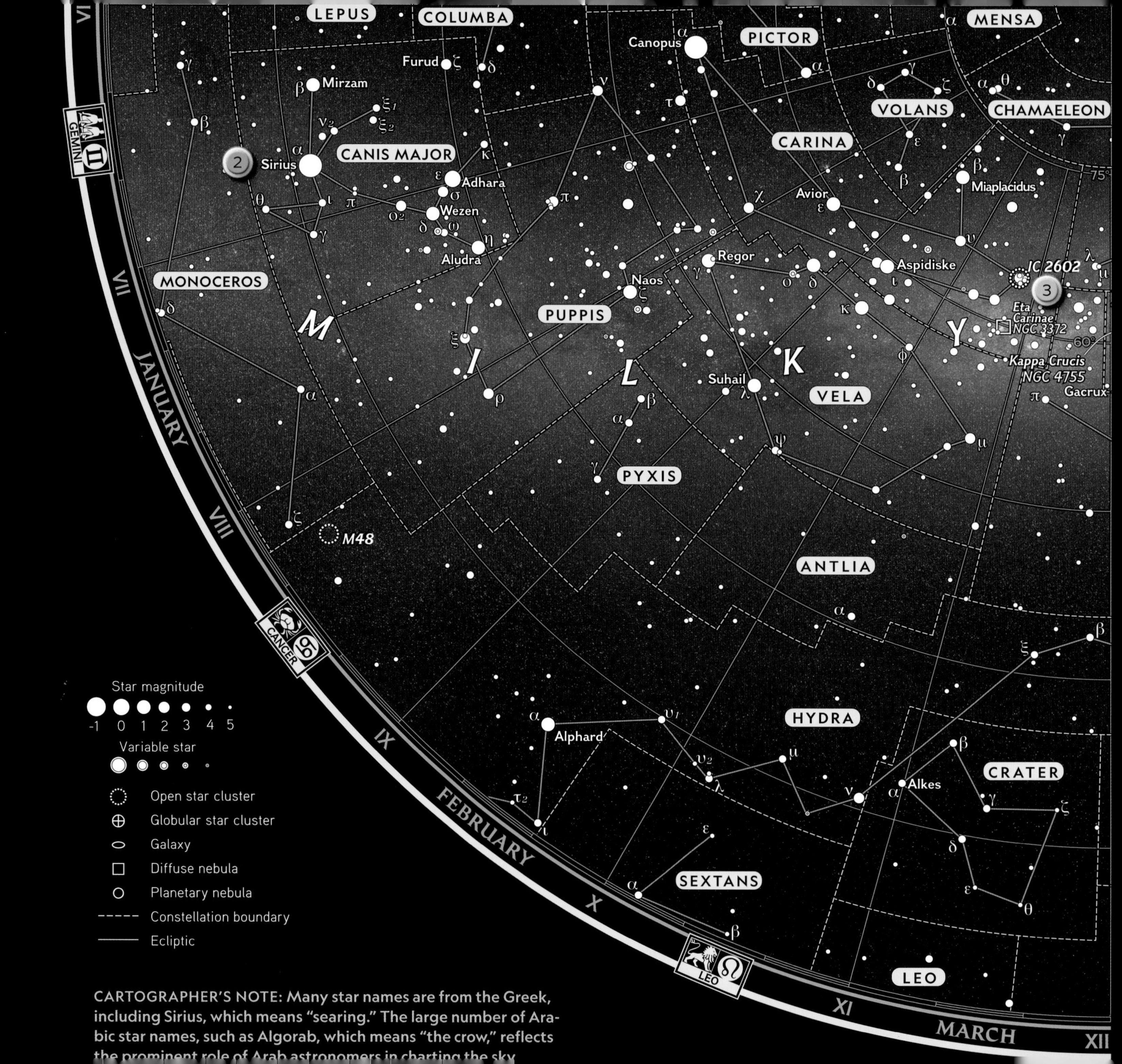

CARTOGRAPHER'S NOTE: Many star names are from the Greek, including Sirius, which means "searing." The large number of Arabic star names, such as Algorab, which means "the crow," reflects the prominent role of Arab astronomers in charting the sky

KEY FEATURES

1. **SOUTHERN CROSS (CRUX):** A prominent constellation visible only from the southern hemisphere
2. **SIRIUS:** Brightest star in the sky (after the sun)
3. **ETA CARINAE:** An unstable massive star, likely to end in a supernova

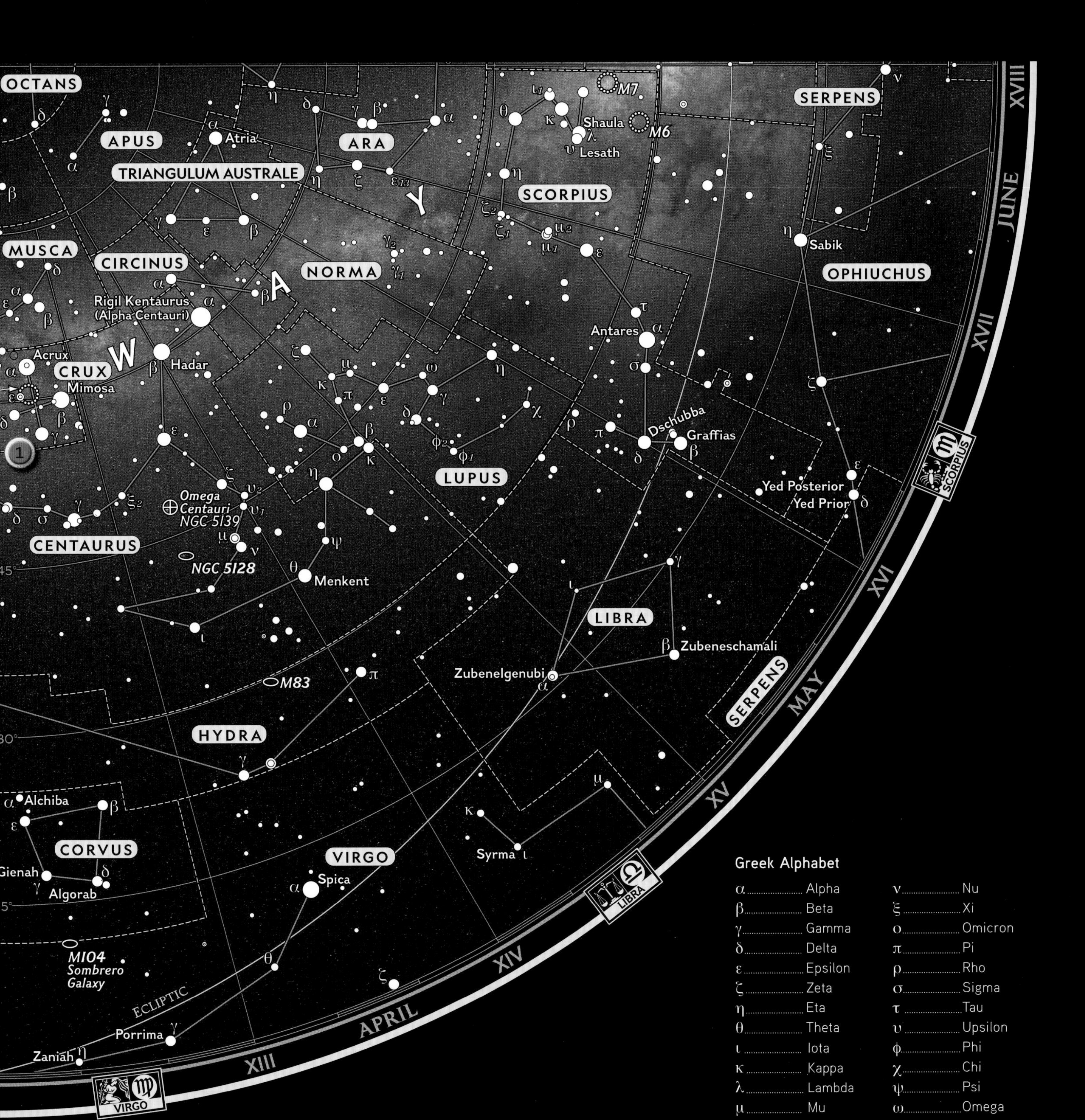

Greek Alphabet

α	Alpha	ν	Nu
β	Beta	ξ	Xi
γ	Gamma	ο	Omicron
δ	Delta	π	Pi
ε	Epsilon	ρ	Rho
ζ	Zeta	σ	Sigma
η	Eta	τ	Tau
θ	Theta	υ	Upsilon
ι	Iota	φ	Phi
κ	Kappa	χ	Chi
λ	Lambda	ψ	Psi
μ	Mu	ω	Omega

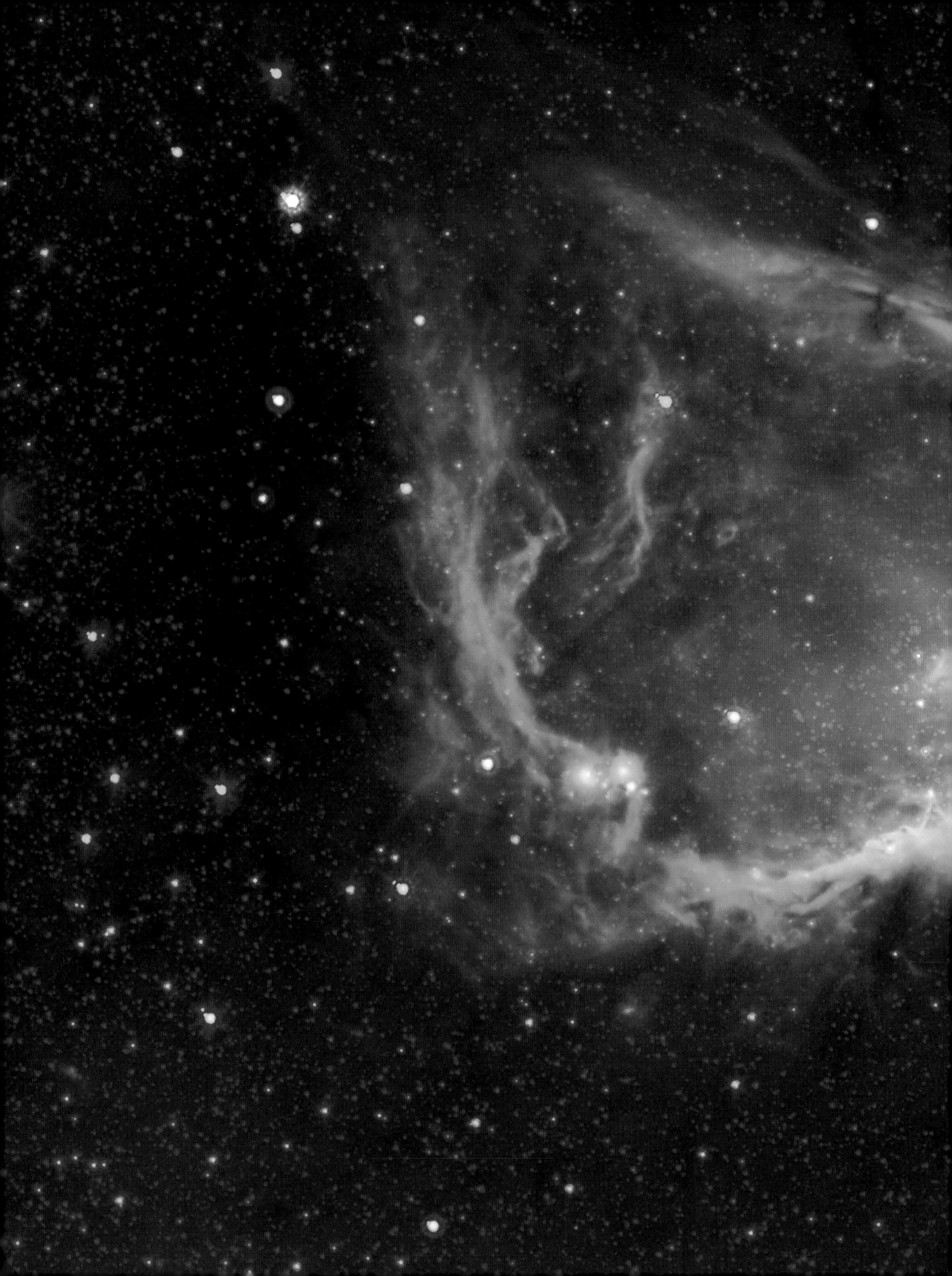

SPACE

Glowing in infrared light, the nebula RCW 120 is illuminated by the radiation from two supermassive stars in its center (not visible in this image). The stars' light and solar winds heat dust inside the nebula (colored red), as well as tiny dust grains at its edges (colored green).

The Chinese say that a journey of a thousand miles begins with a single step. It is fitting, then, that we begin our exploration of the universe in our own astronomical backyard—our own solar system. In this section you will find amazing maps of our nearest neighbors, with a level of detail that would have been unimaginable a generation ago. They are the result of a new method of exploration: the space probe. Every planet in our system has been visited by one or more spacecraft, some of them actually landing on the planet (as on Mars and Venus) and some simply sending back images (as for Jupiter and Saturn). We have explored not only the planets but also their moons. We have come to realize that every world in our system has its own unique

THE SOLAR

story to tell. The old notion that Mars and Venus are places where life could have developed (and might possibly still exist) has been replaced by a focus on the frigid but possibly life-sustaining moons of Jupiter and Saturn, and this change is reflected in the amount of attention this section gives to these moons.

Finally, modern research has extended our idea of the limits of the solar system beyond the orbit of Pluto to what are called the Kuiper belt and Oort cloud. Planet-size worlds have been discovered out there, and we are now seeing the inner planets as just a small part of the entire system. The well-publicized "demotion" of Pluto is a result of this new way of looking at our home system.

SYSTEM

I SOLAR SYSTEM

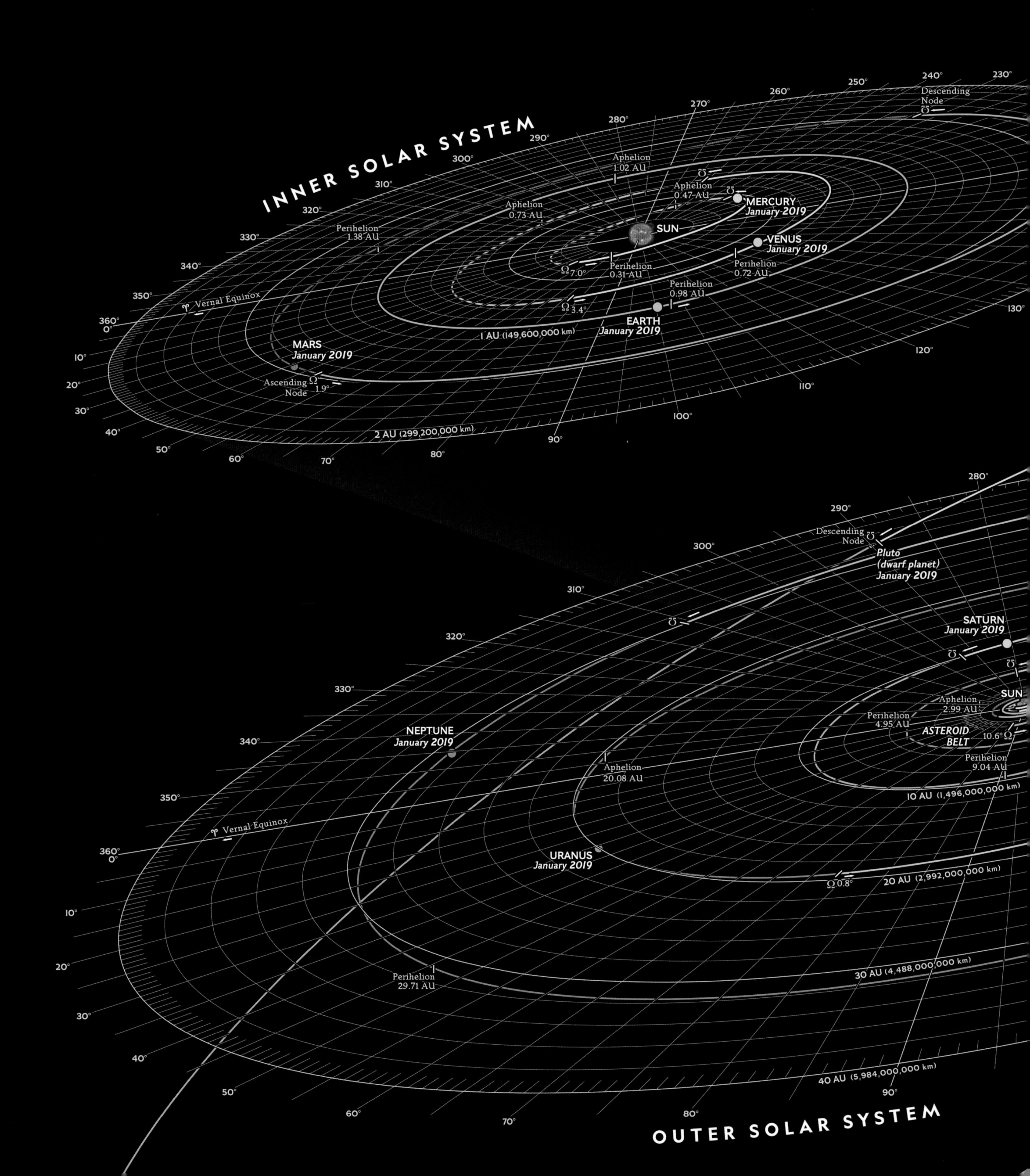

Eight planets, five or more dwarf planets, well over a hundred moons, and countless asteroids and comets orbit our massive sun. The four terrestrial planets form a relatively compact family in the inner solar system. Across the asteroid belt, the large, gaseous outer planets grow increasingly remote from the sun's warmth. All the planets orbit on roughly the same plane as Earth, known as the ecliptic. Pluto, now designated a dwarf planet, is an exception.

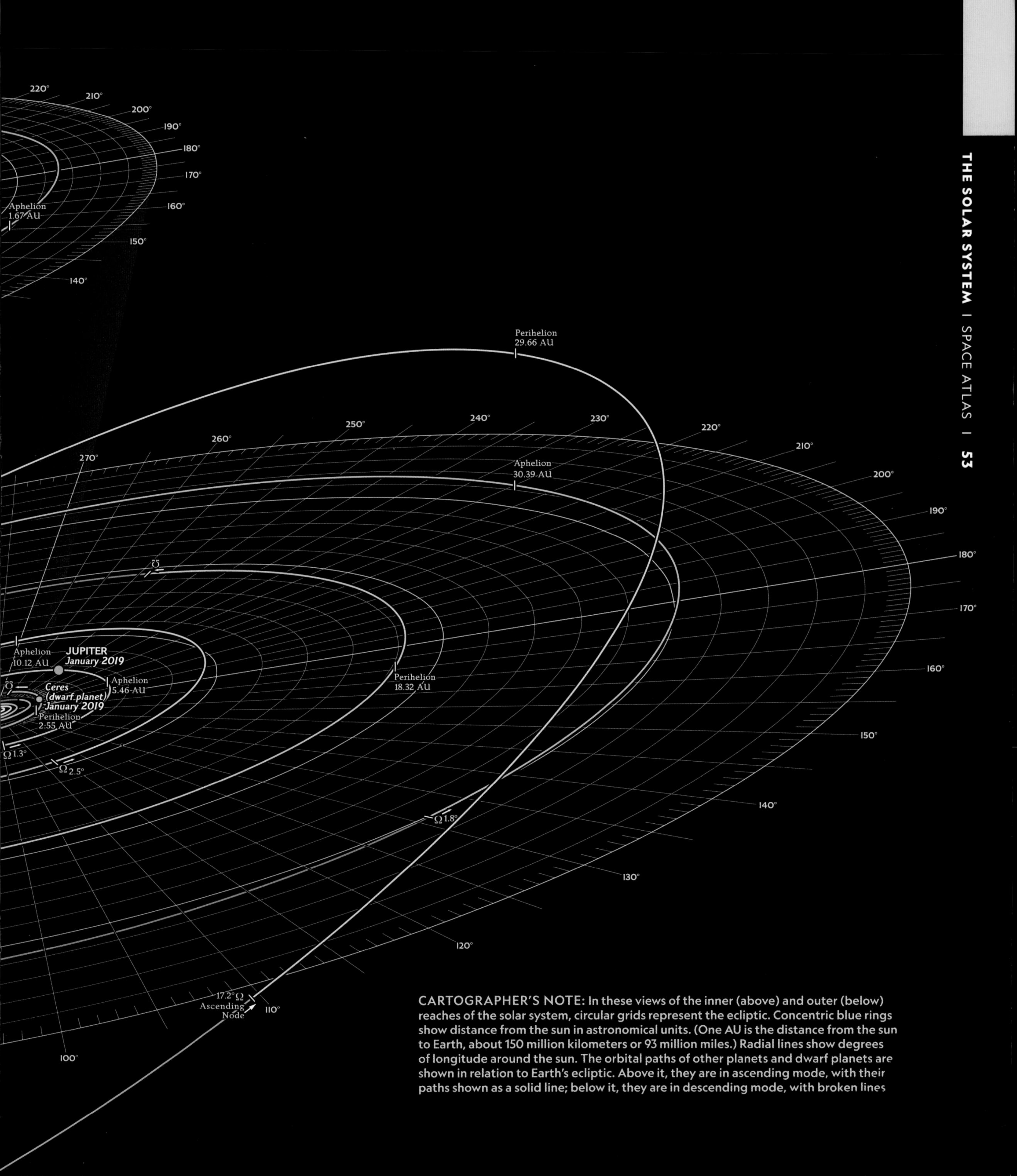

CARTOGRAPHER'S NOTE: In these views of the inner (above) and outer (below) reaches of the solar system, circular grids represent the ecliptic. Concentric blue rings show distance from the sun in astronomical units. (One AU is the distance from the sun to Earth, about 150 million kilometers or 93 million miles.) Radial lines show degrees of longitude around the sun. The orbital paths of other planets and dwarf planets are shown in relation to Earth's ecliptic. Above it, they are in ascending mode, with their paths shown as a solid line; below it, they are in descending mode, with broken lines

It all started about 4.5 billion years ago with a huge interstellar cloud floating in space. Today we see an orderly collection of planets circling a rather ordinary star. The question: How did we get here from there? • Looking at our familiar solar system, we can see regularities that give us hints about the answer to this question: First, all the planets orbit in the same plane; second, all the planets orbit in the same direction; and third, the planets closest to the sun are small and rocky, whereas those farther out are gas giants. The explanation for these regularities—and many others—begins with work by scientists in the 18th century, most notably the French physicist Pierre-Simon Laplace (1749–1827).

FORMATION

BIRTH OF THE SOLAR SYSTEM

AGE: **4.5-4.6 BILLION YEARS**
DISTANCE FROM CENTER OF MILKY WAY: **28,000 LIGHT-YEARS**
TYPE OF STAR: **MAIN SEQUENCE, G2-V**
MAIN ELEMENTS: **HYDROGEN, HELIUM, OXYGEN, CARBON, NITROGEN**

NUMBER OF PLANETS: **8**
TERRESTRIAL PLANETS: **MERCURY, VENUS, EARTH, MARS**
GAS AND ICE GIANTS: **JUPITER, SATURN, URANUS, NEPTUNE**
NUMBER OF DWARF PLANETS: **5 OR MORE**
NUMBER OF MOONS: **169**
DISTANCE FROM SUN TO NEPTUNE'S ORBIT: **30 AU**
DISTANCE FROM SUN TO EDGE OF OORT CLOUD: **~100,000 AU**

Artist's conception of a planetary system forming around the star Epsilon Eridani. (Inset) Art of young solar system.

Laplace reasoned that the ordinary laws of gravity would produce something like our solar system if they operated on diffuse interstellar clouds—the objects in the sky that scientists at the time christened nebulae (singular: nebula), Latin for "clouds." Easily observed with telescopes, these cloudy patches of light can be seen everywhere in the night sky, and Laplace's theory of how they could evolve into a solar system was duly named the nebular hypothesis. With many details filled in, it is essentially our modern theory.

Our understanding of how the solar system formed begins with an examination of its interstellar cloud. Like the others of its kind, it was made up primarily of the primordial gases created in the big bang—hydrogen and helium—with a small admixture of heavier elements produced in stars (see pages 304–11). Modern work indicates that one or more large stars exploded in our interstellar cloud, producing regions where the mass was more concentrated. These mass concentrations exerted a strong gravitational force, pulling in surrounding material. Eventually, the cloud (originally some 10s of light-years across) began to break up and collapse around the places where these mass concentrations started, and one of those accretions of matter, which we now call the presolar nebula, eventually became our own solar system. As the gases collapsed, the nebula began to rotate. Laplace's picture of how this nebula became the solar system depicts an orderly, rather placid process. As we shall see, notions about how orderly the process actually was have changed drastically in the last few years.

SUN AND FROST

Of course, gravity never quits, and it continued its work once the presolar nebula formed. Two important things happened as the inward collapse progressed: First, most of the mass of the presolar nebula became concentrated at the center, where it eventually became the star we call the sun. Second, as the cloud contracted, its spin increased, much as a skater's spin increases when she pulls in her arms. The various forces acting on the spinning, contracting cloud—gravity, pressure, centrifugal force, and even magnetism—caused the small amount of material that hadn't been taken into the nascent sun to flatten out into a rotating disk surrounding the central

CLUES TO EARTH'S FORMATION

People often ask how scientists can know about events like the formation of Earth, which took place billions of years ago. Let's look at one part of that story—the differentiation of the planet into core, mantle, and crust described on page 88. Unraveling this process is a fascinating scientific detective story.

The story starts in the dust cloud from which the solar system formed. In that cloud were a certain number of hafnium-182 atoms. (Hafnium is a relatively rare material, usually seen as a silvery gray metal.) The nuclei of these atoms are unstable, decaying with a half-life of about nine million years into atoms of tungsten-182, which is stable. (Tungsten is the metal customarily used as the filament in incandescent lightbulbs.) What makes these atoms interesting is that hafnium is chemically attracted to the kinds of materials found in Earth's mantle, while tungsten is attracted to the iron and nickel found in the core. This means that if the iron materials sank into the core quickly, before the hafnium had time to decay, most of the tungsten-182 would be found in the mantle. If, on the other hand, the differentiation happened after the hafnium had all decayed, the tungsten-182 would be in the core.

By comparing the amount of tungsten-182 in mantle rocks to the amount found in meteorites (which never underwent differentiation), scientists have concluded that Earth's core formed about 30 million years after the solar system started to condense from its gas cloud.

One small piece of the puzzle in place

sphere. With the formation of this disk, the solar system was starting to take shape.

As the disk was forming, the sun was also firing up, heating nearby material. Out to a place between the present orbits of Mars and Jupiter, the temperature was hot enough that volatile materials like water couldn't exist in solid form. Beyond this boundary, which astronomers refer to loosely as the "frost line," these materials remained as solid ices. Thus, the building blocks available for the inner planets were different from those available for planets beyond the frost line. It's no wonder, then, that the so-called terrestrial (Earthlike) planets near the sun are different from the Jovian (Jupiter-like) planets farther out.

Our main tools for understanding what happened from that point on are massive computer simulations of the early solar system. The descriptions that follow are largely a summary of these calculations.

TERRESTRIAL PLANETS

Let's start with the terrestrial planets. Because the light-weight, volatile materials had mostly been removed from the inner disk, these planets formed primarily from materials with high melting points (think iron, nickel, and the rocky compounds of silicon). As they circled the sun, grains of these materials collided and stuck together, eventually forming boulder-size objects that then aggregated into mountain-size bodies called planetesimals. It is these bodies that eventually came together to form the planets.

Until the last decade of the 20th century, it was assumed that the planets formed in pretty much their present orbits and states. This scenario is not, however, how the computers tell us things happened. In fact, the picture they give us is truly staggering. The end result of the process described above was an inner solar system with dozens of moon-size planetary embryos zipping around. What followed was an impossible game of planetary billiards, with the embryos colliding, melding, breaking apart, and, occasionally, getting kicked out of the solar system completely. By the time the game of planetary billiards was over, the inner solar system was left with the four planets—Mercury, Venus, Earth, and Mars—that we see today.

PLANET-BUILDING DUST AROUND A YOUNG STAR

THE SOLAR SYSTEM GREW OVER

50

MILLION YEARS

GIANT PLANETS

Meanwhile, a different scenario was playing out beyond the frost line. Because there was so much undisturbed material out there, the planetesimals grew more quickly and to a larger size for the terrestrial planets. The great mass of these bodies allowed them to start capturing the abundant hydrogen and helium around them. These planets are known as gas giants, particularly Jupiter and Saturn, and they are the largest planets in our system.

The subsequent story of the outer planets is somewhat more complicated than it is for the terrestrials. The gas giants Jupiter and Saturn formed quickly, as described above. Apparently the next two planets—Uranus and Neptune—formed later and much closer to the sun than they are now. They also formed at a time when the sun was emitting huge streams of particles into space, streams that blew much of the primordial hydrogen and helium out of the solar system. Consequently, these two planets wound up being smaller and different in composition from Jupiter and Saturn. In fact, they are often referred to as ice giants rather than gas giants to emphasize this difference.

The four giant planets, as well as all the leftover planetesimals and other materials, continued in their orbits, interacting with each other through the force of gravity. Our models tell us that Jupiter formed at the outer edge of what is now the asteroid belt. A series of complex gravitational interactions between Jupiter, Saturn, and the remaining material in the planetary disk set in motion a series of events that astronomers call the Grand Tack. (The name refers to the change of directions a sailing ship takes when it moves into the wind).

An illustration depicts the violent formation of Earth during the solar system's early years, as the inner planets were bombarded by countless planetesimals and heated by the collisions.

The Grand Tack began with Jupiter moving inward to a spot between the present orbits of Mars and Earth. During this journey, the nascent planet scattered material from the planetary disk, kicking some material out of the solar system and pushing some into the sun. At this point, the gravitational interaction between Jupiter and Saturn (whose orbit has also moved inward) reversed the direction of the giant planets and moved them outward to their present positions.

Another result of these maneuvers was that Neptune's orbit was pushed outward, sending it careering into the remnants of the protoplanetary disk like a bowling ball among tenpins. At that time, the disk extended to about the present orbit of Uranus, but by the time the planetary migrations were over, the system was clear out to well past the present orbit of Pluto.

The Grand Tack explains several features of the inner solar system. For example, the fact that so much material from the planetary disk was removed explains why Mars is so much smaller than Earth and Venus—its supply of building material was simply removed. The argument can also be used to explain why there is so little material left in the asteroid belt today.

As a result of all these moves, there was a period of a couple of hundred million years when every body in the inner solar system suffered intense collisions. Scientists have taken to calling this period the Late Heavy Bombardment. Its scars can be seen in the craters that still survive on airless worlds like Mercury and the moon.

It has become clear to astronomers over the last few decades that the early life of the solar system was far from the placid, orderly collapse that Laplace had in mind in the 18th century. But once the initial fireworks died down, the solar system became a much more orderly and predictable kind of place—just the sort of place we need in order to begin our tour of the first of our "universes."

I INNER PLANETS

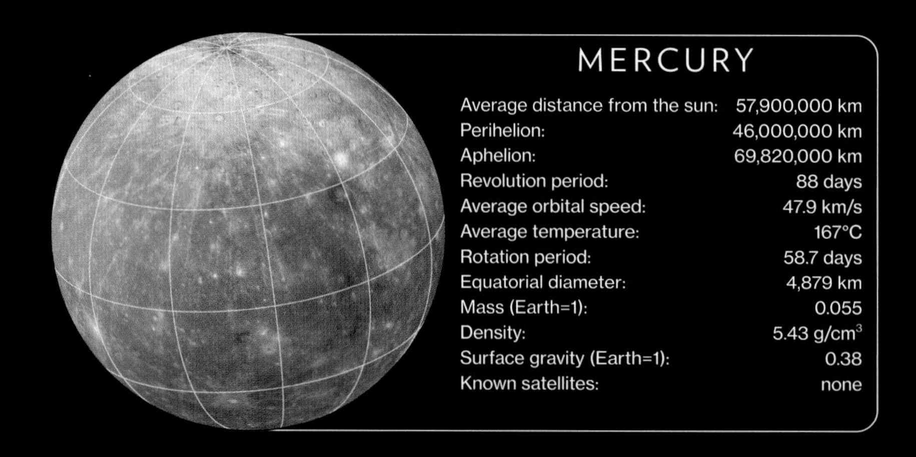

MERCURY

Average distance from the sun:	57,900,000 km
Perihelion:	46,000,000 km
Aphelion:	69,820,000 km
Revolution period:	88 days
Average orbital speed:	47.9 km/s
Average temperature:	167°C
Rotation period:	58.7 days
Equatorial diameter:	4,879 km
Mass (Earth=1):	0.055
Density:	5.43 g/cm^3
Surface gravity (Earth=1):	0.38
Known satellites:	none

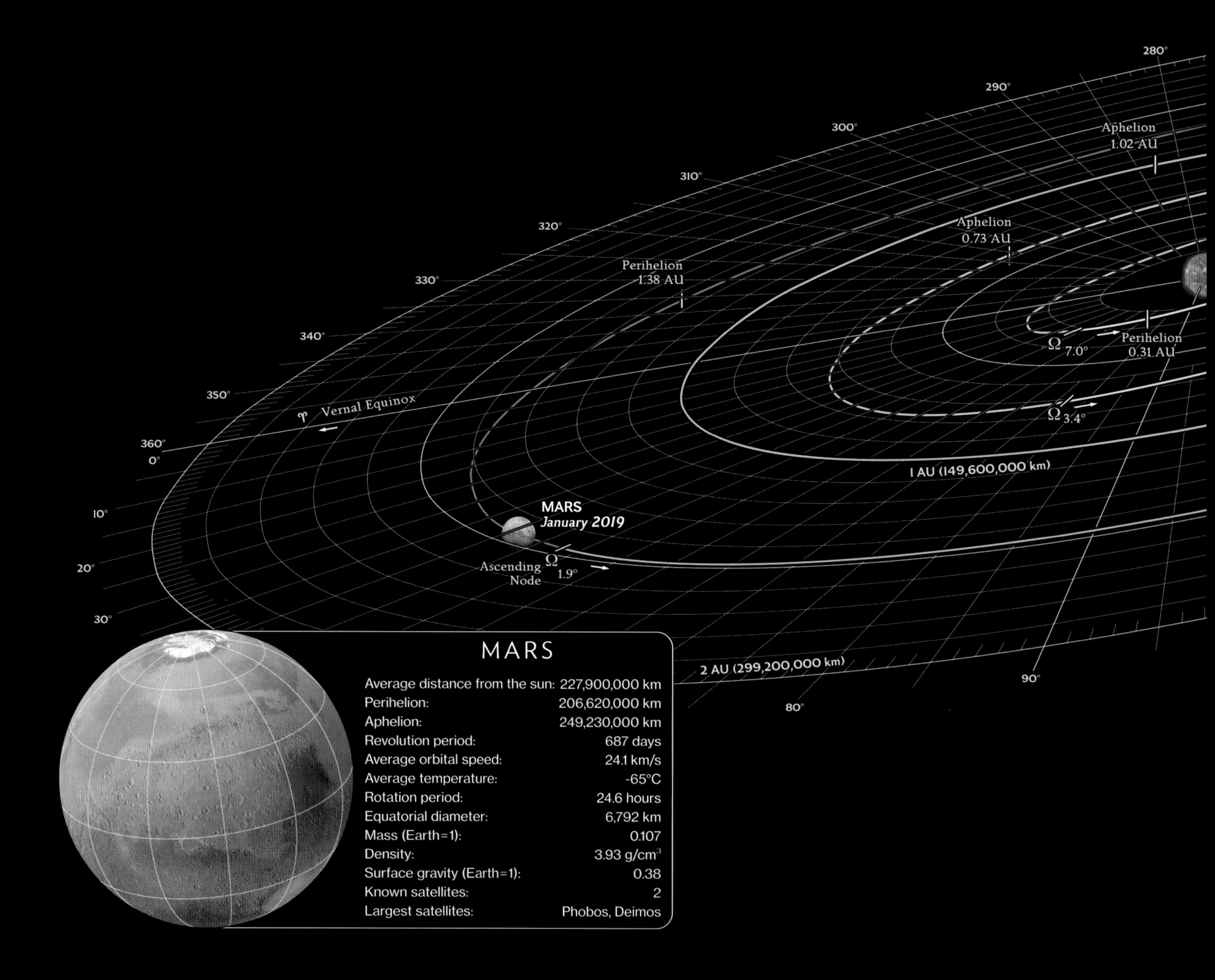

MARS

Average distance from the sun:	227,900,000 km
Perihelion:	206,620,000 km
Aphelion:	249,230,000 km
Revolution period:	687 days
Average orbital speed:	24.1 km/s
Average temperature:	-65°C
Rotation period:	24.6 hours
Equatorial diameter:	6,792 km
Mass (Earth=1):	0.107
Density:	3.93 g/cm^3
Surface gravity (Earth=1):	0.38
Known satellites:	2
Largest satellites:	Phobos, Deimos

The terrestrial planets, small and rocky, form the inner core of the solar system. All possess a secondary atmosphere (appearing after their formation), though Mercury's is barely detectable. Craters from the era of early bombardment mark Mercury and Earth's moon; atmospheric weathering and the effects of volcanism or plate tectonics have erased many of these early traces from Venus and Earth.

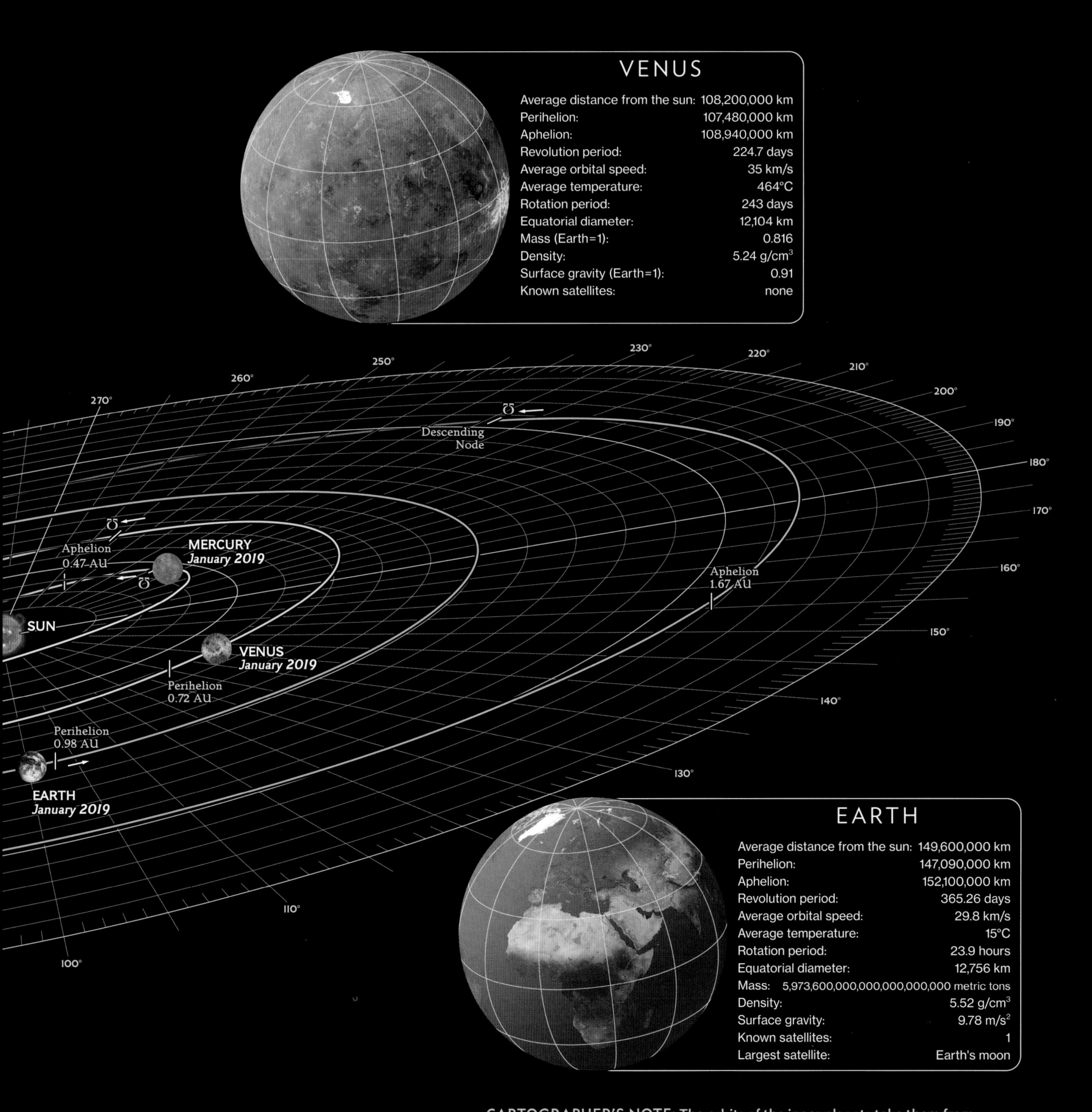

VENUS

Average distance from the sun:	108,200,000 km
Perihelion:	107,480,000 km
Aphelion:	108,940,000 km
Revolution period:	224.7 days
Average orbital speed:	35 km/s
Average temperature:	464°C
Rotation period:	243 days
Equatorial diameter:	12,104 km
Mass (Earth=1):	0.816
Density:	5.24 g/cm^3
Surface gravity (Earth=1):	0.91
Known satellites:	none

EARTH

Average distance from the sun:	149,600,000 km
Perihelion:	147,090,000 km
Aphelion:	152,100,000 km
Revolution period:	365.26 days
Average orbital speed:	29.8 km/s
Average temperature:	15°C
Rotation period:	23.9 hours
Equatorial diameter:	12,756 km
Mass:	5,973,600,000,000,000,000,000 metric tons
Density:	5.52 g/cm^3
Surface gravity:	9.78 m/s^2
Known satellites:	1
Largest satellite:	Earth's moon

CARTOGRAPHER'S NOTE: The orbits of the inner planets take them from the scorching temperatures of Mercury to the deep winter chill of Mars. Swift Mercury races around the sun, its hemispheres burning and freezing. Torrid Venus bakes under an atmosphere that holds in most of the sun's energy, the greenhouse effect writ large. Uniquely sited, Earth is in the habitable zone, where water can exist as a liquid. Frigid Mars offers tantalizing evidence of a warmer, wetter past. Well studied in the space age, each terrestrial planet has been visited and mapped in detail by spacecraft.

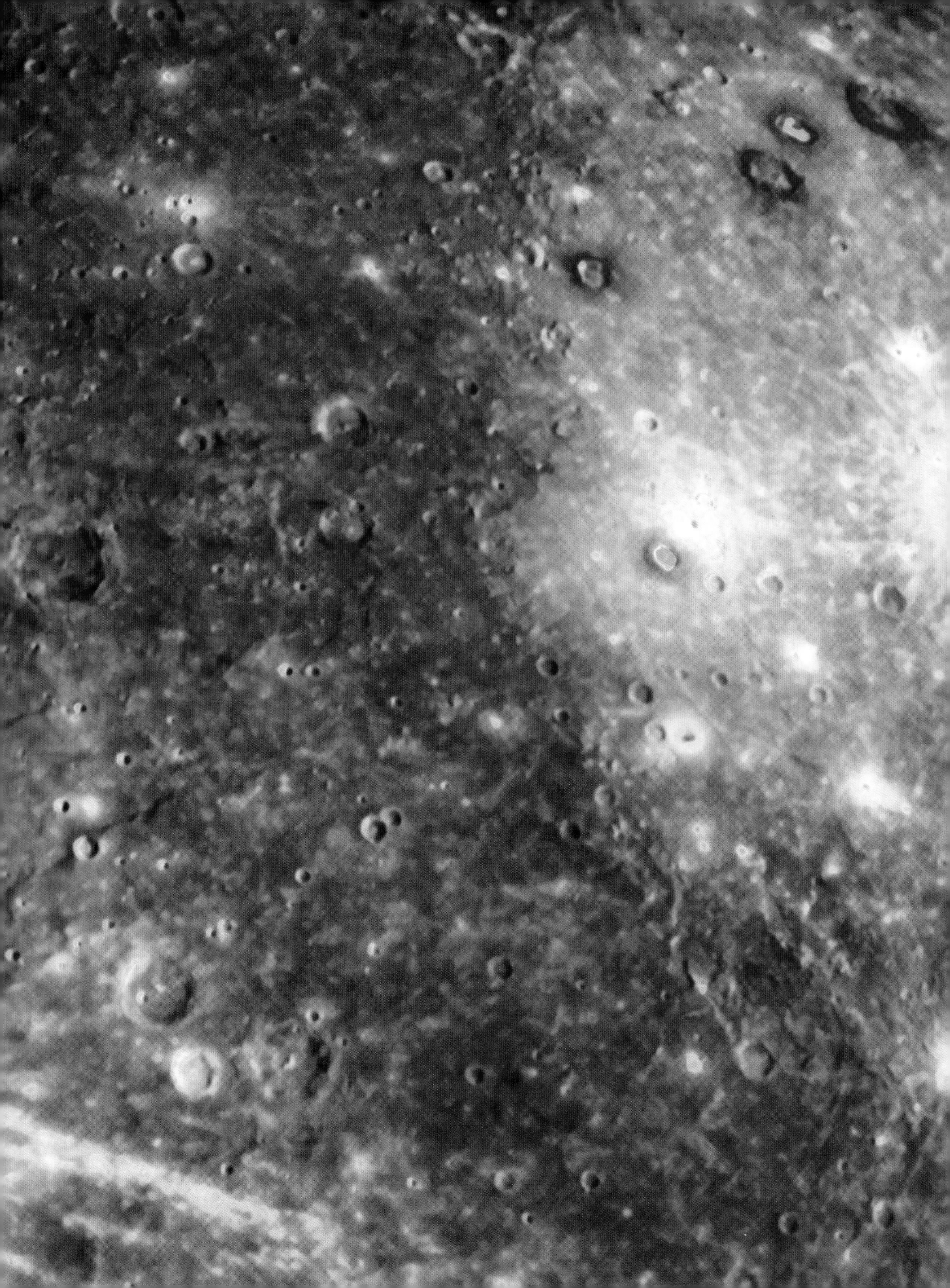

Mercury is the closest planet to the sun, which means that observers on Earth will always see it near the sun in the sky. During the day, light from the planet is overwhelmed by light from the sun, so we can see it only at dawn or sunset when the sun is safely below the horizon. The planet can never appear high in the night sky, for the simple reason that in that situation it will always be on the other side of Earth from the observer. Like Venus, then, Mercury always appears as a morning or evening "star" to observers. Records of naked-eye sightings of Mercury by Assyrian astronomers go back to the 14th century B.C., and by the 4th century B.C. the Greeks had realized that their morning and evening stars were actually a single body.

MERCURY

| SWIFT, SUN-BLASTED WORLD |

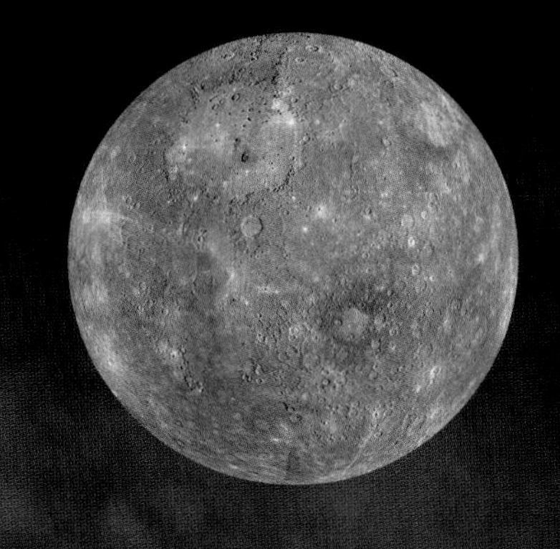

DISCOVERER: **UNKNOWN**
DISCOVERY DATE: **PREHISTORIC**
NAMED FOR: **ROMAN MESSENGER GOD**

MASS: **0.06 × EARTH'S**
VOLUME: **0.06 × EARTH'S**
MEAN RADIUS: **2,440 KM (1,516 MI)**
MIN./MAX. TEMPERATURE: **−173/427°C (−279/801°F)**
LENGTH OF DAY: **58.65 EARTH DAYS**
LENGTH OF YEAR: **87.97 EARTH DAYS**
NUMBER OF MOONS: **0**
PLANETARY RING SYSTEM: **NO**

False-color image of Caloris Basin on Mercury's surface. (Inset) Mercury.

I MERCURY
CALORIS ANTIPODE

Mercury is heavily cratered, but at least 40 percent of the surface consists of smooth plains, hinting at a volcanic past.

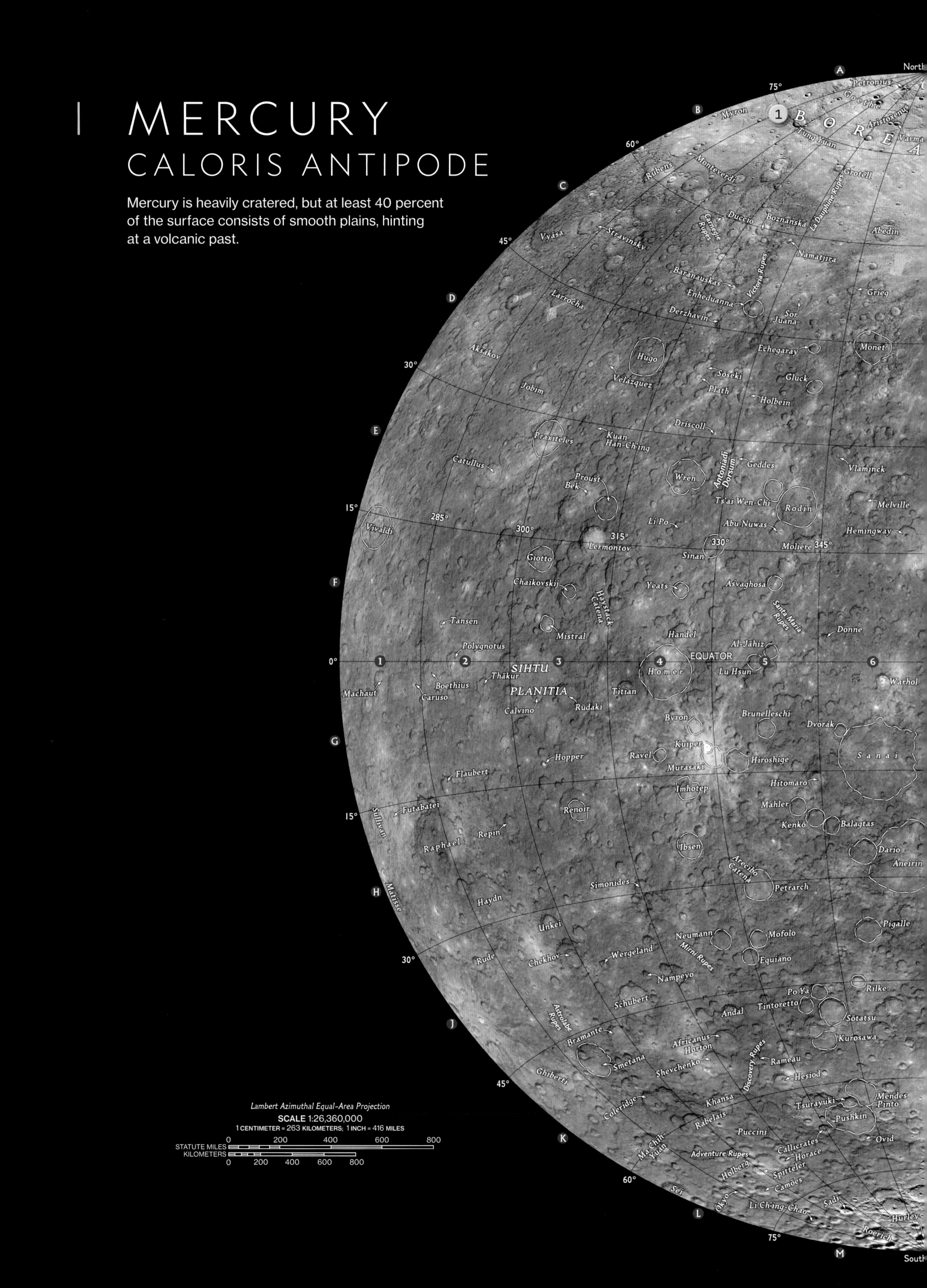

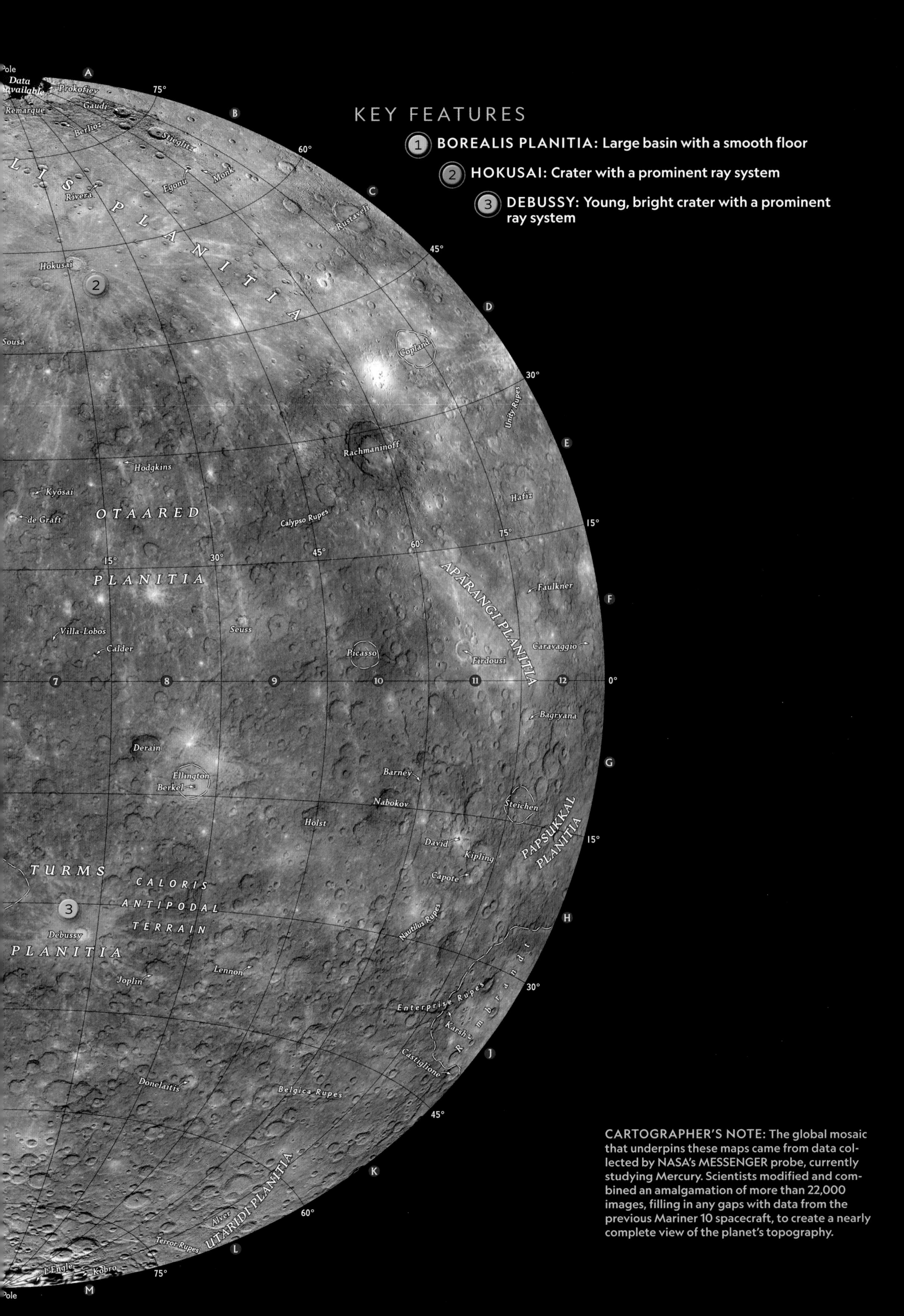

KEY FEATURES

1. **BOREALIS PLANITIA: Large basin with a smooth floor**
2. **HOKUSAI: Crater with a prominent ray system**
3. **DEBUSSY: Young, bright crater with a prominent ray system**

CARTOGRAPHER'S NOTE: The global mosaic that underpins these maps came from data collected by NASA's MESSENGER probe, currently studying Mercury. Scientists modified and combined an amalgamation of more than 22,000 images, filling in any gaps with data from the previous Mariner 10 spacecraft, to create a nearly complete view of the planet's topography.

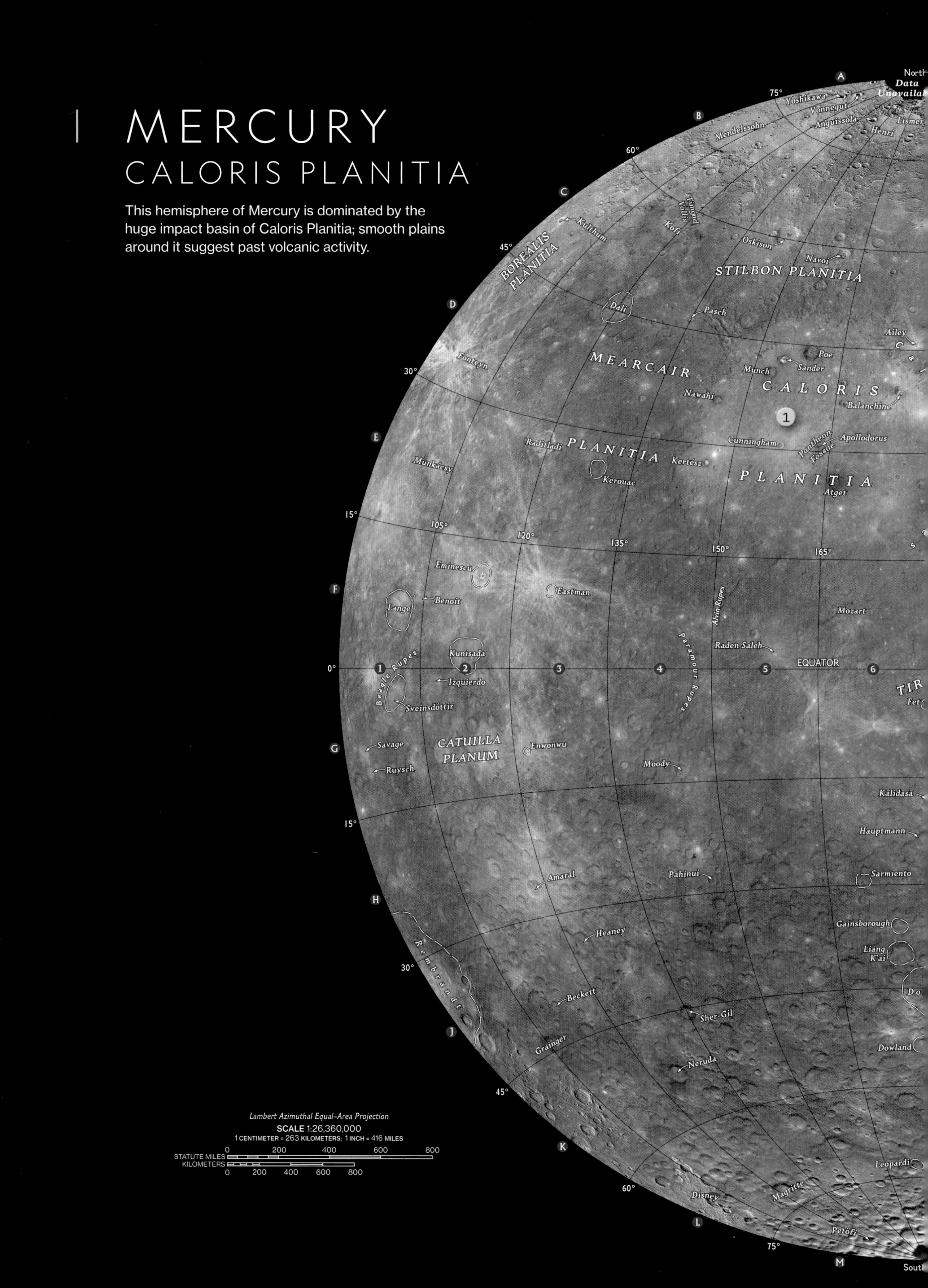

MERCURY
CALORIS PLANITIA
This hemisphere of Mercury is dominated by the huge impact basin of Caloris Planitia; smooth plains around it suggest past volcanic activity.
North
Data Unavailable
Yoshikawa
Vonnegut
Anguissola
Henri
Lismer
Mendelssohn
Tingzud Vallis
Kofi
Kulthum
BOREALIS PLANITIA
Oskison
Navoi
STILBON PLANITIA
Dali
Pasch
Aitey
Poe
Sander
Munch
Fonteyn
MEARCAIR
Nawahi
CALORIS
Balanchine
Cunningham
Pantheon Fossae
Apollodorus
Raditladi
PLANITIA
Kertész
Kerouac
Munkácsy
PLANITIA
Atget
EQUATOR
Eminescu
Eastman
Lange
Benoit
Alvin Rupes
Mozart
Paramour Rupes
Raden Saleh
Kunisada
Beagle Rupes
Izquierdo
Sveinsdóttir
Fet
Savage
CATUILLA PLANUM
Enwonwu
Ruysch
Moody
Kālidāsā
Hauptmann
Amaral
Pahinui
Sarmiento
Gainsborough
Liang K'ai
Heaney
Rembrandt
Beckett
Sher-Gil
Grainger
Dowland
Neruda
Leopardi
Disney
Magritte
Petőfi
South
75°
60°
45°
30°
15°
0°
15°
30°
45°
60°
75°
105°
120°
135°
150°
165°
A
B
C
D
E
F
G
H
J
K
L
M
1
2
3
4
5
6
Lambert Azimuthal Equal-Area Projection
SCALE 1:26,360,000
1 CENTIMETER = 263 KILOMETERS; 1 INCH = 416 MILES
STATUTE MILES 0 200 400 600 800
KILOMETERS 0 200 400 600 800

KEY FEATURES

1. **CALORIS PLANITIA: The plains of Caloris Basin, one of the largest craters in the solar system**
2. **BEETHOVEN: An old impact basin whose rim is covered by ejected material**
3. **HERO RUPES: An escarpment 300 kilometers (190 mi) long**

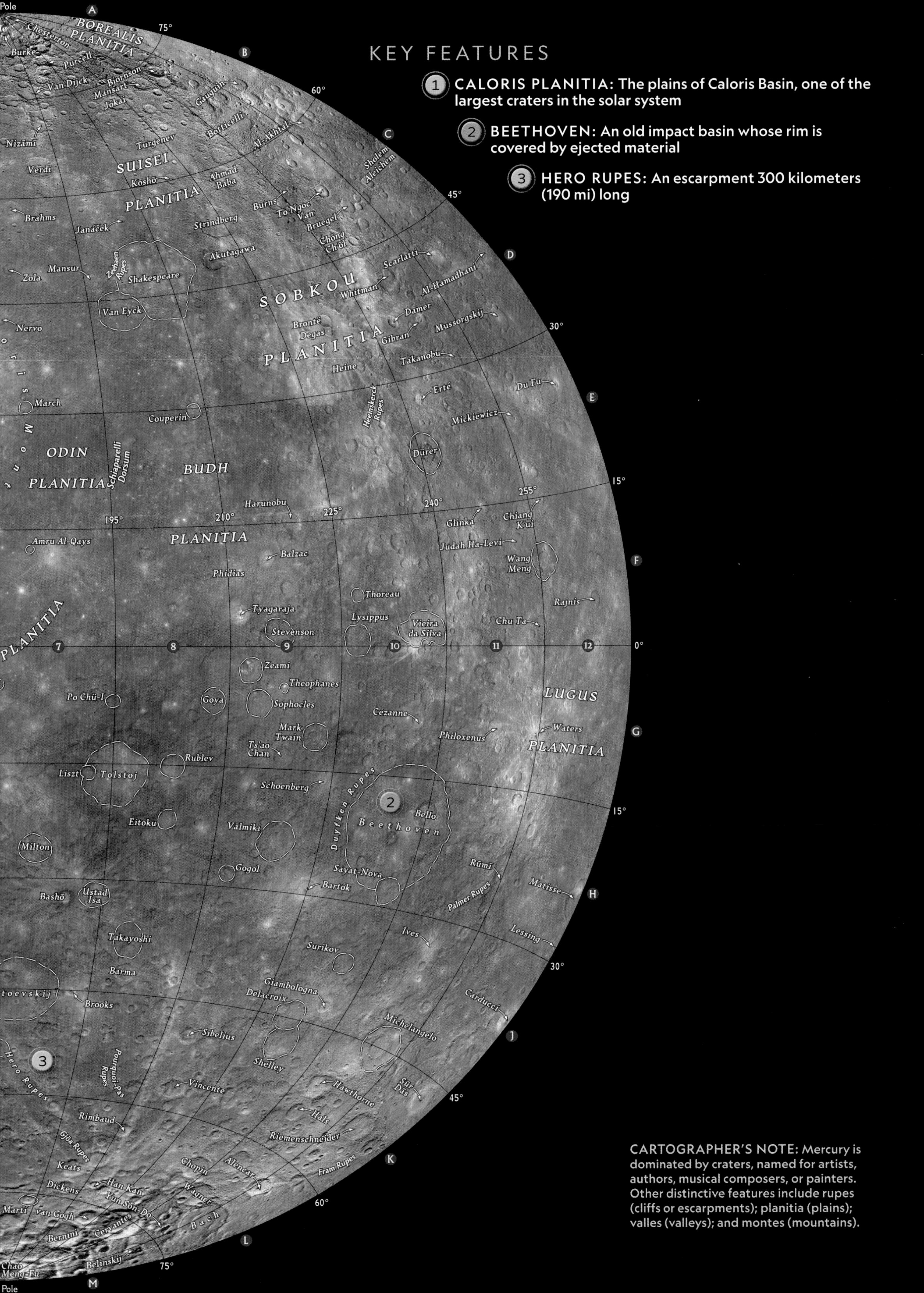

CARTOGRAPHER'S NOTE: Mercury is dominated by craters, named for artists, authors, musical composers, or painters. Other distinctive features include rupes (cliffs or escarpments); planitia (plains); valles (valleys); and montes (mountains).

The Romans gave Mercury its current title, naming it after the swift-footed messenger of the gods (presumably because its rapid motion in the sky imitated the god's swift flight). It is interesting that the Babylonians named the planet Nabu, after the messenger of the gods in their pantheon, probably for the same reason.

HOT AND COLD

Mercury is small—it has only about 5 percent the mass of Earth—and it has long since lost whatever atmosphere it ever had to space and the blistering radiation from the sun. Like our own moon, it is a dead world, with no geological activity to raise mountains or atmosphere to erode its surface. The main features of the planet, like those of the moon, are craters that bear mute testimony to long-ago impacts. The planet rotates on its axis every 176 Earth days and circles the sun every 88 days or so. Thus, every part of Mercury's surface is exposed to both the direct rays of the sun (during its day) and the cold of space (at night).

As you might expect for a planet so close to the sun, Mercury's surface temperatures can get quite hot—427°C (801°F) at the equator at high noon. This is hotter than the melting point of lead. What you might not expect is that it can also get quite cold: minus 173°C (–279°F) at midnight. The reason for this is that once a part of the planet moves into the night side, there is no atmosphere to act as a blanket and keep it warm. The energy accumulated during the day is quickly radiated into space and the temperature falls quickly. (To be exact, there is a thin mist of atoms above Mercury's surface—scientists call it an exosphere. It's made up of atoms boiled off the surface and eventually lost to space.)

CRATERED WORLD

Because the planet has no atmosphere, the craters on Mercury, like those on the moon, last for a long time. The largest of these, the Caloris Basin, is almost 1,600 kilometers (1,000 mi) across and must be the scar of a very large impact. In fact, on the exact opposite side of the planet from Caloris is a region of jumbled hills that goes by the descriptive name Weird Terrain. Some scientists think this area was created by a shock wave from the impact that created Caloris. Many of the larger craters on Mercury appear to have a smooth surface similar to the maria on the moon. It is thought that these smooth regions are caused by lava outflows, probably as a result of the impact itself. Between the craters are regions of rolling

A cutaway view shows Mercury's structure. The planet has a large, molten core surrounded by a solid mantle 500 to 700 kilometers (300–400 mi) thick. This, in turn, is covered by 100 to 300 kilometers (60–180 mi) of crust.

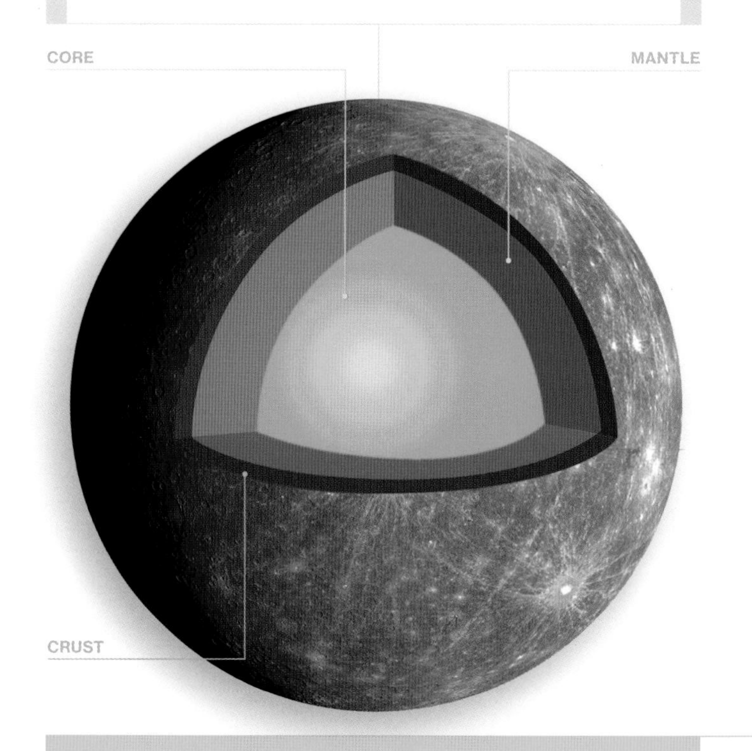

Half in sunlight, Mercury's south pole displays the planet's characteristic craters. Like the moon, the planet has almost no atmosphere, so its craters do not erode; thus, some of these structures are billions of years old.

MERCURY'S SURFACE IS

ANCIENT

AND HEAVILY CRUSTED

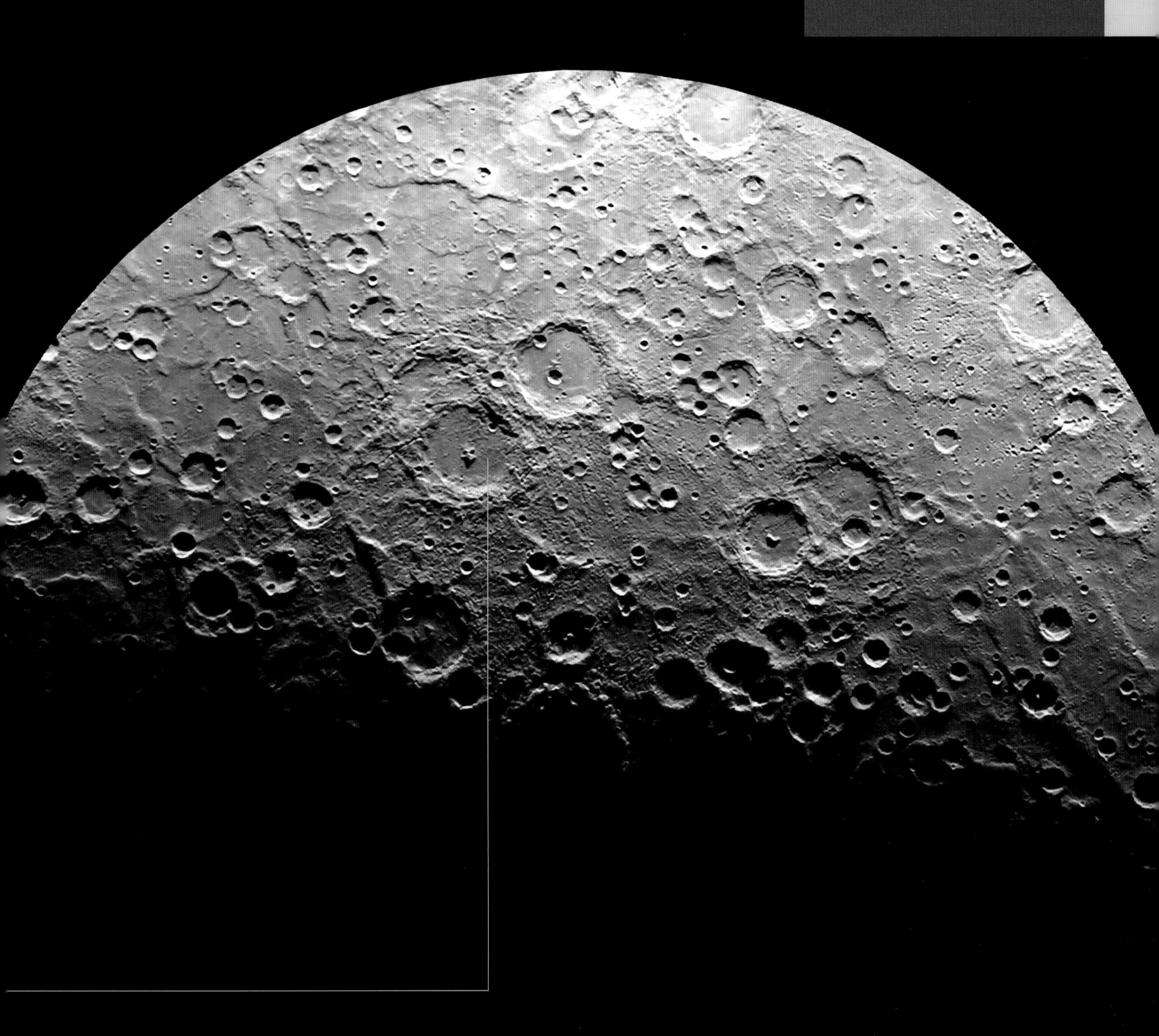

hills, representing Mercury's oldest surviving surface. These plains are crisscrossed with long ridges that may have been created by surface wrinkling as Mercury cooled (think of an apple shriveling as it dries out). Scientists believe that this shrinking process is still going on. In 2016 scientists examining photos taken by the MESSENGER spacecraft (see page 71) found an extraordinary canyon on Mercury's surface. Three kilometers (2 mi) deep, 400 kilometers (250 mi) across and 966 kilometers (600 mi) long, the canyon's geology lent further support to the "shriveled apple" theory.

Like the other terrestrial planets, Mercury is a rocky world. It has a small magnetic field, about one percent as strong as the one on Earth, which supports the idea that, like Earth, it has a core composed mainly of iron—you can think of the planet as a giant permanent magnet.

In fact, based largely on data gathered from the space probes described below, scientists believe that the planet has an unusually large iron core, constituting greater than 42 percent of its volume. Several hypotheses have been advanced to explain this unusual composition. The most popular explanation is that after it had

MERCURY'S WEIRD TERRAIN

Although Mercury is visible to the naked eye, only since the launch of the MESSENGER spacecraft have we been able to explore it in detail. Mercury is an airless world, so geological features, once formed, do not erode. This means that the history of the planet can still be read on its unchanging surface.

By far the most striking feature on Mercury's surface is the Caloris Basin, a crater almost 1,600 kilometers (1,000 mi) across. The collision that formed it is thought to have occurred about 3.8 billion years ago, at about the same time as the impacts that created the maria on Earth's moon. It is one of the largest craters in the solar system.

The power of that impact can be seen in the fact that the walls thrown up around the edges of the crater are over 1.6 kilometers (1 mi) high. More interesting, though, is a jumbled region of irregular hills on the opposite side of Mercury that scientists have given the name Weird Terrain. This region was clearly created by the impact that formed the Caloris Basin. There are two theories to explain how this happened. In one, seismic waves created by the impact traveled around the planet to converge at the antipode, breaking up the previously smooth surface. The other theory is that material thrown up by the impact traveled around the planet and fell back to the surface, creating the irregular overlay we see today.

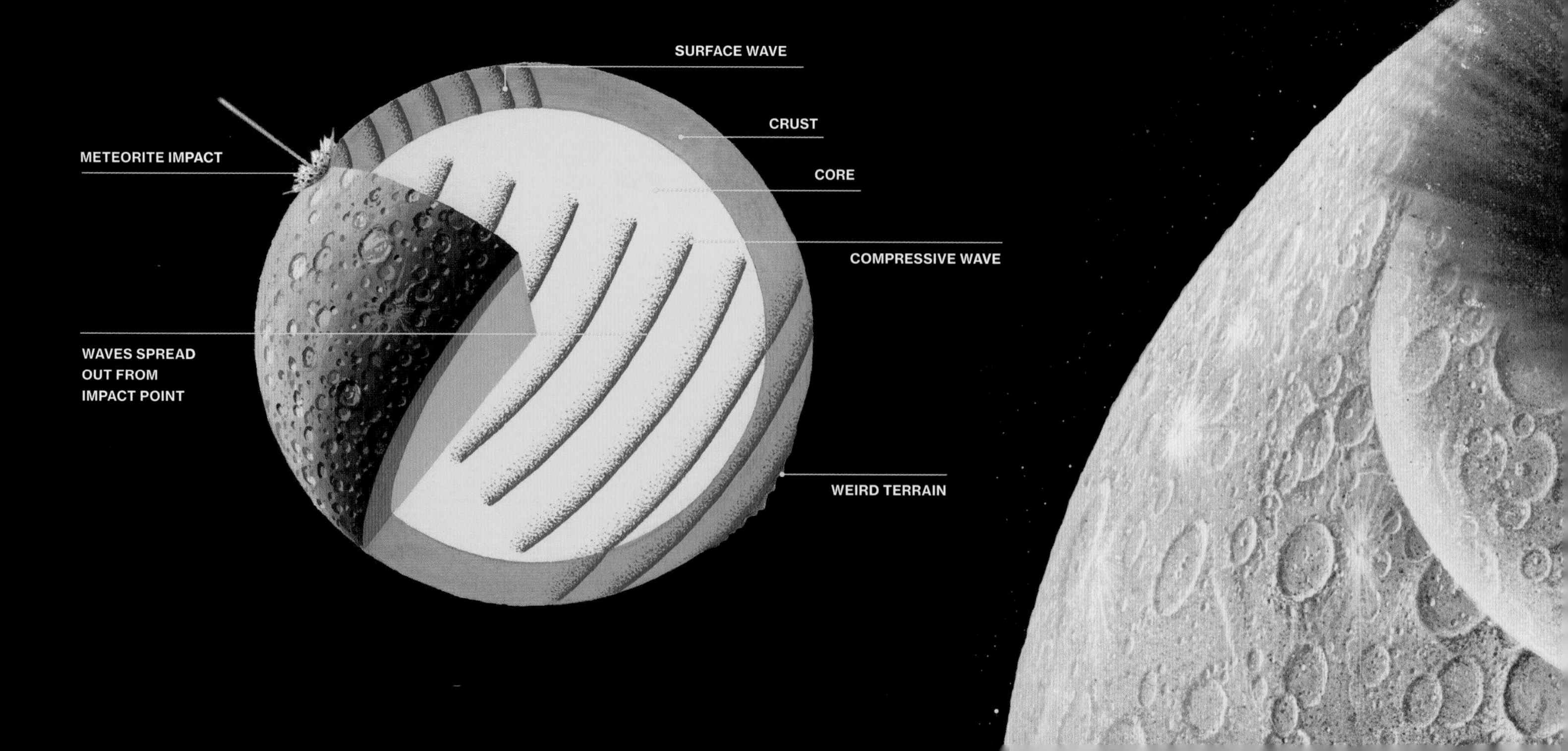

gone through the process of differentiation (see page 88), Mercury was struck by a large planetesimal during the Late Heavy Bombardment about four billion years ago. The collision blew off a lot of the lighter outer layer of the planet, leaving behind the iron in the core.

SPACECRAFT TO MERCURY

Because it is so close to the sun, observing Mercury with ground-based telescopes is difficult. In fact, most of our information about the details of the planet's structure comes from two space probes. The first of these, Mariner 10, arrived at the planet in 1974 and made three close approaches before it ran out of fuel. Most likely the craft is still in orbit around the sun, making undocumented close approaches to the planet as you read this.

The MESSENGER (MErcury Surface, Space ENvironment, GEochemistry, and Ranging) spacecraft was launched from Cape Canaveral in 2004, and, after visiting Venus and Earth, it made its first flyby of Mercury on January 14, 2008. On March 18, 2011, it went into orbit around the planet, providing detailed data on Mercury's surface and magnetic field and confirming the presence of ice in shaded polar craters, On April 30, 2015, after 4,105 orbits, MESSENGER, its fuel expended, was crashed into the planet's surface. A European Space Agency probe called Bepi Columbo is expected to arrive at Mercury in 2024 and will represent the next wave of exploration of the planet.

We can't leave the planet without mentioning what may be its greatest contribution to the advance of science. Like the other planets, Mercury orbits the sun in an elliptical path, with the point of closest approach being called the perihelion. Because of the gravitational pull of the other planets, scientists expected that each time the planet came around, the perihelion would shift a little—think of the ellipse rotating a little in space each time around the sun. In the late 19th century, calculations showed that there was more shift than could be explained by simple gravitational effects—about 43 seconds of arc per century, in fact. When Albert Einstein published the theory of general relativity in 1915, it turned out that he was able to explain exactly this much greater shift in Mercury's perihelion. Thus, the planet provided one of the initial tests of our best current theory of gravity.

A MASSIVE IMPACT CREATES CALORIS BASIN.

The second world from the sun, Venus has often been called Earth's sister planet. Indeed, it is the planet closest in mass to our own, weighing in at 85 percent of Earth's mass. Like Mercury, Venus is seen only as a morning or evening "star," and it also is named for a god in the Roman pantheon. Venus was the goddess of love, and a number of ancient civilizations saw it in the same way—the Babylonians, for example, named it Ishtar, after their own goddess of desire. • Aside from the moon, Venus is the brightest object in the night sky, and it can be seen even through the glare of modern cities. Perhaps because of its prominence, Venus also has the dubious distinction of being the most frequently reported UFO.

VENUS

A BEAUTIFUL INFERNO

DISCOVERER: **UNKNOWN**
DISCOVERY DATE: **PREHISTORIC**
NAMED FOR: **ROMAN GODDESS OF LOVE**

MASS: **0.82 × EARTH'S**
VOLUME: **0.86 × EARTH'S**
MEAN RADIUS: **6,052 KM (3,760 MI)**
SURFACE TEMPERATURE: **462°C (864°F)**
LENGTH OF DAY: **243 EARTH DAYS (RETROGRADE)**
LENGTH OF YEAR: **224.7 EARTH DAYS**
NUMBER OF MOONS: **0**
PLANETARY RING SYSTEM: **NO**

A computer-generated view of volcano Gula Mons, Venus. (Inset) Eastern hemisphere of Venus.

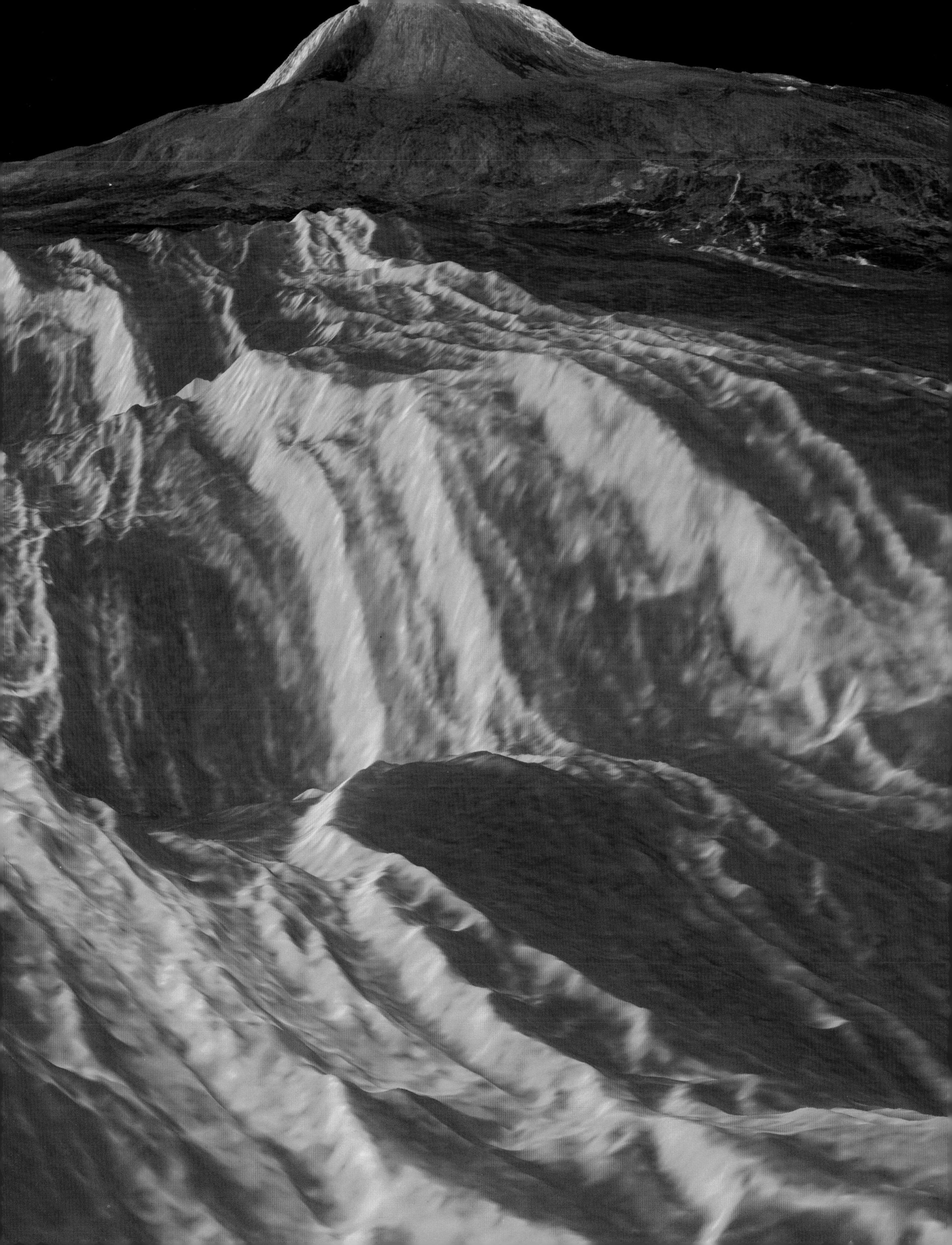

I VENUS WEST

Under a smothering atmosphere, Venus has a varied, volcanic landscape.

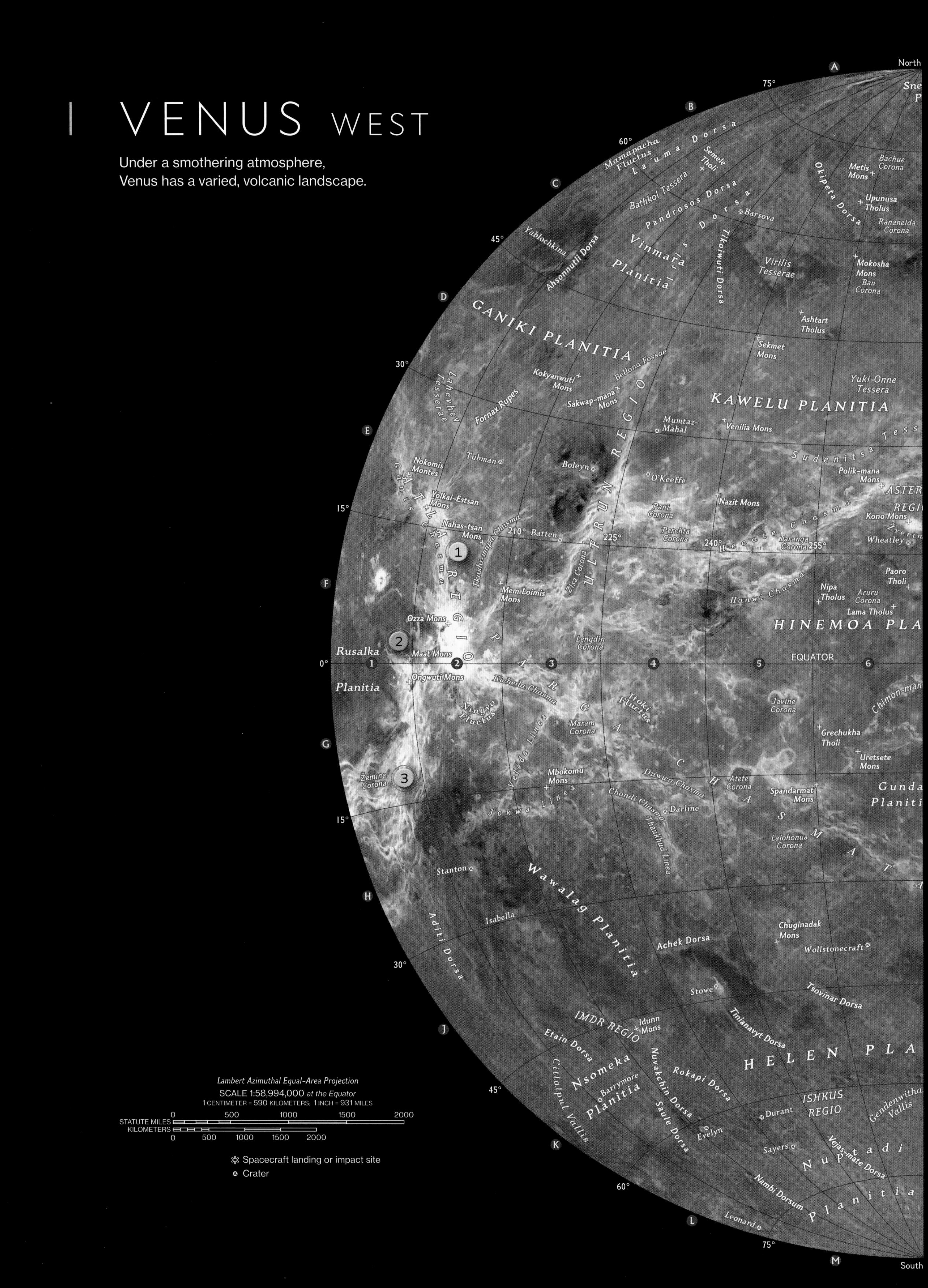

KEY FEATURES

1. **ATLA REGIO:** A region of old volcanic flows
2. **MAAT MONS:** Highest volcano on Venus
3. **ŽEMINA CORONA:** A domed feature in the steepest terrain on Venus

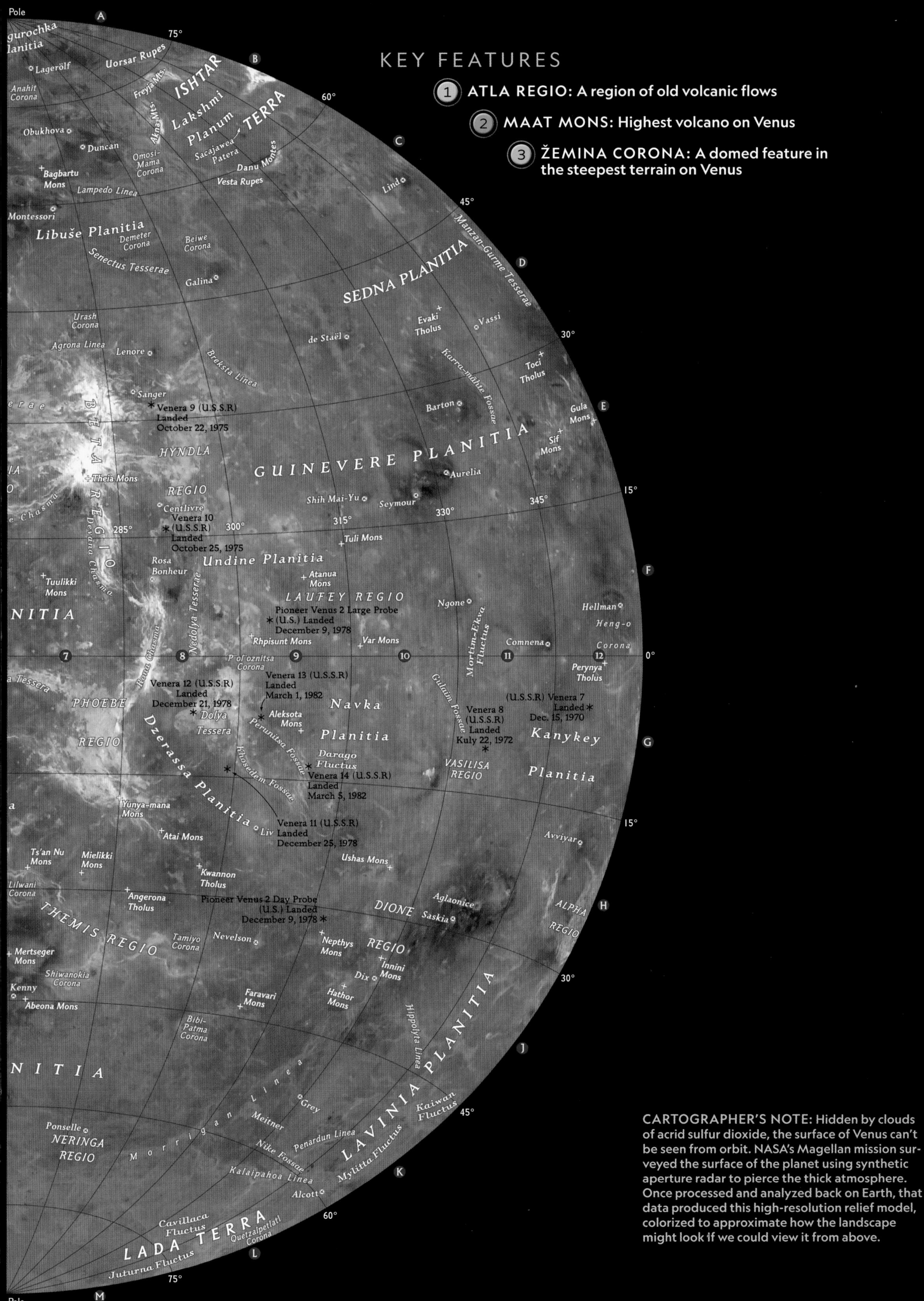

CARTOGRAPHER'S NOTE: Hidden by clouds of acrid sulfur dioxide, the surface of Venus can't be seen from orbit. NASA's Magellan mission surveyed the surface of the planet using synthetic aperture radar to pierce the thick atmosphere. Once processed and analyzed back on Earth, that data produced this high-resolution relief model, colorized to approximate how the landscape might look if we could view it from above.

VENUS EAST

The highlands of Aphrodite Terra, about half the size of Africa, dominate the equatorial region.

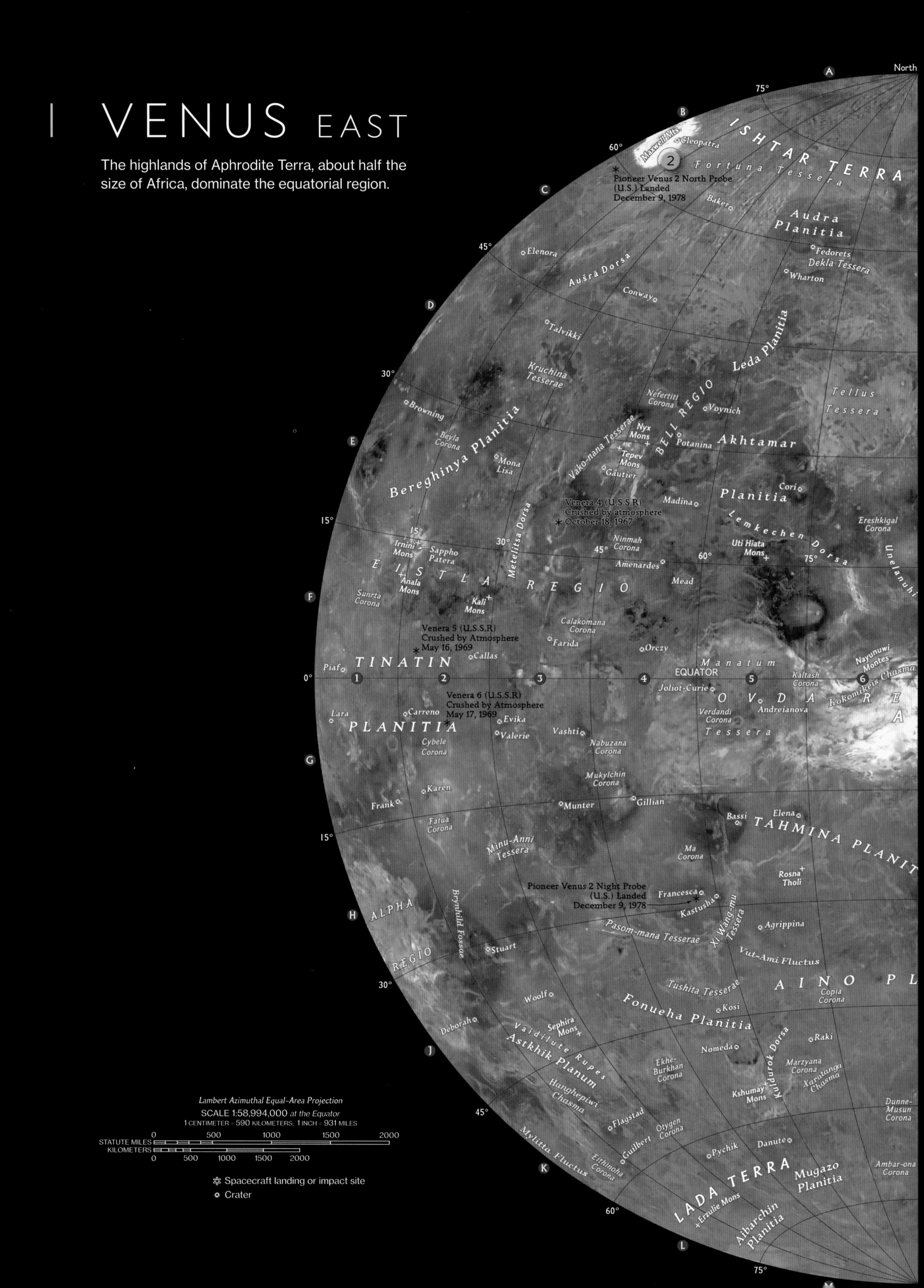

KEY FEATURES

1. **APHRODITE TERRA:** Huge highland region
2. **MAXWELL MONTES:** Mountain massif 11 km (7 mi) high
3. **ARTEMIS CORONA:** Largest of these domed features on Venus

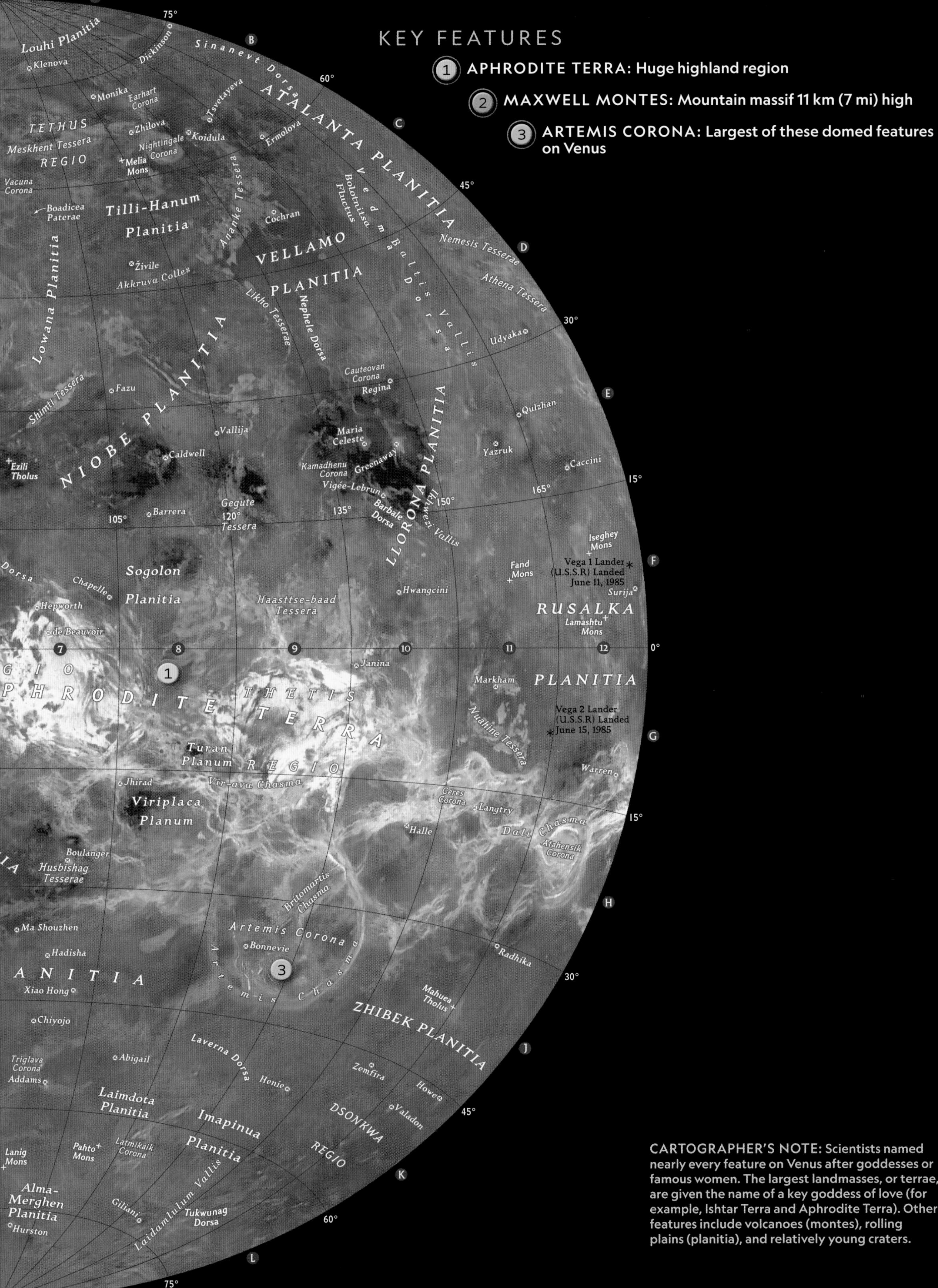

CARTOGRAPHER'S NOTE: Scientists named nearly every feature on Venus after goddesses or famous women. The largest landmasses, or terrae, are given the name of a key goddess of love (for example, Ishtar Terra and Aphrodite Terra). Other features include volcanoes (montes), rolling plains (planitia), and relatively young craters.

Venus completes an orbit around the sun in about 225 days, but its rotation is somewhat unusual. Viewed from above, the planets in the solar system all move around the sun in a counterclockwise direction, and most of them rotate on their axes in the same direction, counterclockwise as seen from above. This situation is a holdover from the spinning disk of matter from which the planets formed. Venus, however, rotates backward and completes a "day" in 243 Earth days, the slowest rate of rotation of all the planets. Scientists suggest that these anomalies are the result of a collision during the violent early days of planetary formation (see pages 55–8).

Despite the planet's proximity to Earth, astronomers learned little about Venus until the latter part of the 20th century. This is because a thick layer of white clouds perpetually obscures the surface of the planet. Then, in the 1960s, the United States and the Soviet Union began systematic programs of sending space probes to the planet, first to observe it from orbit or in a flyby and then to land probes on the surface itself. In 1962 Mariner 2 flew by the planet and probed it with microwave and infrared sensors. This was when we learned that the surface of Venus is extremely hot—around 462°C (864°F), which is hotter than the surface of Mercury, despite Venus's greater distance from the sun.

MISSIONS TO VENUS

In 1966 the Soviet probe Venera 3 crash-landed on the Venusian surface. The mission's goal was to land a probe and return data, but the intense pressure of the atmosphere crushed the capsule on the way down. In 1967 the strengthened Venera 4 spacecraft entered the atmosphere and sent back data, but the parachute that slowed its velocity was too effective—the craft took so long to descend that its batteries gave out before it reached its destination. Finally, in 1970, Venera 7, bolstered against the pressure and equipped with a smaller parachute, actually made it to the surface and sent back pictures. This was followed by several more spacecraft that landed successfully, typically sending back data for less than an hour before succumbing to the extreme conditions on the planetary surface.

In 1978 the American Pioneer Venus mission produced the first detailed maps of Venus, using radar signals to penetrate the clouds. Subsequent missions from the United States and the Soviet Union continued the exploration, with radar maps from the Magellan probe in 1989 revealing the surface in unprecedented, three-dimensional form. In 2005 the European Space Agency launched

A cutaway view of the interior of Venus. The core is mainly solid iron, the mantle is thick, and the crust is thin—about half as thick as Earth's. Venus's thick atmosphere supports a greenhouse effect that makes it the hottest planet in the solar system.

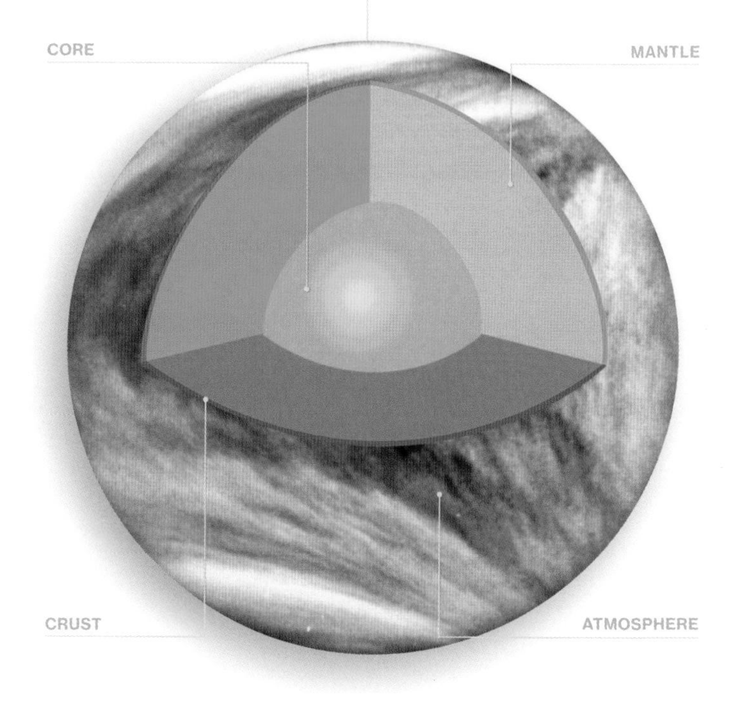

These images, taken 24 hours apart by the European Space Agency's probe Venus Express, show a rapidly evolving storm, or vortex, over Venus's south pole (marked by a yellow dot).

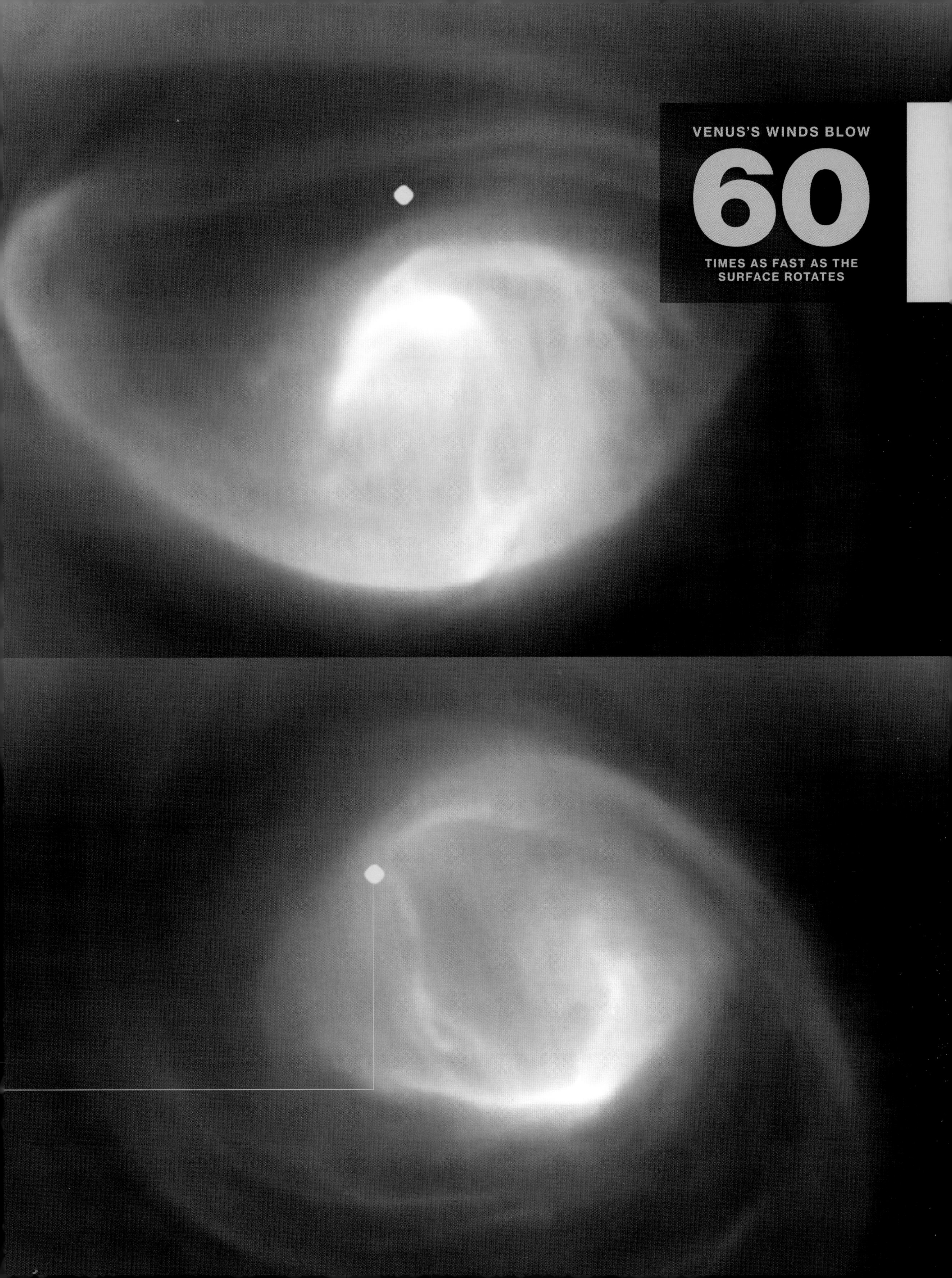

VENUS'S WINDS BLOW

60

TIMES AS FAST AS THE SURFACE ROTATES

AT VENUS'S SURFACE, AIR WEIGHS

90

TIMES MORE THAN ON EARTH

Venus Express, which entered a polar orbit around the planet in 2006 and has since been sending back data about Venus's dramatic atmosphere.

HOT, TOXIC, AND VOLCANIC

Venus's atmosphere is almost pure carbon dioxide—more than 95 percent—with most of the rest being nitrogen. Atmospheric pressure at the surface is 92 times that at sea level on Earth. This is about the same pressure that you would experience at a depth of one kilometer (~3,000 ft) below the surface of the ocean on Earth. No wonder those first space probes were crushed! Scientists suggest that early in its history, Venus may have had oceans, but that they were lost because the sun's intense radiation evaporated them. Without oceans to pull carbon dioxide out of the atmosphere, the concentration of this gas grew as volcanoes spewed it out. The planet experienced a runaway greenhouse effect that produced the intense temperatures we see today.

Because of Venus's dense atmosphere, the surface temperature is essentially constant everywhere. The surface winds are slow (only a few miles an hour), but it would be difficult to stand up to them on the surface because of the high density of the atmosphere—think of the slow winds as being more like a tidal wave than a gentle breeze.

Venus's clouds are composed mainly of sulfur dioxide and sulfuric acid. Intense winds at these higher altitudes have speeds clocking at several hundred miles an hour; we do not yet understand what causes them. It can actually rain sulfuric acid on Venus, but the drops evaporate as they fall through the thick atmosphere, never reaching the surface. The clouds also generate the planet's small magnetic field by means of a complex interaction with particles streaming from the sun. Venus's slow rotation precludes it from generating a magnetic field like Earth's, which depends on the rotation of its liquid iron core.

Radar mapping reveals a Venusian surface shaped primarily by volcanic activity. About 80 percent of the surface consists of smooth plains, with two highland "continents" making up the rest. One hundred sixty-seven volcanoes on the Venusian surface are bigger than the one forming the Big Island in Hawaii, the largest volcano on Earth.

The plains are dotted with impact craters, most of which show no evidence of weathering. Scientists argue that this fact implies that about 500 million years ago Venus underwent a "resurfacing" event, during which lava flows covered the old surface (with its craters), creating the smooth plains we see now and presenting a new surface for incoming meteorites. Models suggest that over time the temperature of the mantle rises, the crust weakens, and, every 100 million years or so, a new surface is created.

NASA has two Venus probes on the drawing boards, scheduled for launch in the 2020s. The VERITAS (Venus Emissivity, Radio Science, InSAR, Topography and Spectroscopy) mission will go into orbit around the planet and make detailed maps of its surface. It will seek, among other things, evidence that water once flowed on the surface and that there are active volcanoes, which would indicate that the planet is still geologically active.

The second mission is DAVINCI (Deep Atmosphere Venus Investigation of Noble gases, Chemistry, and Imaging). Once it arrives at Venus, this spacecraft will descend into the atmosphere, where it is expected to transmit data for a little more than an hour before it succumbs to the hellish conditions on the planet.

An artist's depiction of a lightning bolt on Venus. With an atmosphere made mainly of carbon dioxide, a crushing surface pressure, and sulfuric acid clouds, Venus is not the abode of life so often imagined in early science fiction.

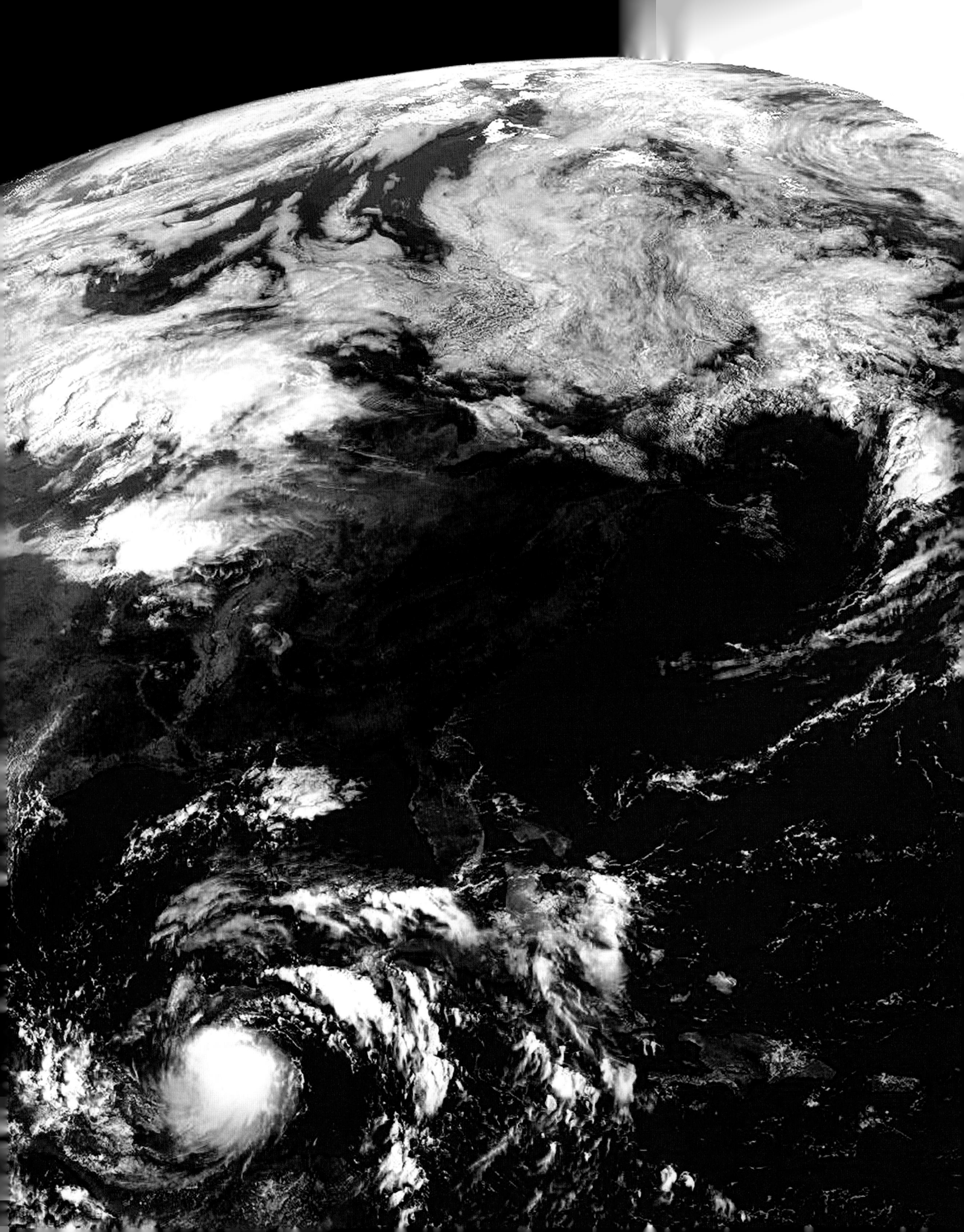

For obvious reasons, we know a lot more about Earth than we do about any other object in the solar system. But there is a lot to be learned by looking at Earth as just one more object in a solar system full of planets and moons. • So what distinguishes Earth from other worlds? There are two important distinctions: First, it is the largest terrestrial planet. As we shall see below, Earth's size is related to the fact that its surface is constantly changing, constantly moving. And second, Earth's orbit lies in a narrow band around the sun known as the habitable zone, which is defined as the region in which liquid water can stay on the planetary surface for long periods. Because of this, Earth is the only object in the solar system where we know life exists. • Let's look at these distinctions separately.

EARTH

| OCEAN PLANET |

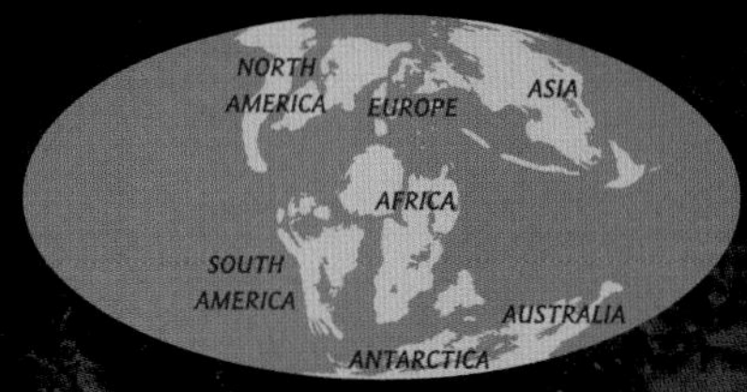

DISCOVERER: **UNKNOWN**
DISCOVERY DATE: **PREHISTORIC**
NAMED FOR: **OLD ENGLISH WORD *ERTHA*, MEANING "GROUND"**

MASS: **5,972,190,000,000,000,000,000,000 KG**
VOLUME: **1,083,206,916,846 KM³ (259,875,159,532 MI³)**
MEAN RADIUS: **6,371 KM (3,959 MI)**
MIN./MAX. TEMPERATURE: **−88/58°C (−126/136°F)**
LENGTH OF DAY: **23.93 HOURS**
LENGTH OF YEAR: **365.26 DAYS**
NUMBER OF MOONS: **1**
PLANETARY RING SYSTEM: **NO**

Cyclones cross the Atlantic. (Inset) Earth's continents 200 million years ago (left) and 100 million years ago (right).

I EARTH WEST

Earth is the only planet in the solar system whose surface is constantly changing.

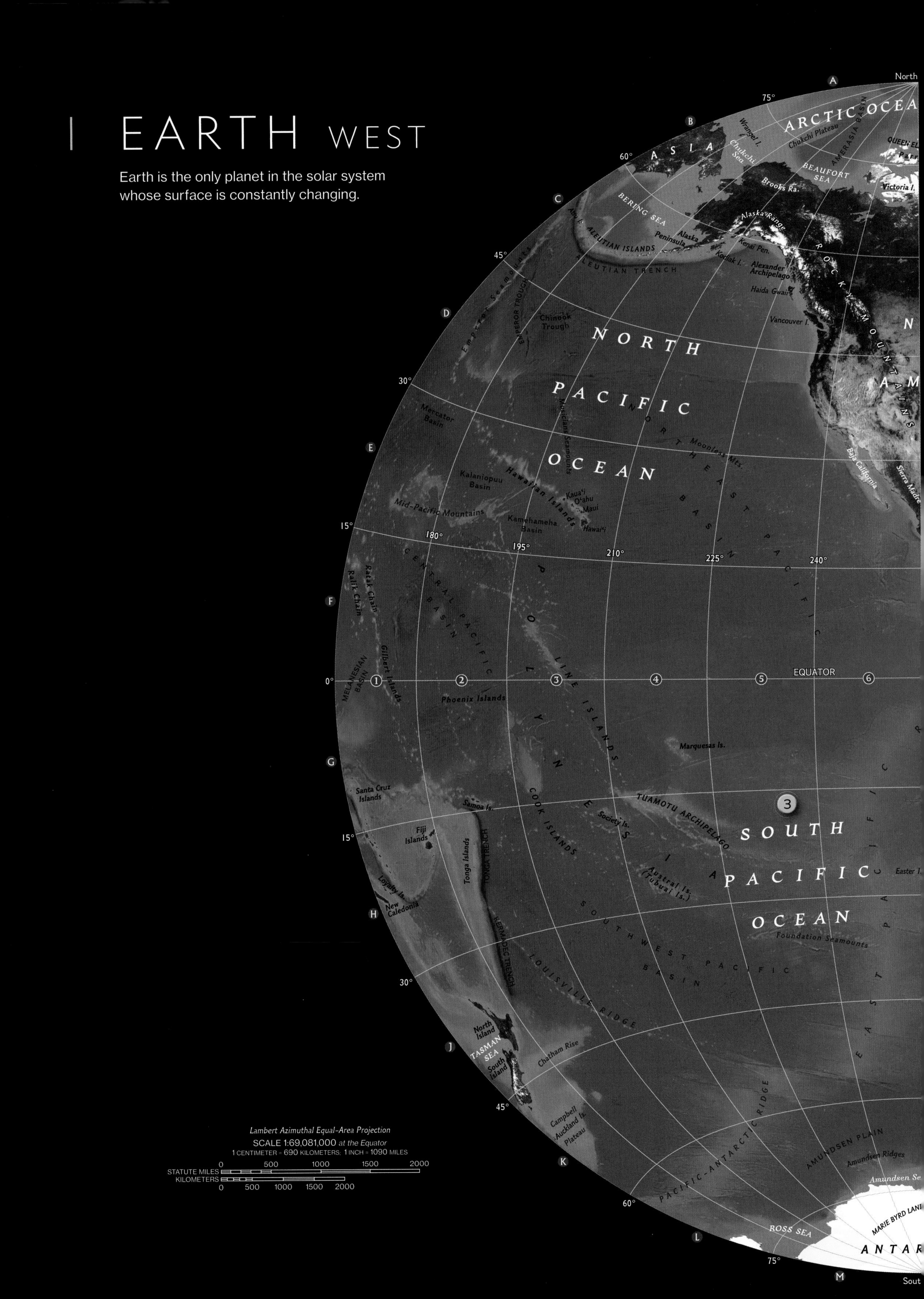

KEY FEATURES

1. **ANDES MOUNTAINS:** Created by the subduction of tectonic plates

2. **ICELAND:** The northernmost reach of the Mid-Atlantic Ridge

3. **PACIFIC OCEAN:** Part of a liquid, surface ocean, a feature unique to Earth

CARTOGRAPHER'S NOTE: Satellite imagery provides the detail of lake and mountain, desert and forest, showing every corner of our living planet. Peering beneath the watery ocean depths, global seafloor topographic data were used to create a representation of the ocean crust invisible from the surface.

EARTH EAST

Arid regions contrast vividly with the green of forests and river valleys in Earth's eastern hemisphere.

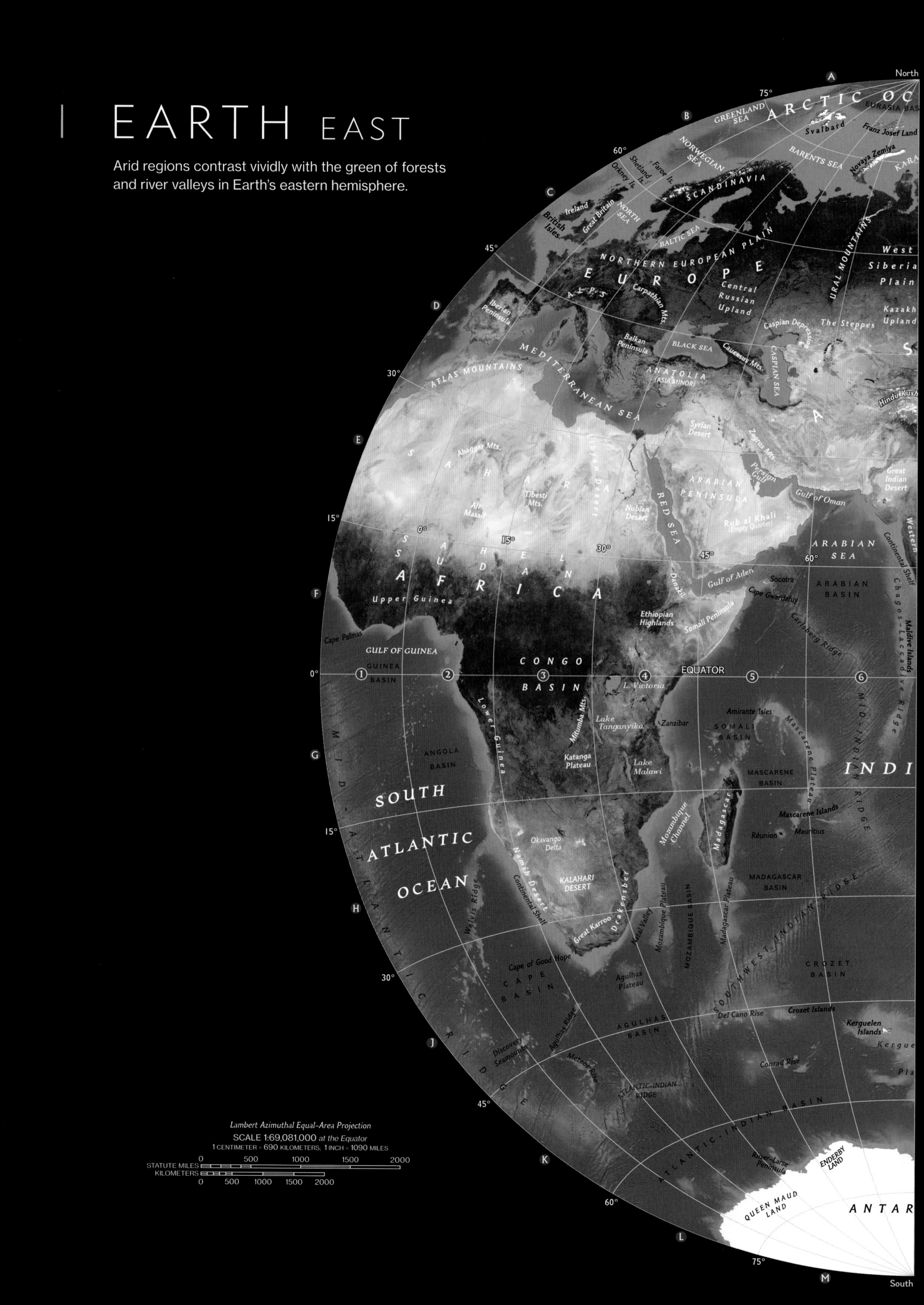

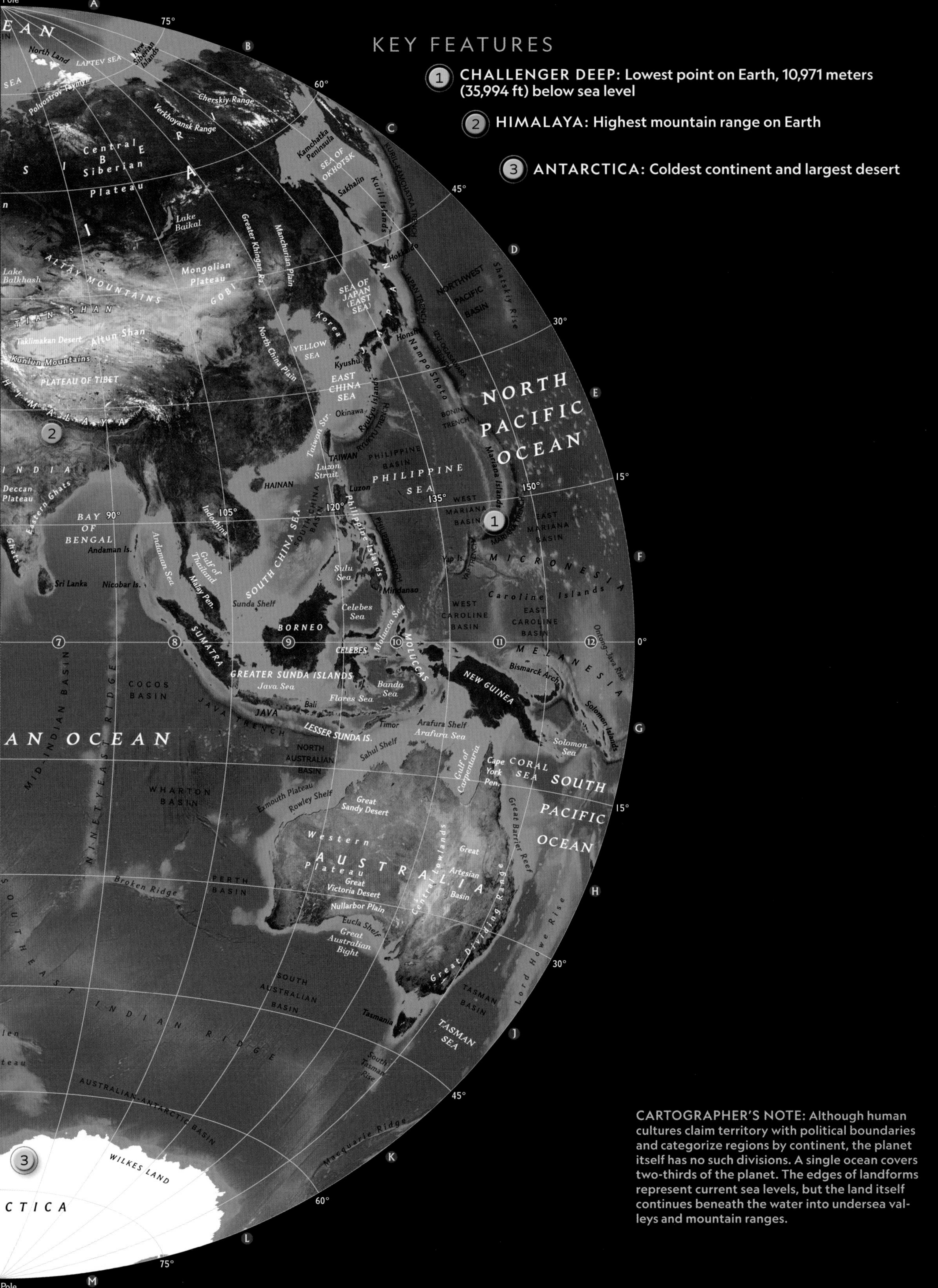

KEY FEATURES

1. **CHALLENGER DEEP:** Lowest point on Earth, 10,971 meters (35,994 ft) below sea level
2. **HIMALAYA:** Highest mountain range on Earth
3. **ANTARCTICA:** Coldest continent and largest desert

CARTOGRAPHER'S NOTE: Although human cultures claim territory with political boundaries and categorize regions by continent, the planet itself has no such divisions. A single ocean covers two-thirds of the planet. The edges of landforms represent current sea levels, but the land itself continues beneath the water into undersea valleys and mountain ranges.

For the first half billion years of its existence, give or take a few hundred million years, Earth swept around its orbit collecting debris from the process of planetary formation. Had anyone been standing on the surface, he or she would have seen the fiery impact of large meteorites all around. Scientists refer to this period as the Great Bombardment. (Please note that the Great Bombardment happened early in the formation of the solar system and is different from the Late Heavy Bombardment, which occurred when the solar system was half a billion years old.) Each impact added a certain amount of energy to the newly forming planet, energy that was converted to heat. Eventually, Earth either melted all the way through or heated up to the point that it became soft enough for materials to flow easily—scientists are still debating the details. What is not under debate, however, is that as a result of this early heating the planet went through a process known as differentiation. The heaviest materials—mainly iron—sank to the center to form Earth's core, while lighter materials formed the mantle and crust. Like a salad dressing left in one place too long, the materials within Earth separated themselves out under the influence of gravity.

Earth's differentiation created the magnetic field in which we all live. The heavy iron and nickel sank down, and at the very center of the planet the pressure was high enough to force those atoms into a solid. Above this solid core, however, is a region where the temperatures and pressure are only high enough to form a liquid layer. It is the rotation of this fluid metal core that ultimately produces the planet's magnetic field.

THE BOILING EARTH

The interior of the nascent planet had two sources of heat—the leftover heat from the Great Bombardment and the heat generated by the decay of radioactive elements in the rock. Like a pot of water on a stove, the interior of the planet had to find a way of getting this heat to the surface, where it could radiate away. And like a pot on the stove, Earth "boiled." Over hundreds of millions of years, the solid rocks in the mantle circulated, with hot material rising in one place, cooling, and sinking somewhere else. As this boiling went on, the lighter material that had risen to just under Earth's surface during differentiation was carried along, like leaves on a stream. And on top of this, a collection of the lightest material of all—the stuff we call the "solid ground" of continents—rode along like passengers on a raft.

The best way to picture the working of the planet, then, is to imagine a thin layer of oil on a pot of boiling water. The action of the water will break the layer of oil

A cutaway view of Earth shows a solid iron-nickel core at the center, surrounded by a liquid layer of the same material. This two-layer core is overlain by a thick mantle and an outer crust.

A volcanic eruption on the Big Island of Hawaii brings many different kinds of materials to Earth's surface—particles lofted into the atmosphere, gases (not visible here), and red-hot lava flowing into the sea.

EARTH HAS
12
MAJOR TECTONIC PLATES

into pieces, creating a kind of jigsaw puzzle. In just the same way, the boiling of the mantle breaks the planet's surface into pieces called plates.

The theory of plate tectonics is based on the idea that the surface of the planet is made from plates that respond to the movement of rocks in the mantle. ("Tectonics" comes from the same Greek root as "architect," and refers to the process of building.) Some plates carry continents, some do not, but because of the constant boiling, the plates and everything on them are always in motion, reshaping the face of the planet. There have been times, for example, when all of the continents were strung around the Equator, like some titanic necklace, and times when they were stuck together in a single land mass called Pangaea ("all Earth"). Other planets don't operate this way. Mercury and Mars (as well as our moon) are small enough that they got rid of their heat early on and are now "frozen." Although there is probably geological activity on Venus, the planet is apparently just a little too small to foster anything like plate tectonics.

HABITABLE ZONE

Earth is what astronomers call a Goldilocks planet—not too cold, not too hot, but juuuust right. For the last four billion years, while the luminosity of the sun increased by a third (see page 223), processes on our planet adjusted so that the atmospheric temperature stayed between the freezing and boiling points of water. This meant that liquid water could always be found at the surface. As we saw on page 83, the presence of liquid water is thought to be a necessary prerequisite for the development of life. Had Earth been closer to the sun than it is, it might have followed the same path as our sister planet Venus, with a runaway greenhouse effect extinguishing whatever life-forms had developed. Had it been farther from the sun, it might have frozen solid. In either case, there would be no life on our planet.

There is a narrow band around any star in which the surface temperature of a planet will remain below the boiling point and above the freezing point of water.

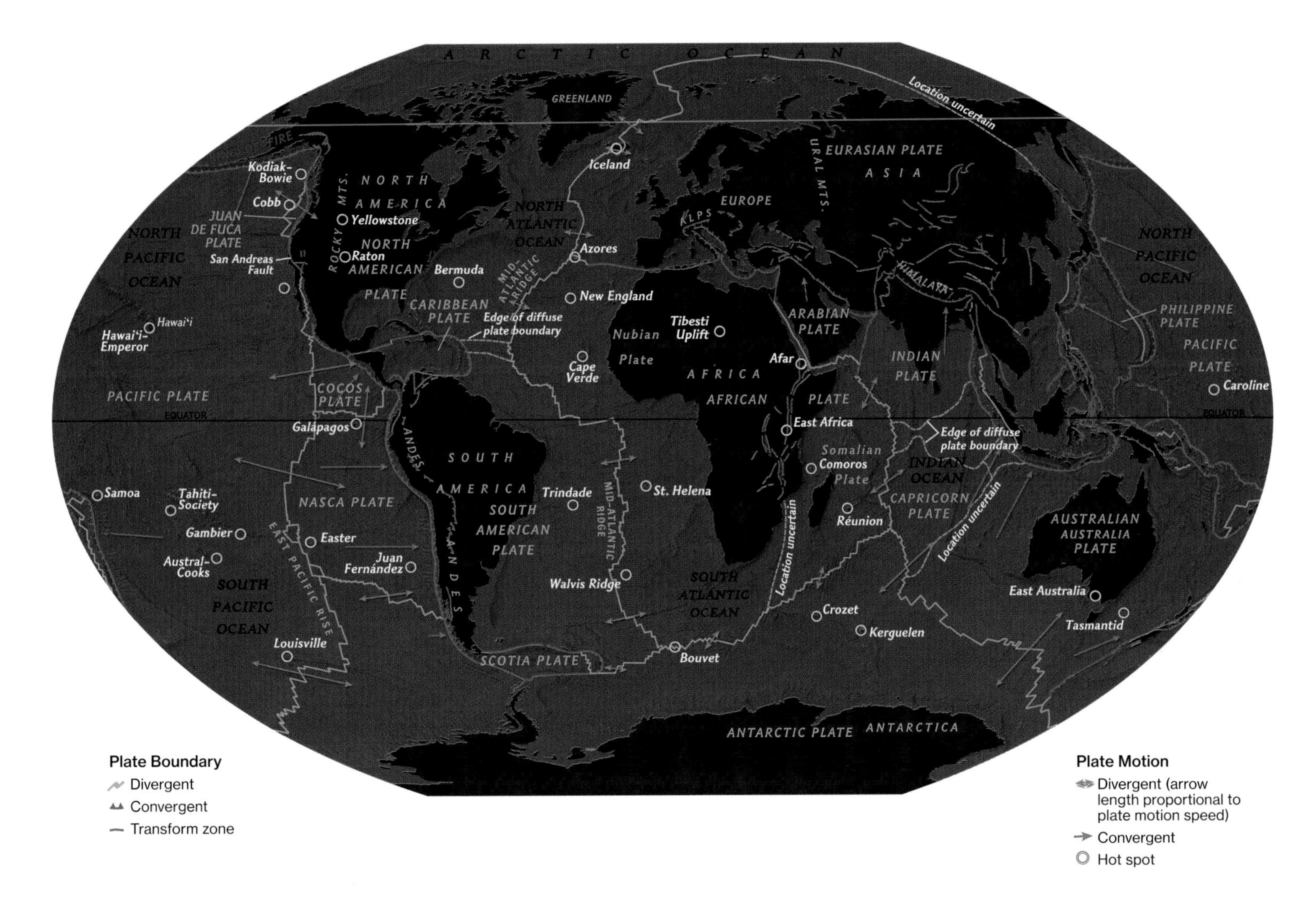

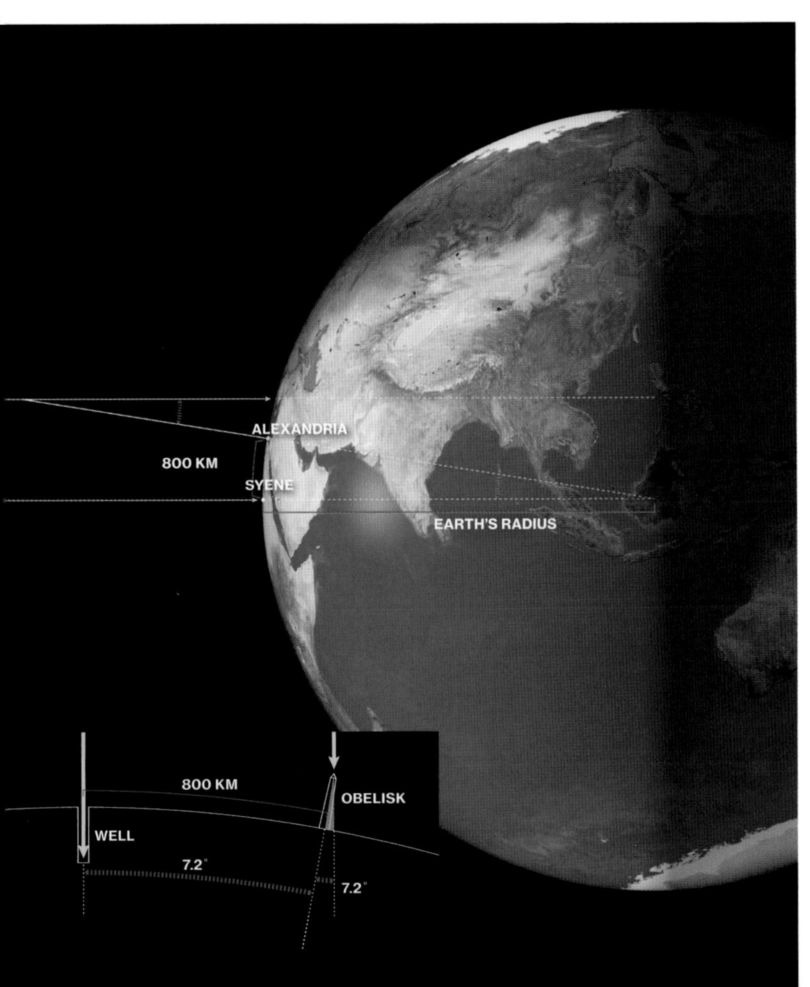

A map shows Earth's tectonic plates. The plates move in response to the slow churning of Earth's mantle, a churning driven in part by heat from radioactive decay. Alone among the planets in the solar system, Earth features a surface that is constantly changing because of this motion.

This narrow band is called the continuously habitable zone (CHZ) of the star. Earth is in the CHZ of the sun, which is why life exists here.

The presence of life changes a planet dramatically. On Earth, for example, the inclusion of the corrosive element oxygen in our atmosphere is a result of the metabolic activity of life, as are many of the organic processes that break up rocks and create soil. Astronomers look for these chemical signs of life as they search for an Earth-type planet in the CHZ of another star.

The moon, familiar to all of us from childhood, is the brightest object in the night sky. It completes an orbit around Earth in a little more than 27 days and turns on its axis in exactly the same amount of time. This means that it always keeps the same face toward us. In the jargon of astronomers, we say that the moon has been de-spun because of a complex series of gravitational interactions between its mass and that of Earth. We shall see that this is a common phenomenon for moons in the solar system. These gravitational interplays are also pulling the moon away from Earth at the rate of about four centimeters (1.5 in) per century.

EARTH'S MOON

OUR STEADY NEIGHBOR

DISCOVERER: **UNKNOWN**
DISCOVERY DATE: **PREHISTORIC**
NAMED FOR: **OLD ENGLISH WORDS FOR "MOON" AND "MONTH"**
DISTANCE FROM EARTH: **384,400 KM (238,855 MI)**

MASS: **0.012 × EARTH'S**
VOLUME: **0.02 × EARTH'S**
MEAN RADIUS: **1,738 KM (1,080 MI)**
SURFACE GRAVITY: **0.17 × EARTH'S**
MIN./MAX. TEMPERATURE: **−233/123°C (−387/253°F)**

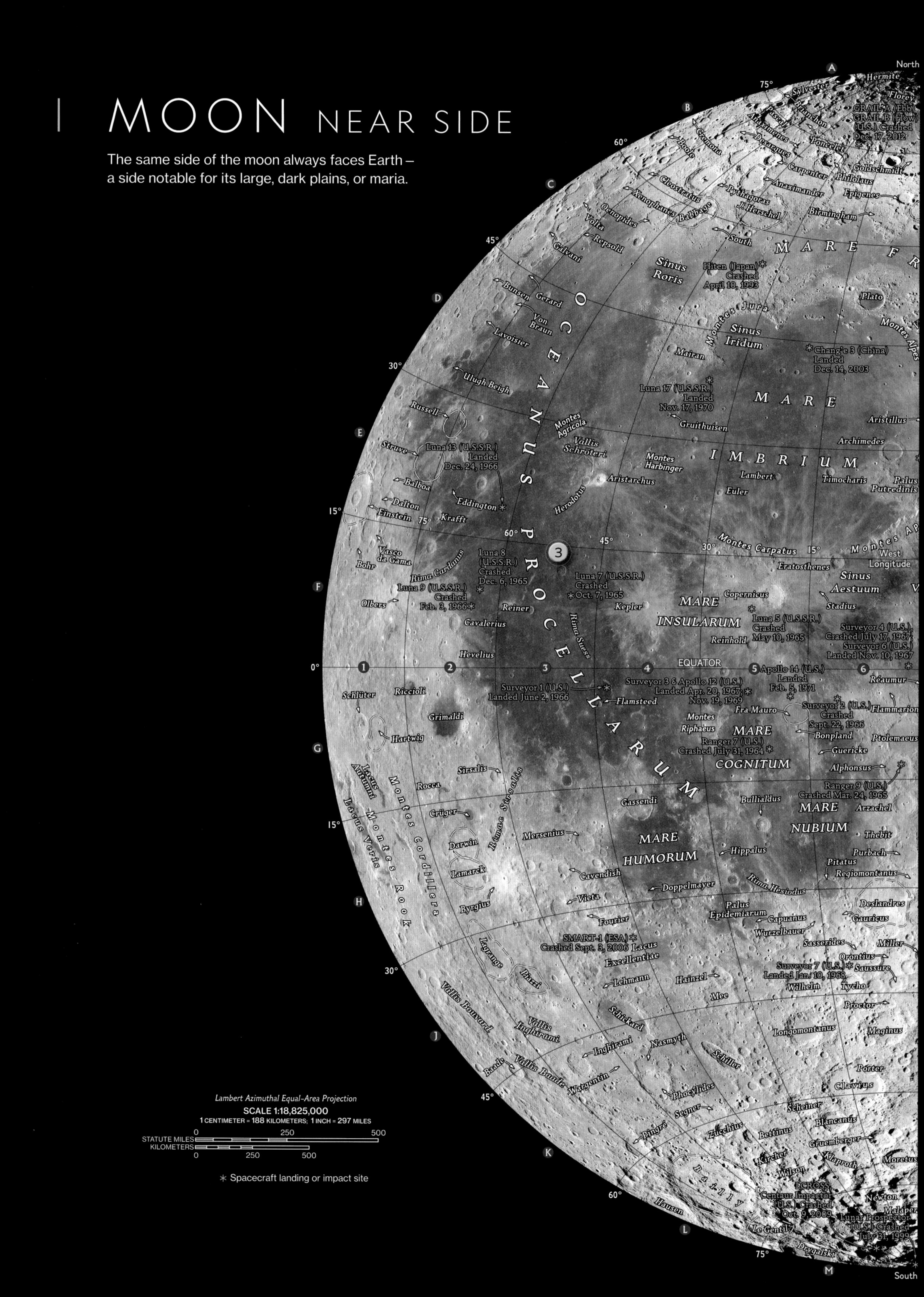

1 MOON NEAR SIDE
The same side of the moon always faces Earth – a side notable for its large, dark plains, or maria.
North
South
A
B
C
D
E
F
G
H
J
K
L
M
1
2
3
4
5
6
EQUATOR
0°
15°
30°
45°
60°
75°
West Longitude
Hermite
Sylvester
Florey
GRAIL A (Ebb) GRAIL B (Flow) (U.S.) Crashed Dec. 17, 2012
Cremona
Desargues
Poncelet
Goldschmidt
Philolaus
Anaximander
Carpenter
Epigenes
Boole
Cleostratus
Pythagoras
J. Herschel
Birmingham
Xenophanes
Oenopides
Babbage
Volta
Repsold
South
Galvani
MARE FR
Sinus Roris
Hiten (Japan) Crashed April 10, 1993
Plato
Bunsen
Gerard
Von Braun
Lavoisier
Montes Jura
Sinus Iridum
Montes Alpes
Mairan
Chang'e 3 (China) Landed Dec. 14, 2003
Ulugh Beigh
Luna 17 (U.S.S.R.) Landed Nov. 17, 1970
MARE IMBRIUM
Russell
Gruithuisen
Aristillus
Montes Agricola
Vallis Schröteri
Archimedes
Struve
Luna 13 (U.S.S.R.) Landed Dec. 24, 1966
Montes Harbinger
Lambert
Timocharis
Palus Putredinis
Balboa
Aristarchus
Euler
Eddington
Herodotus
Dalton
Einstein
Krafft
OCEANUS PROCELLARUM
Montes Carpatus
Montes Ap
Vasco da Gama
Rima Cardanus
Luna 8 (U.S.S.R.) Crashed Dec. 6, 1965
Eratosthenes
Sinus Aestuum
Bohr
Luna 7 (U.S.S.R.) Crashed Oct. 7, 1965
Luna 9 (U.S.S.R.) Crashed Feb. 3, 1966
Copernicus
Stadius
Olbers
Reiner
Kepler
MARE INSULARUM
Cavalerius
Rima Suess
Luna 5 (U.S.S.R.) Crashed May 10, 1965
Surveyor 4 (U.S.) Crashed July 17, 1967
Reinhold
Surveyor 6 (U.S.) Landed Nov. 10, 1967
Hevelius
Apollo 14 (U.S.) Landed Feb. 5, 1971
Surveyor 1 (U.S.) Landed June 2, 1966
Surveyor 3 & Apollo 12 (U.S.) Landed Apr. 20, 1967; Nov. 19, 1969
Réaumur
Schlüter
Riccioli
Flamsteed
Fra Mauro
Surveyor 2 (U.S.) Crashed Sept. 22, 1966
Flammarion
Grimaldi
Montes Riphaeus
MARE COGNITUM
Bonpland
Ptolemaeus
Hartwig
Ranger 7 (U.S.) Crashed July 31, 1964
Guericke
Alphonsus
Lacus Autumni
Montes Rook
Sirsalis
Rocca
Rimae Sirsalis
Ranger 9 (U.S.) Crashed Mar. 24, 1965
Gassendi
Bullialdus
MARE NUBIUM
Arzachel
Crüger
Lacus Veris
Montes Cordillera
Mersenius
Thebit
Darwin
MARE HUMORUM
Hippalus
Purbach
Lamarck
Pitatus
Regiomontanus
Cavendish
Rima Hesiodus
Doppelmayer
Deslandres
Byrgius
Vieta
Palus Epidemiarum
Gauricus
Fourier
Capuanus
Wurzelbauer
SMART-1 (ESA) Crashed Sept. 3, 2006
Lacus Excellentiae
Sasserides
Miller
Lagrange
Orontius
Piazzi
Surveyor 7 (U.S.) Landed Jan. 10, 1968
Saussure
Lehmann
Hainzel
Wilhelm
Tycho
Mee
Proctor
Vallis Bouvard
Schickard
Vallis Inghirami
Longomontanus
Maginus
Inghirami
Nasmyth
Schiller
Baade
Vallis Baade
Wargentin
Porter
Phocylides
Clavius
Segner
Scheiner
Pingré
Zucchius
Bettinus
Blancanus
Gruemberger
Kircher
Klaproth
Moretus
Wilson
Bailly
LCROSS Centaur Impactor (U.S.) Crashed Oct. 9, 2009
Newton
Hausen
Malapert
Lunar Prospector (U.S.) Crashed July 31, 1999
Le Gentil
Drygalski
Lambert Azimuthal Equal-Area Projection
SCALE 1:18,825,000
1 CENTIMETER = 188 KILOMETERS; 1 INCH = 297 MILES
STATUTE MILES 0 250 500
KILOMETERS 0 250 500
* Spacecraft landing or impact site

KEY FEATURES

1. **APOLLO 11 LANDING SITE**
2. **MONTES APENNINUS:** High lunar mountain range
3. **OCEANUS PROCELLARUM:** Ocean of storms, largest of the maria

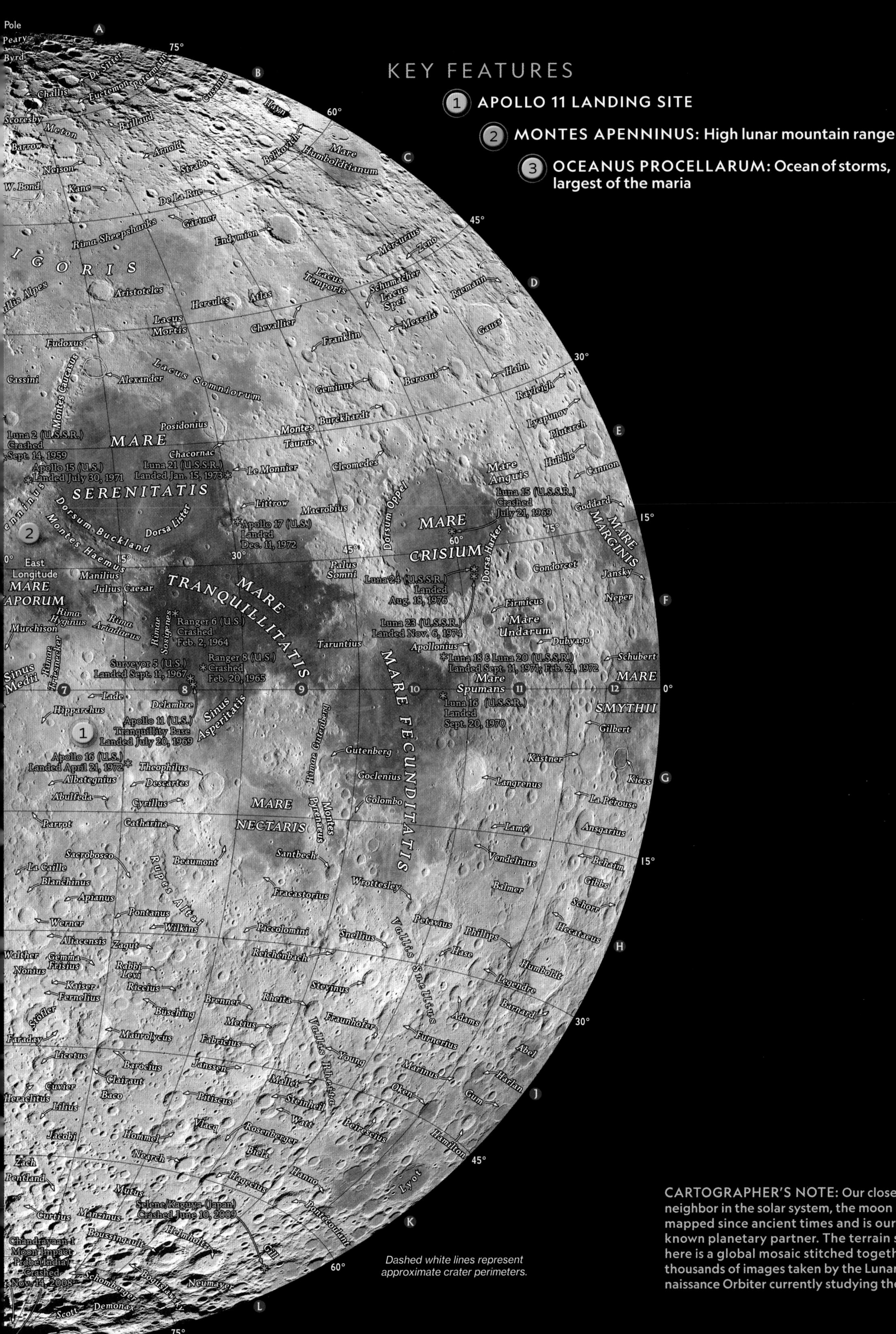

Dashed white lines represent approximate crater perimeters.

CARTOGRAPHER'S NOTE: Our closest neighbor in the solar system, the moon has been mapped since ancient times and is our best known planetary partner. The terrain shown here is a global mosaic stitched together from thousands of images taken by the Lunar Reconnaissance Orbiter currently studying the moon.

I MOON FAR SIDE

The far side of the moon, facing perpetually away from Earth, was not seen until the space age.

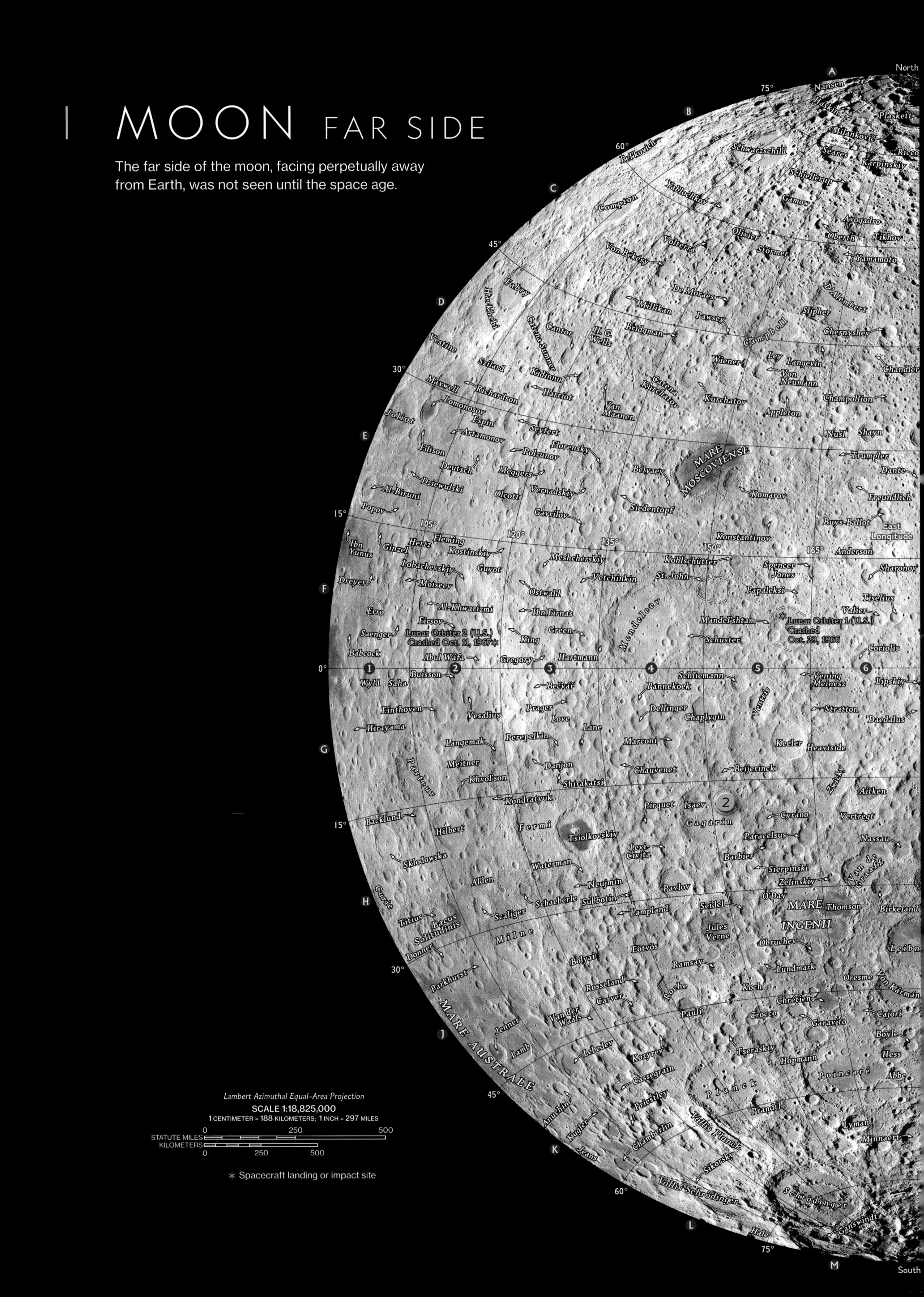

KEY FEATURES

1. **APOLLO:** Large-impact basin
2. **GAGARIN:** Named for first man in space
3. **SOUTH POLE:** Possible site of water ice

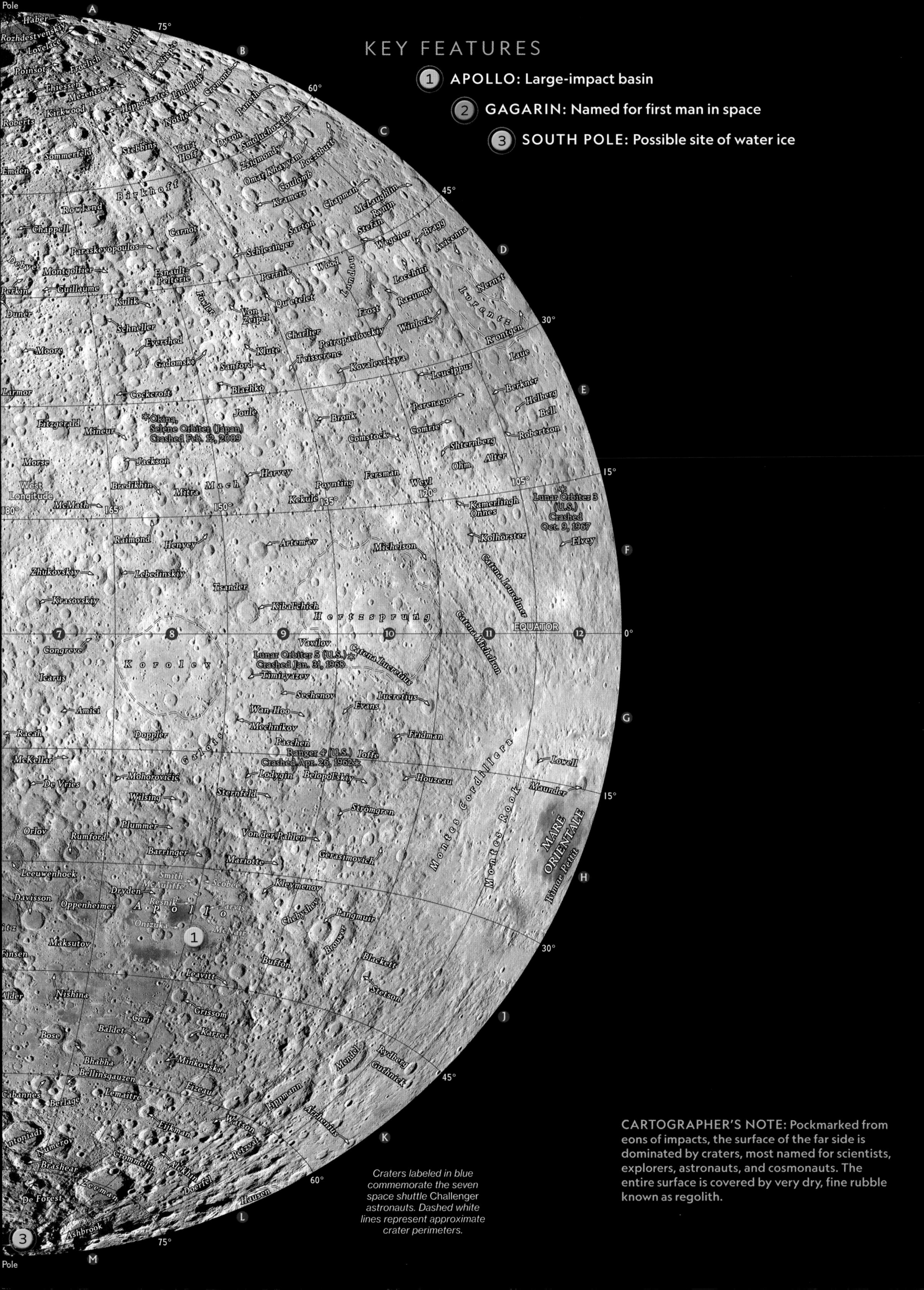

Craters labeled in blue commemorate the seven space shuttle Challenger *astronauts. Dashed white lines represent approximate crater perimeters.*

CARTOGRAPHER'S NOTE: Pockmarked from eons of impacts, the surface of the far side is dominated by craters, most named for scientists, explorers, astronauts, and cosmonauts. The entire surface is covered by very dry, fine rubble known as regolith.

There is a long history of scientific studies of the moon. Greek, Chinese, and Indian astronomers all realized that the moon shone because it reflected light from the sun, and Aristotle taught that the moon marked the boundary between the earthly and heavenly spheres. The astronomer Claudius Ptolemy (ca 100 A.D.), expanding upon earlier work by Greek astronomers, estimated both the distance to the moon and the satellite's size to within a few percent of the currently accepted values.

In 1609 Galileo used his new telescope to produce drawings of the lunar surface, showing mountains, plains, and craters. The realization that the moon had geological features similar to those on Earth played a role in discrediting the old, sun-centered theories of the universe, since those theories had held that the moon was a smooth, featureless sphere. The far side of the moon—never visible from Earth—was first photographed by the Soviet probe Luna 3 in 1959, and both American and Soviet unmanned spacecraft landed on the lunar surface in the same year.

ONE SMALL STEP

The space race, fueled by the Cold War, resulted in the first human beings setting foot on the moon in 1969, when astronaut Neil Armstrong uttered his famous line "That's one small step for a man, one giant leap for mankind" as he stepped off of the ladder of the Apollo 11 spacecraft. More important from a scientific viewpoint, astronauts on the six lunar landing missions brought back 380 kilograms (840 lb) of lunar rock samples for scientific study, some of which date back to the solar system's earliest years. In the foreword, astronaut Buzz Aldrin, one of a dozen human beings who have walked on the moon, gives a detailed history of the Apollo program.

Since the end of the Apollo program in 1972, only unmanned probes have gone to the moon. In addition to the United States, the European Space Agency, India, Japan, and China have engaged in lunar exploration over the past few decades.

THE MOON'S ANATOMY

We now know that the moon, like Earth, was formed about 4.5 billion years ago. Scientists have long debated how this happened. The basic problem is that the moon is significantly less dense than Earth, primarily because the moon has such a small iron core. How could Earth and the moon, both of which apparently formed in the same part of the planetary cloud, end up looking so different?

The currently favored theory is that early in the formation of Earth (but after differentiation had occurred)

A cutaway view of the moon. The moon is a dead world with a small, solid iron core, a thick mantle, and a crust full of craters. Because the moon has no atmosphere, the craters, once formed, never disappear.

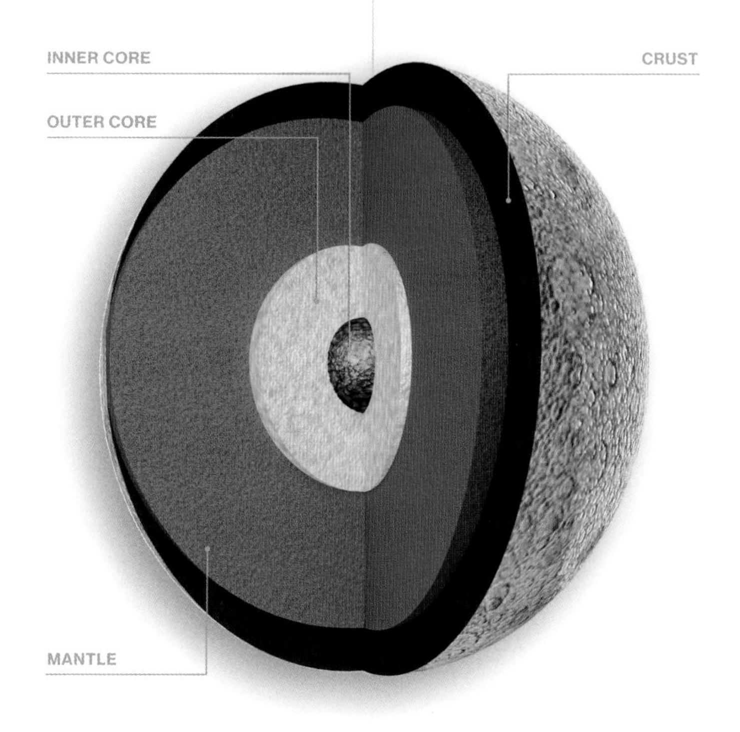

Apollo 17 astronaut plants the American flag on the moon. The lunar vehicle was driven to places near the landing site to collect geological samples. Much of what we know about the formation of the moon resulted from the study of those rocks. The footprints you see are still there.

Earth and a Mars-size object (on which differentiation had also occurred) collided. This collision blew off a large chunk of Earth's lower-density mantle, and some of this ejected material, along with material from the other object, went into orbit around Earth. At this point, the same process of accretion that built the terrestrial planets came into play, and the moon formed from that orbiting material. Although the moon is only the fifth largest satellite in the solar system, it is the largest in relation to its planet. Its radius is a quarter that of Earth's, and it weighs in at 1/81 of Earth's mass.

Earth's densest material (its iron core) did not contribute to the makeup of the moon, and this explains the density difference between the two. The young moon went through the same kind of differentiation process as Earth (see page 88), so its interior has a layered structure, although, for the reasons we discussed above, it has a much smaller core. The main features on the near side (the side visible from Earth) are the large dark plains that cover about a third of its surface (the far side does not have many of these features). These plains are called maria (singular: mare), Latin for "oceans," because early astronomers thought they were seas. They are actually massive outflows of lava, with the biggest ones dating back to between 3 and 3.5 billion years ago. The lighter areas on the moon, usually referred to as highlands, are older—perhaps 4.4 billion years old—and represent the first materials to crystallize from molten material as the moon cooled. The maria and highlands together produce the familiar "man in the moon" image that you can see on any clear night.

THE SURFACE

Craters, the result of meteorite impacts over the eons, dot the lunar surface. Since the moon has no atmosphere to speak of, and since it is now a frozen world with no

TIME AND TIDES

Everybody knows that tides are caused by the moon's gravitational pull on the oceans, but two things about tides make the story a little complicated: First, there are two tides a day rather than one; and second, high tide occurs when the moon is on the horizon, not when it is directly overhead. Thus, the tides are not simply a matter of the moon pulling the ocean water toward it. In that case, there would be one high tide every day, and it would occur when the moon is directly overhead.

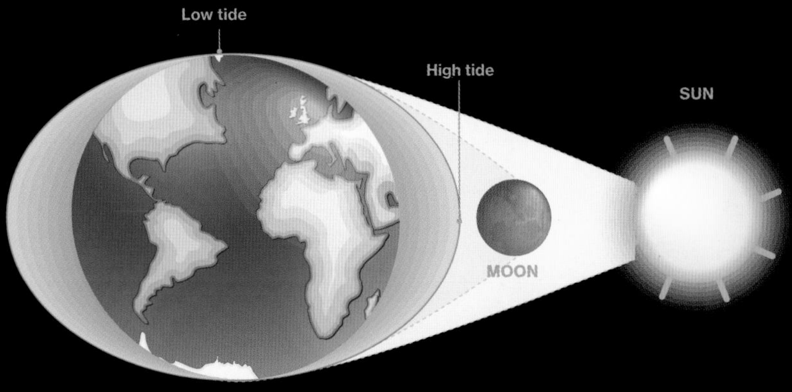

We usually think of the moon as going around Earth, but in fact the moon and Earth rotate around a spot in between them known as the center of mass. As Earth rotates around the center of mass, its motion creates a centrifugal force that raises a second tidal bulge in the oceans directly opposite to the one raised by the moon's gravity. This is why there are two tides every day.

The fact that these high tides occur when the moon is on the horizon rather than overhead is related to the fact that Earth's oceans are relatively shallow—their average depth is only 5 kilometers (3 mi). This means that the tidal bulges can't keep up with the spot underneath the moon as Earth rotates; they lag behind. If the oceans were 97 kilometers (60 mi) deep, high tides would occur 12 hours later, when the moon is overhead.

geological activity, there is nothing to remove these craters once they form. Consequently, there are literally hundreds of thousands of them visible on the moon today.

In fact, the only change agent on the moon's surface is the arrival of new meteorites. Small impacts break up the surface rock, creating small, glassy pieces that weld themselves together (think of damp Rice Krispies). This material, called the lunar regolith, is found everywhere on the moon except on steep surfaces. It is about 10 to 20 meters (30 to 60 ft) deep on the older highlands and about 3 to 5 meters (9 to 15 ft) deep in the maria.

Because people often talk about building permanent bases or colonies on the moon, the question of whether or not there is water at the surface has garnered a lot of scientific attention over the years. The best place to look for water is in deep craters at the moon's poles, since these areas are never exposed to direct sunlight. In 2009 the first Indian moon mission, named Chandrayaan-1, found evidence for water in light reflected from the lunar surface. A few weeks later the American spacecraft LCROSS dropped an impactor the size of a pickup truck into one of the shadowed craters and saw in the resulting debris indications of enough water to fill a small wading pool.

We can't leave the subject of the moon without dealing with some popular misconceptions:

- There is no statistically significant evidence that more people are admitted to psychiatric wards during a full moon than at any other time.
- There is no evidence of an alien UFO base on the far side of the moon.
- The fact that the moon looks bigger on the horizon than when it is overhead is an optical illusion. You can verify this yourself by marking the apparent size of the moon on a stick, first when the moon is on the horizon, then a few hours later when it is overhead. You'll find the same result in both cases.

LOW TIDE, BAY OF FUNDY

Mars, the fourth planet from the sun, is the most thoroughly explored of all the planets except Earth and has probably figured in more science fiction than any other astronomical object. Named for the Roman god of war, it often appears to have a reddish cast owing to the iron oxide (rust) on its surface. Mars is smaller than Earth, having about half the radius and only about 11 percent the mass of our home planet. Because of its small mass, Mars lost most of its atmosphere to space long ago and today has only a thin atmosphere composed mostly of carbon dioxide. The average air pressure on Mars is roughly equivalent to the pressure 35 kilometers (24 mi) above sea level on Earth.

MARS

THE RED DESERT

DISCOVERER: **UNKNOWN**
DISCOVERY DATE: **PREHISTORIC**
NAMED FOR: **ROMAN GOD OF WAR**

MASS: **0.11 × EARTH'S**
VOLUME: **0.15 × EARTH'S**
MEAN RADIUS: **3,390 KM (2,106 MI)**
MIN./MAX. TEMPERATURE: **−87/−5°C (−125/23°F)**
LENGTH OF DAY: **1.03 EARTH DAYS**
LENGTH OF YEAR: **1.88 EARTH YEARS**
NUMBER OF MOONS: **2**
PLANETARY RING SYSTEM: **NO**

A composite view of pictures of the Martian landscape taken by the Curiosity rover. (Inset) Icy clouds drifting over planet Mars.

MARS WEST

Mars is a planet of extremes: towering volcanoes, plunging valleys, smooth plains, and ragged craters.

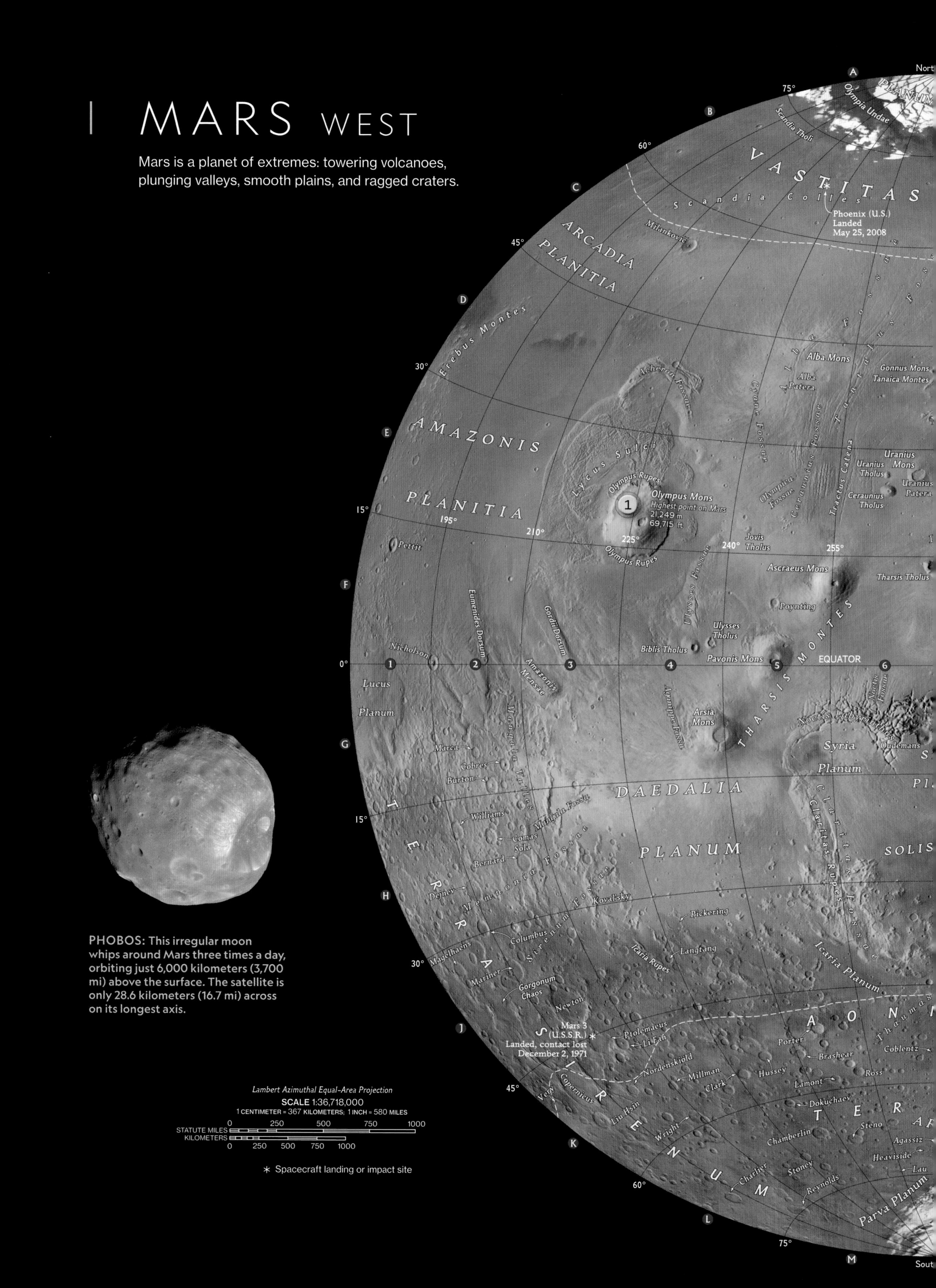

PHOBOS: This irregular moon whips around Mars three times a day, orbiting just 6,000 kilometers (3,700 mi) above the surface. The satellite is only 28.6 kilometers (16.7 mi) across on its longest axis.

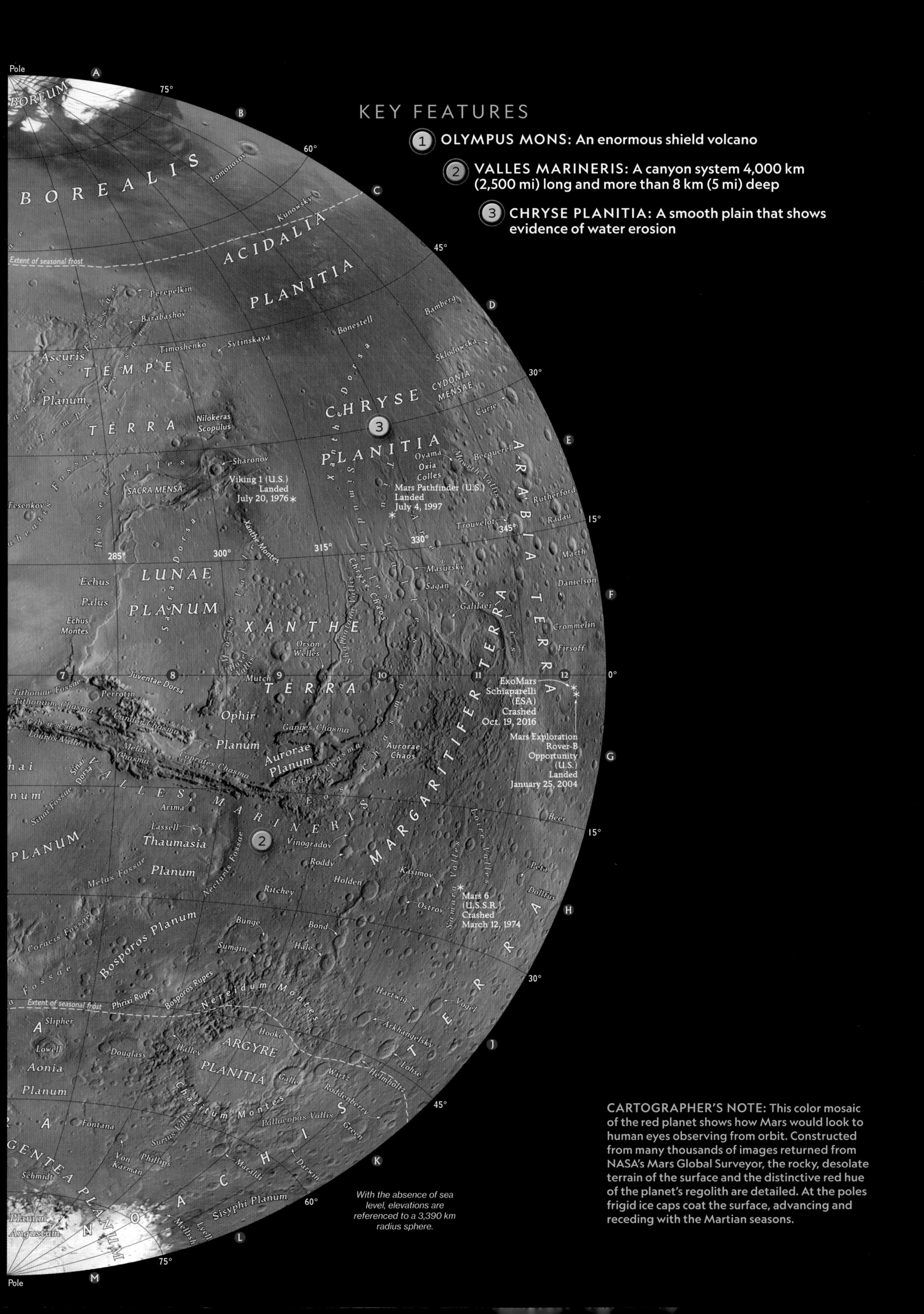

With the absence of sea level, elevations are referenced to a 3,390 km radius sphere.

KEY FEATURES

1. **OLYMPUS MONS:** An enormous shield volcano
2. **VALLES MARINERIS:** A canyon system 4,000 km (2,500 mi) long and more than 8 km (5 mi) deep
3. **CHRYSE PLANITIA:** A smooth plain that shows evidence of water erosion

CARTOGRAPHER'S NOTE: This color mosaic of the red planet shows how Mars would look to human eyes observing from orbit. Constructed from many thousands of images returned from NASA's Mars Global Surveyor, the rocky, desolate terrain of the surface and the distinctive red hue of the planet's regolith are detailed. At the poles frigid ice caps coat the surface, advancing and receding with the Martian seasons.

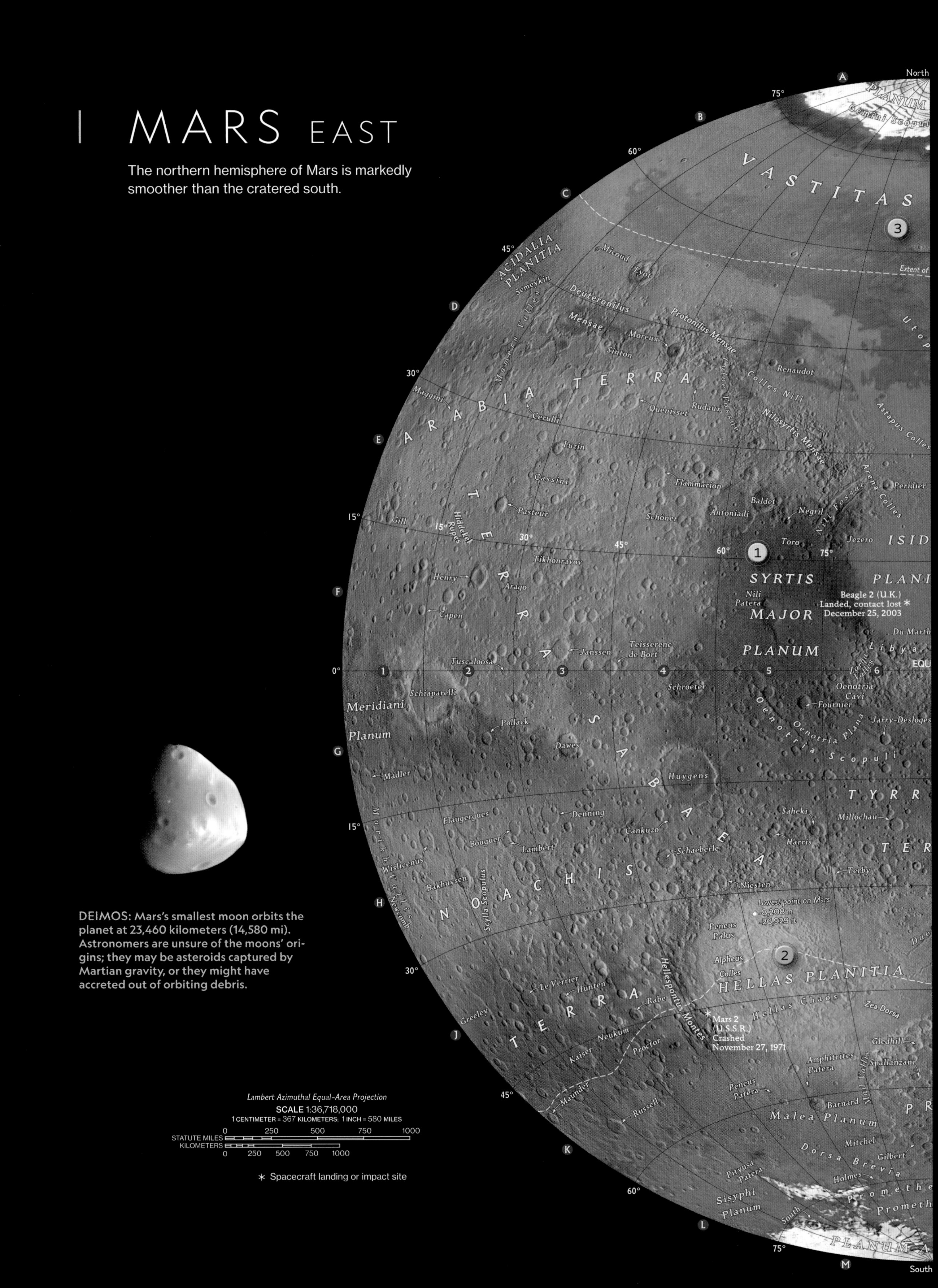

DEIMOS: Mars's smallest moon orbits the planet at 23,460 kilometers (14,580 mi). Astronomers are unsure of the moons' origins; they may be asteroids captured by Martian gravity, or they might have accreted out of orbiting debris.

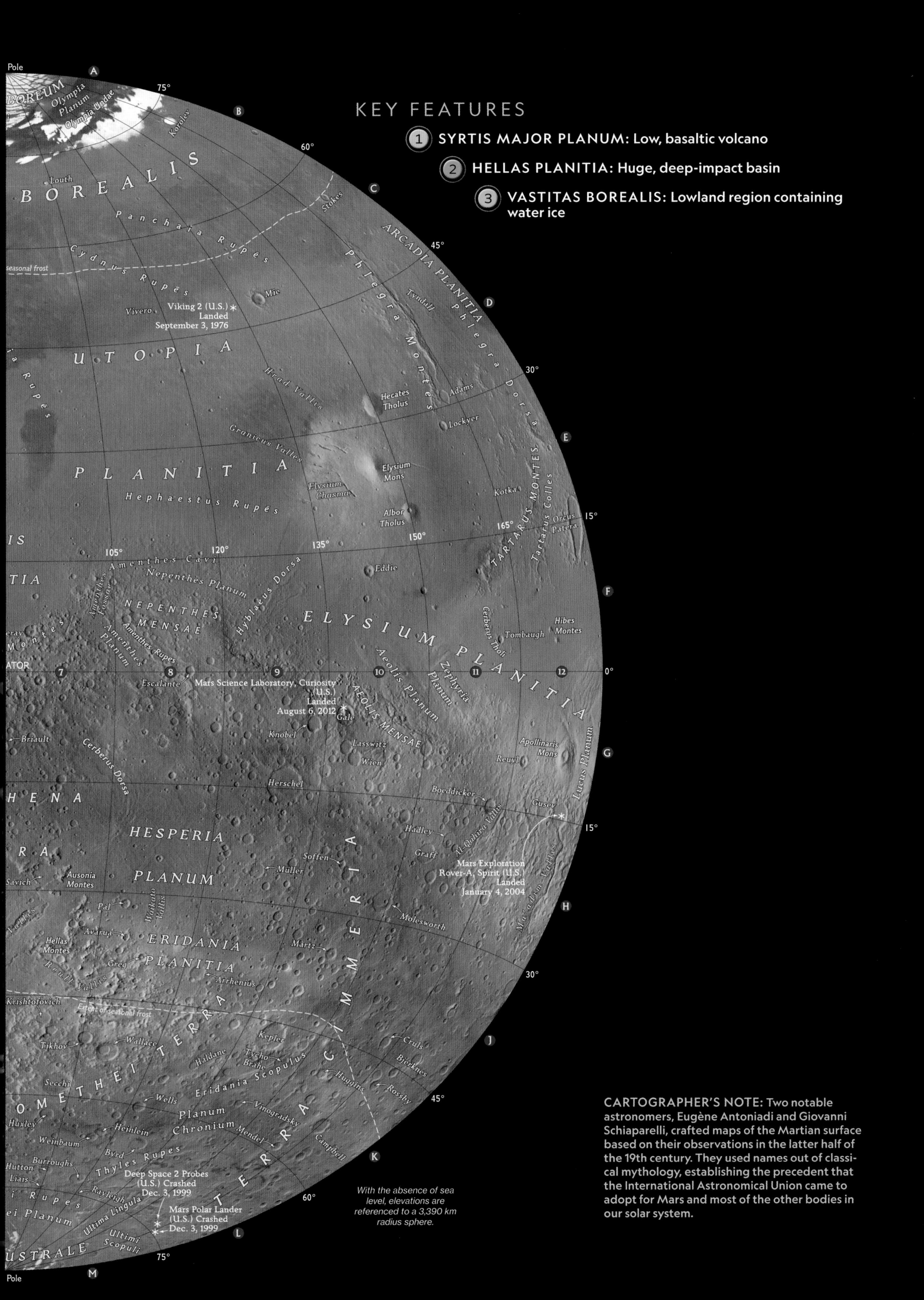

KEY FEATURES

1. **SYRTIS MAJOR PLANUM:** Low, basaltic volcano
2. **HELLAS PLANITIA:** Huge, deep-impact basin
3. **VASTITAS BOREALIS:** Lowland region containing water ice

With the absence of sea level, elevations are referenced to a 3,390 km radius sphere.

CARTOGRAPHER'S NOTE: Two notable astronomers, Eugène Antoniadi and Giovanni Schiaparelli, crafted maps of the Martian surface based on their observations in the latter half of the 19th century. They used names out of classical mythology, establishing the precedent that the International Astronomical Union came to adopt for Mars and most of the other bodies in our solar system.

The axis of rotation for Mars is tilted at approximately the same angle as that of Earth, so the red planet has seasons, just as we do. Its "year" is about twice as long as ours, however, so each of its seasons lasts twice as long as those on Earth. During the winter in each hemisphere no sunlight reaches the pole, as on Earth, but on Mars large amounts of carbon dioxide freeze out of the atmosphere to produce a thick layer of frozen carbon dioxide—the stuff we know as dry ice on Earth. This dry ice layer at the poles disappears when the sun returns. Under the dry ice are large, permanent polar caps of water ice. The water ice in Mars's northern ice cap contains a little less than half the ice found in the Greenland ice sheet on Earth.

Mars has a relatively smooth northern hemisphere formed by lava flows and a more complex southern hemisphere displaying old impact craters. Current hypotheses suggest that both hemispheres were covered by oceans early in Mars's history—presumably the water evaporated and, because of the planet's small size, was eventually lost to space. Scientists believe that the most recent ocean was located in the northern hemisphere. Moving southward, we pass through a transitional terrain (one author refers to it as "beachfront property") before we get to the rough, cratered southern hemisphere.

MOUNTAINS AND VALLEYS

Many of the remarkable features on the Martian surface, most visible only in data sent back by spacecraft, are eerily reminiscent of familiar formations on our own planet—only more so. Two deserve special attention. Olympus Mons (Mount Olympus), an extinct volcano, is the largest mountain yet discovered in the solar system. With a height of about 27 kilometers (18 mi), it is over three times as tall as Mount Everest. Valles Marineris (Mariner Valleys) is a canyon system about 4,000 kilometers (2,600 mi) long and up to 7 kilometers (5 mi) deep. (For reference, the Grand Canyon is about 450 kilometers/300 miles long and up to 2 kilometers/1.3 miles deep.)

THE SEARCH FOR WATER

In 1877 the Italian astronomer Giovanni Schiaparelli produced the first detailed map of the Martian surface. Using a telescope, he saw lines on the surface, which he called *canali* (channels). Unfortunately, this was translated as "canals" in English, which suggested the presence of intelligent life on the planet. Schiaparelli was followed by the American astronomer Percival Lowell, whose book *Mars as the Abode of Life* brought the concept of an inhabited Mars to the attention of the general public. Lowell not only

A cutaway view of Mars. Like the moon, Mars has no tectonic activity. It has a solid core, mostly iron, and a mantle. The crust, with an average thickness of about 50 kilometers (30 mi), is slightly thicker than Earth's.

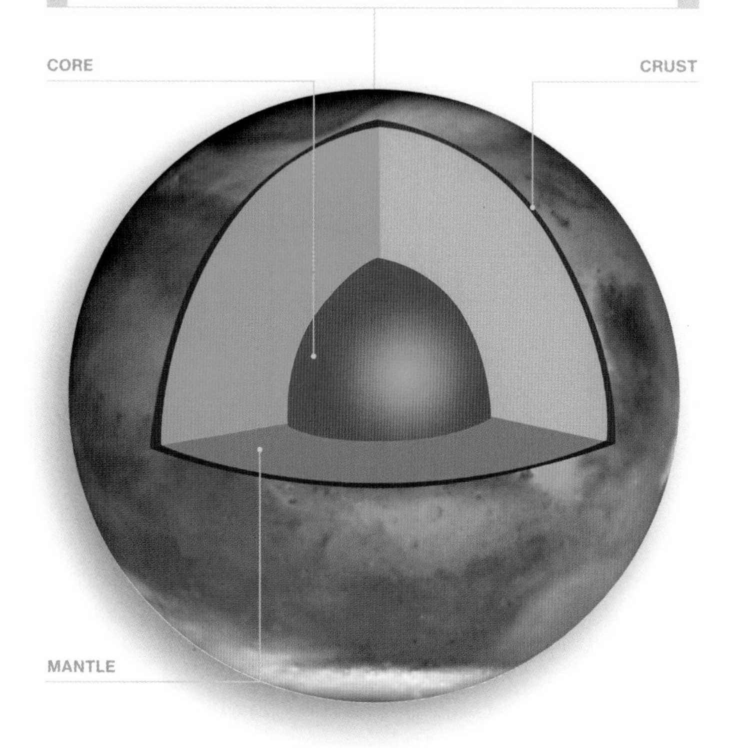

The first intimation that water might once have flowed on the Martian surface was the discovery of gullies like these in the southern highlands, resembling water-formed channels on Earth.

claimed to see canals but also reported on how they filled with water or drained according to the seasons.

The hypothesis promulgated at the time was one of those elegant, beautiful, and wrong ideas that pop up now and again. Mars was the home of a dying civilization, Earth the home of a flourishing civilization, while hot, swampy Venus represented the future. We now know that Lowell's canals were optical illusions, and his conclusions were based on the well-known tendency of humans to see patterns in random assortments of images (think Rorschach test).

The modern exploration of Mars started in 1964, when Mariner 4 flew by the planet, and gained detail in 1971, when Mariner 9 went into orbit. The biggest surprises that these spacecraft produced for scientists were photographs of gullies on Mars, gullies that looked for all the world like ordinary river watersheds on Earth. This was the first genuine intimation we had that there was once liquid water on the Martian surface, an idea that is now well accepted in the scientific community.

The biggest Martian event for the general public, though, was the landing of the Viking 1 and 2 spacecraft in 1976. The resulting photographs from the Martian surface were one of the first big hits on the Internet and were seen in newspapers and magazines around the world.

NASA's Mars Global Surveyor, launched in 1996, was the next important exploration mission. In its 10 years of operation while in orbit around Mars, this spacecraft produced a detailed map of the Martian surface. In 1997

the first robotic vehicle—Sojourner—dropped to the Martian ground. This was the first successful landing of a rover on another planet, and it pioneered the technique of surrounding the rover with airbags to cushion the landing, then deflating the airbags to allow the vehicle to move. Originally scheduled for a month of operation, Sojourner sent back data for three, establishing a precedent of longevity for future Mars rover missions.

MODERN EXPLORATION

The 21st century has seen a veritable flotilla of spacecraft, landers, and orbiters launched toward the red planet. In addition to NASA's projects, European Space Agency, Russian, Chinese, and Finnish missions are either in progress or on the drawing boards. The most dramatic missions were those of the Mars exploration rovers Spirit and Opportunity, which landed successfully in 2004. These rovers examined Martian rocks and minerals, establishing in short order that liquid water once existed on the Martian surface.

Expected to operate for only a few months, both rovers functioned for an incredible six *years*—an amazing tribute to the engineers that built them. Scientists link the long lifetime of these vehicles to the fact that storms and dust devils on the Martian surface kept the solar panels free of grit, allowing both rovers to operate at full power. In 2010 Spirit got stuck in deep sand and, after prolonged attempts to free her, was converted to a stationary observation station. By this time Opportunity had traveled no less than 20 kilometers (14 mi) over the Martian surface.

The next stage of exploration began in 2011, with the launch of the Mars Science Laboratory. A Volkswagen Beetle–size vehicle dubbed Curiosity, this rover car-

THE FACE ON MARS

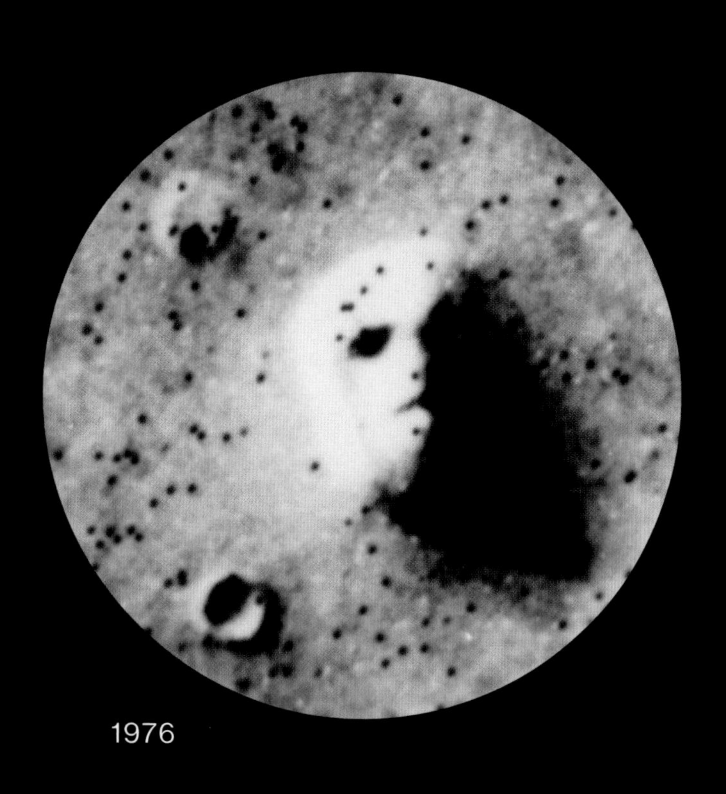

1976

We might as well get down to the only result of the missions to Mars to wind up in tabloids at supermarket checkout counters – the so-called face on Mars. On July 25, 1976, the Viking 1 spacecraft was in orbit around the red planet, photographing possible landing sites for its sister ship, Viking 2. Over a region of the planet's northern hemisphere that marks a kind of coastal zone between the smooth plains of the north and the cratered terrains of the south, a low-resolution photo showed what appeared to be a face staring up at the craft, surrounded by Egyptian-style pyramids. The scientists at control center smiled – clearly it was an optical illusion – but some NASA functionary decided that releasing the photo would be a good way to generate public interest in Martian exploration.

Well, I guess! For years after that decision, the face on Mars became a staple of fringe science. I can recall, for example, a tabloid headline that blared out the news that the face was that of Elvis Presley!

It wasn't until 1998 that the Mars Global Surveyor revisited the site of the face and took photos with much higher resolution (and different lighting). As expected, the face disappeared, a result borne out by several subsequent missions. Although it was fun while it lasted, the face on Mars has joined Lowell's canals in the collection of objects that wishful thinkers have placed on the red planet.

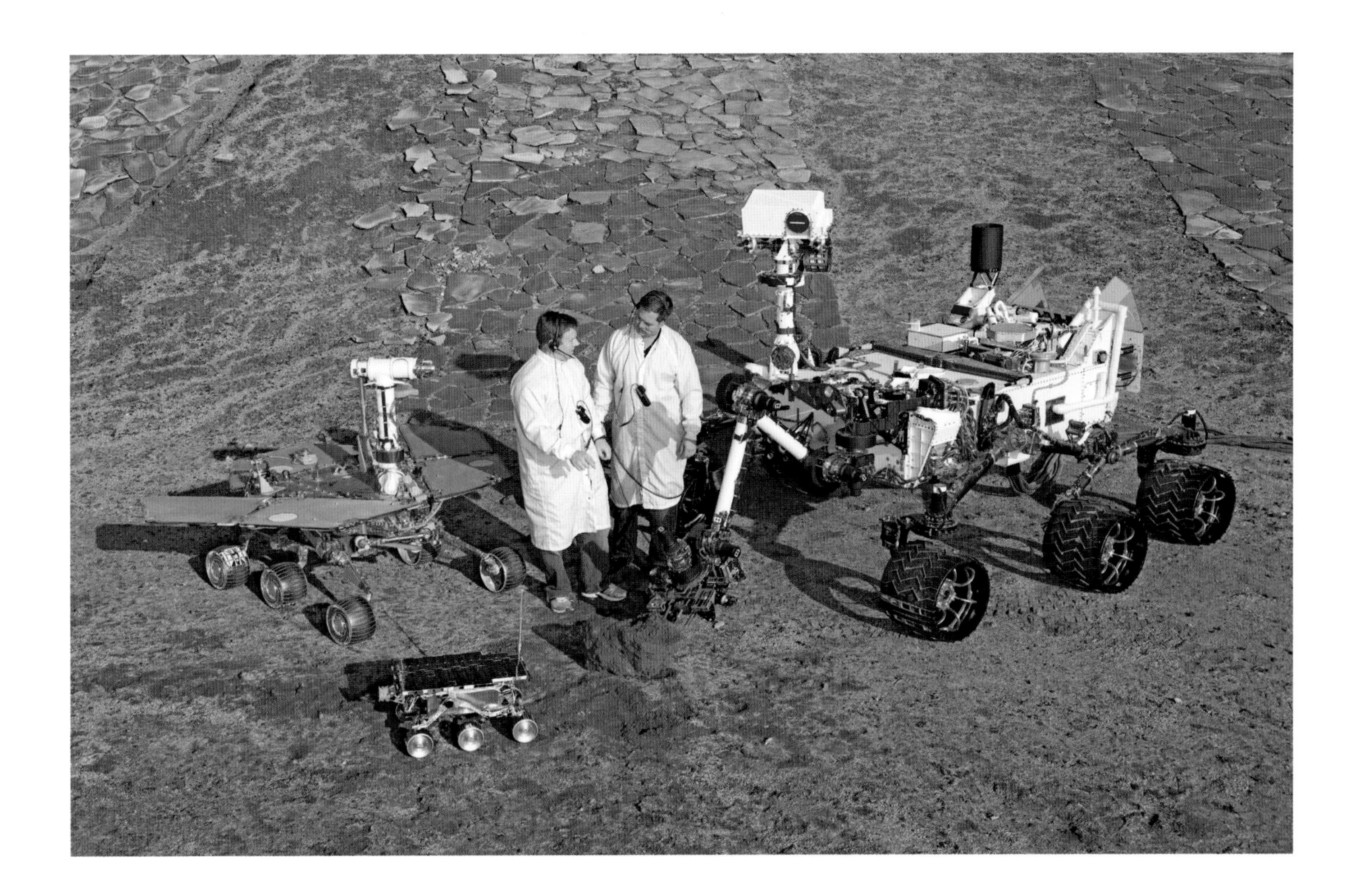

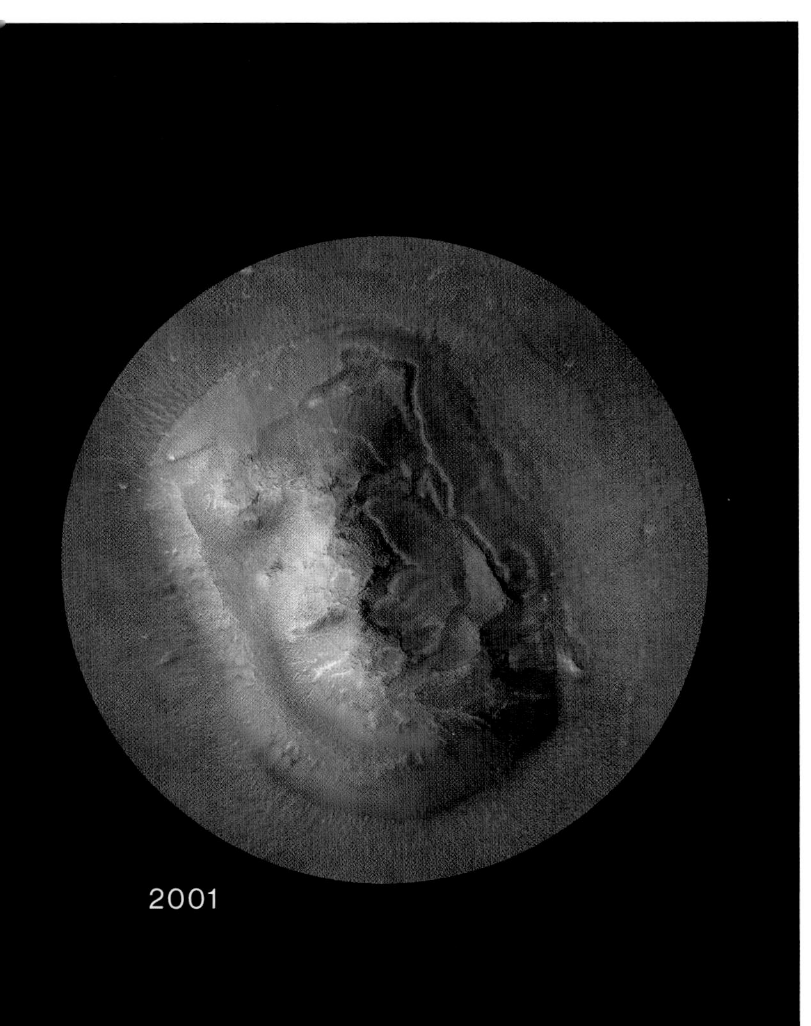

NASA scientists stand among models of rovers that have explored Mars, starting with the small rover Sojourner on the lower left and ending with Curiosity on the right. This illustrates how Mars rovers have increased in size and complexity over the years.

ries scientific instruments from six different countries. Curiosity landed on Mars in 2012 at a place called the Bradbury Landing Site, named in honor of science fiction writer Ray Bradbury. The site is in the middle of a large crater whose geology was expected to yield important information about the planet's history. In 2013, the rover began a year-long trek to a geological formation known as Mount Sharp, some 8 kilometers (5 mi) away. Sampling rocks along its path and on the mountain, Curiosity found further evidence that water had once existed on the Martian surface.

At the same time, the Mars Reconnaissance Orbiter, one of the many spacecraft in orbit around the red planet, reported evidence that could indicate that liquid water may still come to the surface periodically. The Orbiter saw lines on slopes that appeared to darken and lighten with the seasons. One theory is that the lines represent briny water coming to the surface—the idea being that dissolved salts would lower the water's freezing point, much as salt scattered on an icy sidewalk allows the ice to melt. If this is the case, then it suggests there might be liquid water beneath the Martian surface today. Recent studies suggest these lines could be caused by moving sand and dirt instead of water.

NASA's Mars 2020 mission, scheduled for launch in that year, will feature an improved version of Curiosity to search for, among other things, evidence of existing life on the planet.

COLONIZATION AND TERRAFORMING

Even though Mars is the most Earth-like of the planets in our solar system, conditions on its surface are not suitable for human beings. For one thing, the atmosphere is very thin and contains no oxygen. In addition, the atmospheric pressure is so low that exposed liquids (like the fluids that line your lungs) would boil at body temperature. Thus, anyone venturing onto the Martian surface would need to be in a pressure suit and supplied with oxygen, much like those worn by pilots at high altitudes.

In addition, Mars has no magnetic field, and therefore, unlike Earth, offers no protection against cosmic rays. This means that humans spending long periods on the planet would have to have shielded shelters to keep their radiation exposure down.

In the introduction to this book, Apollo astronaut Buzz Aldrin lays out a detailed plan for how the colonization of Mars might be carried out. It is an obvious next step in our exploration of the solar system.

Those who think about a long-term presence on Mars imagine colonies consisting of shielded domes on the surface (or, perhaps, underground). As it happens, an experiment in constructing such a structure was carried out in Arizona in the 1990s. A huge greenhouse-style building called Biosphere 2 was built for the express purpose of exploring the possibility of maintaining a completely

Biosphere 2, located north of Tucson, Arizona, was built as a prototype for a possible Mars colony. In 1994 eight "bionauts" spent a year inside to explore the possibility of human survival in such a structure.

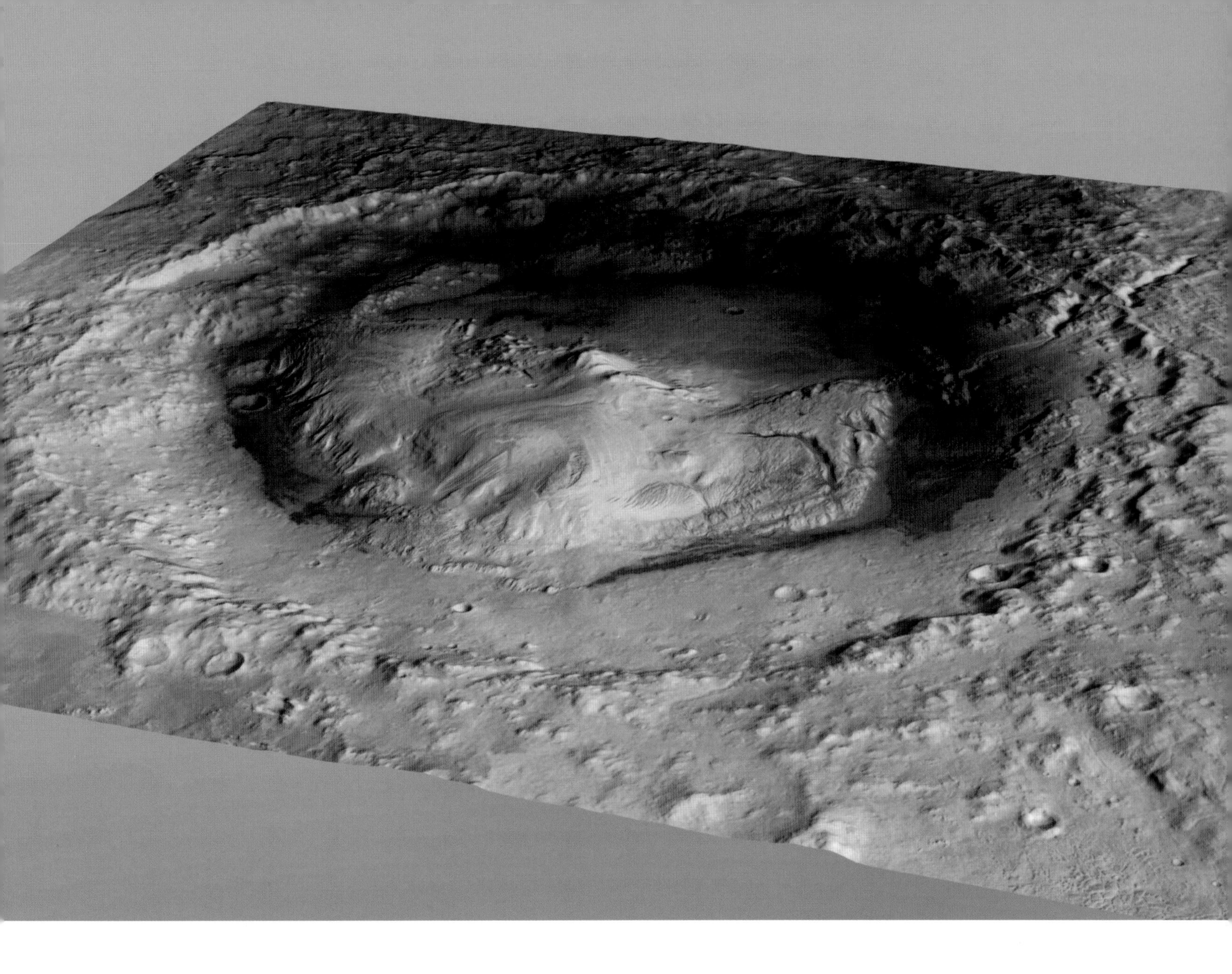

Composite photograph of Gale crater, which the Curiosity rover is exploring. The peak in the center, known as Mount Sharp, contains a geological record of deposits laid down after the crater was formed by an impact more than three billion years ago.

closed ecosystem—the sort of structure you'd need in a Mars colony. In 1994 eight intrepid "bionauts" spent an entire year inside Biosphere 2 as a kind of "proof of concept" operation.

Incidentally, the structure was named "Biosphere 2" because, in the minds of the designers, Earth itself is "Biosphere 1."

Another, longer range strategy for human colonization of Mars lies in the notion of changing the planet to make it more hospitable for a human presence, a process known as terraforming. Some schemes involve introducing gases, such as ammonia or methane, into the Martian atmosphere to trigger a greenhouse effect and raise the surface temperature. Once begun, a process like this would vaporize the large quantities of frozen carbon dioxide (dry ice) at the Martian poles, which would amplify the heating. Other schemes involve massive engineering projects like dismantling the planet's moons and spreading the resulting dark material on the surface to increase the absorption of sunlight or the construction of large mirrors in space to direct sunlight to the planet. At the moment, terraforming remains largely a matter of theoretical speculation rather than serious engineering design.

Dividing the terrestrial and Jovian planets in our solar system is a thin ring of debris known as the asteroid belt. Before we get into a description of the asteroids, we must deal with two popular misconceptions: First, despite the crowded, rock-strewn scenes you may have witnessed in the movies, the asteroid belt is almost completely empty. Spacecraft have flown through the belt without encountering asteroids at all. In fact, the chance of meeting an asteroid during a transit has been estimated to be about one in a billion. Second, the belt is not the remains of an exploded planet—it contains only a tiny fraction of a planet's mass. The "exploded planet" theory was popular in the 19th century and probably served as the inspiration for Krypton, Superman's fictional home planet.

ASTEROID BELT

REMNANTS FROM THE SOLAR SYSTEM'S BIRTH

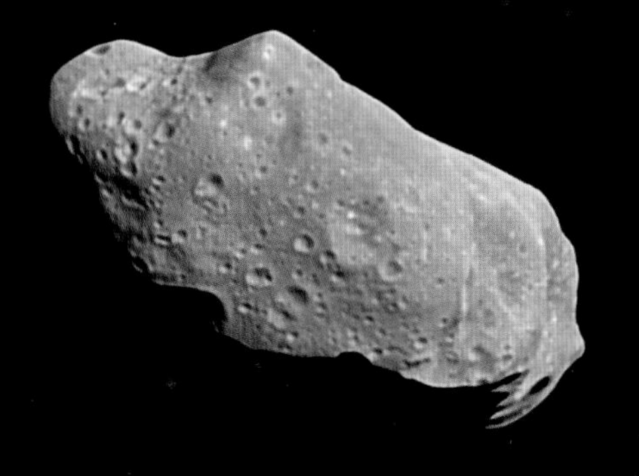

DISCOVERER: **GIUSEPPE PIAZZI**
DISCOVERY DATE: **JANUARY 1, 1801**
NAMED FOR: **GREEK WORD FOR "STARLIKE"**

LOCATION OF MAIN BELT: **BETWEEN 2.1 AND 3.3 AU FROM SUN**
TOTAL NUMBER: **MORE THAN 570,000**
DIAMETER OF LARGEST ASTEROIDS:
CERES: **950 KM (640 MI)**
VESTA: **580 KM (360 MI)**
PALLAS: **540 KM (335 MI)**
HYGIEA: **430 KM (270 MI)**
NEAR-EARTH ASTEROIDS DISCOVERED TO DATE: **8,484**

Artist's conception of Ceres, largest asteroid.
(Inset) Asteroid Ida and its satellite Dactyl

The first asteroid, Ceres, was discovered in 1801. Because the telescopes of that time produced pointlike images of asteroids, the small, dimly reflective bodies resembled stars, hence their name ("asteroid" comes from the Greek for "starlike"). It is no accident that Ceres was the first to be seen: It is the largest body in the asteroid belt at about 950 kilometers (640 mi) across. Ceres is the only object in the belt big enough to be spherical—the mutual gravitational attraction between its constituents pulls it into that shape—and it constitutes about a third of the total mass in the asteroid belt. Technically, this makes it a dwarf planet (see pages 202–203). Everything else in the belt is smaller and irregular in shape. It is estimated that there are more than a million asteroids larger than 1 kilometer (0.6 mi) across.

Not all asteroids are found in the asteroid belt. Some travel in paths that bring them inside the orbits of Mars and Earth—which raises the possibility that they might collide with either planet.

HOW THE BELT TOOK SHAPE

When the solar system was forming, the process of planetesimal accumulation (see pages 56–7) went on in the asteroid belt as it did throughout the inner solar system. Had it not been for the presence of Jupiter, there might well have been a planet where the belt is now. One theory has it that the gravitational pull of the giant planet sped up nearby planetesimals in their orbits, so that they were either knocked out of the asteroid belt by the influence of Jupiter directly or split up when they collided with each other. In either case, it would appear that the influence of Jupiter was able to prevent a planet from forming.

Indeed, computer models suggest that much of the material in the original asteroid belt was ejected in

TARGET: EARTH

On a peaceful day 65 million years ago, dinosaurs were going about their daily business when a piece of rock about 12 kilometers (8 mi) across came hurtling out of the sky. Carrying an energy equivalent to many thousands of times the total modern arsenal of nuclear weapons, the asteroid hit near what is now the Yucatán Peninsula in Mexico, blasting out a crater 180 kilometers (112 mi) across and several kilometers deep. The impact and the pulverized rock thrown up by the impact set in motion a train of events that wiped out fully two-thirds of all the species of plants and animals on Earth, including the dinosaurs. Aptly termed a mass extinction event, the extinction of the dinosaurs and all those other species marks just one of many such events in the history of our planet.

More recently, in 1908, a rock some 10s of meters across (roughly 50 feet or so) fell out of the sky and exploded in the air over the Tunguska River in Siberia. The impact, estimated to have a thousand times the energy of a World War II–era atomic bomb, leveled trees as far as 70 kilometers (50 mi) from the blast center.

These two events highlight an important truth: Earth is part of the larger solar system, and occasionally an asteroid shows up to remind us of that fact. You can verify this for yourself by watching for shooting stars in the sky, pebble-size rocks burning up in the atmosphere.

In fact, collisions with objects like the one that did in the dinosaurs are expected to occur every 100 million years or so, and smaller collisions are expected to be more frequent – a kilometer-size (1,000 yd) object will hit every 70,000 years, and a rock 140 meters (153 yd) across every 30,000 years. A Tunguska-size event could happen every few centuries. Asteroids in the size range of 5 to 10 meters (15 to 30 ft) in diameter enter Earth's atmosphere about once a year but typically explode at high altitude and cause little (if any) damage.

In 2005, recognizing the danger of these impacts, Congress directed NASA to catalog 90 percent of all detectable asteroids, comets, and other potentially dangerous objects near Earth by 2020. (Funding shortfalls make it unlikely that this deadline will be met.) One important part of the search for dangerous asteroids is NASA's Asteroid Grand Challenge program, whose goal is to find all asteroid threats to human beings. This is a large task – we estimate today, for example, that we have cataloged only about 10 percent of asteroids whose diameter is less than 300 meters (900 ft) and one percent of asteroids whose diameter is less than 100 meters (300 ft).

the first million years or so of the solar system's history, with a good portion of the rest being expelled during the Late Heavy Bombardment. During that early period, the minerals in asteroids were subject to influences such as heating from collisions, the radioactive decay of nuclei, and, in some cases, the same process of differentiation that produced the structure of Earth (see page 88). Asteroids may help us understand how.

MISSIONS TO THE ASTEROIDS

Beginning in 1972, a number of spacecraft—the Pioneers, the Voyagers, and Ulysses—passed through the asteroid belt without mishap, but none of them tried to image the asteroids they passed. Since then, we have obtained flyby images from a number of spacecraft bound for other targets—Galileo on its way to Jupiter, for example, and Cassini traveling to Saturn. The probe NEAR (Near Earth Asteroid Rendezvous) went into orbit around the near-Earth asteroid Eros in 2000. In 2010 the Japanese probe Hayabusa returned after a seven-year, five-billion-kilometer (3 billion mi) landing mission to a small asteroid named Itokawa. Although the spacecraft burned up on entry into Earth's atmosphere, the sample return packet, containing small particles of the asteroid, landed safely in Australia. Analysis showed that the asteroid indeed dated to the early days of the solar system.

NASA's DAWN mission, launched in 2007, went into orbit around the massive asteroid Vesta in 2011; in 2015 it went into orbit around Ceres.

There are a number of other missions planned. In 2016 NASA launched the OSIRIS REX (Origins, Spectral Interpretation, Security, Regolith Explorer). The ARM mission (Asteroid Redirect Mission) is slated for launch in the 2020s.

EARTH IMPACT

I CERES

Discovered in 1801 by Italian priest and astronomer Giuseppe Piazzi, Ceres is the largest object in the asteroid belt between Mars and Jupiter. It was reclassified as a dwarf planet in 2006.

WESTERN HEMISPHERE

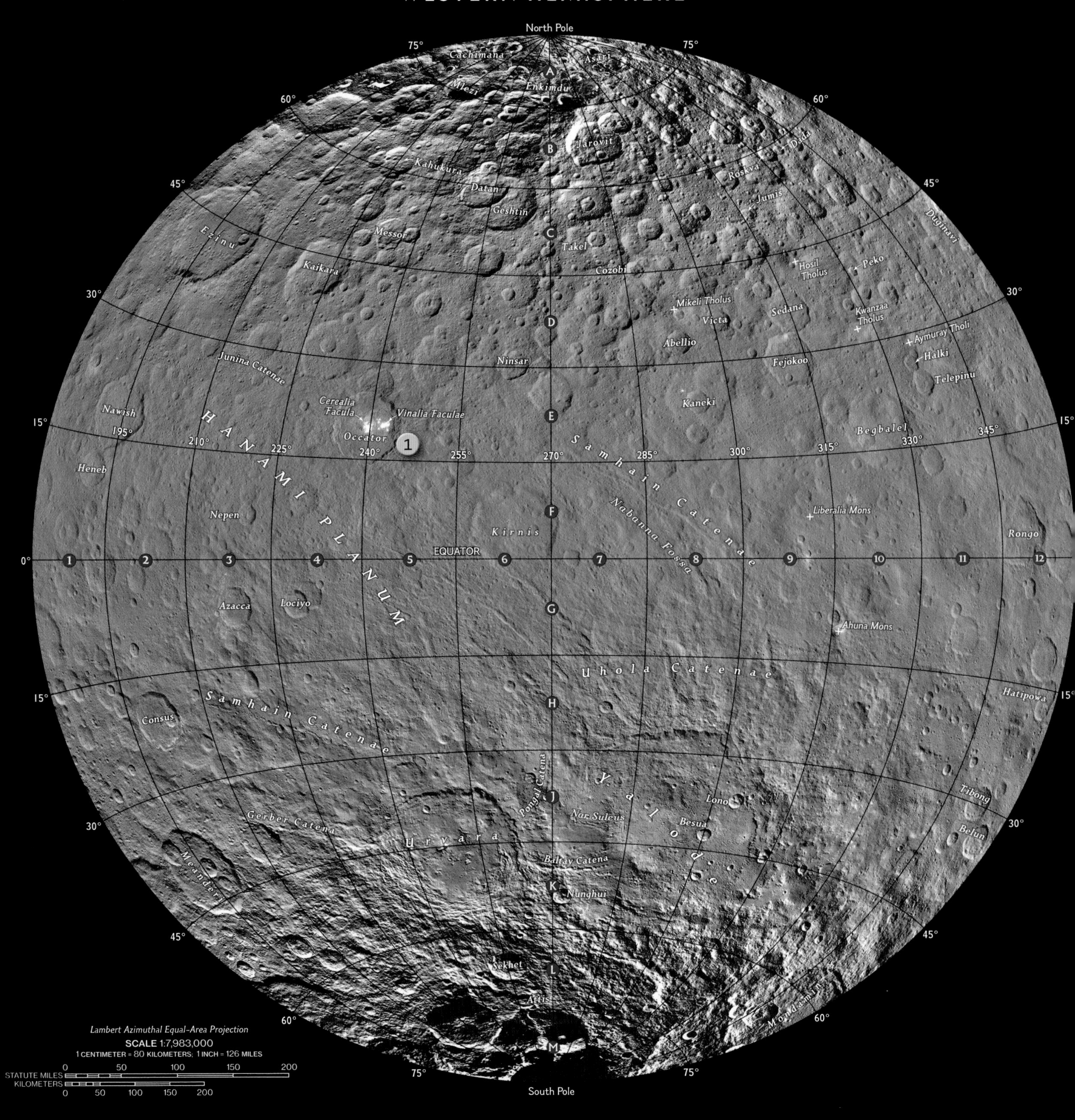

KEY FEATURES

1. **ACCATOR:** This 57-mile-wide crater contains lightly colored deposits.
2. **VENDIMIA PLANITIA:** A rugged plain covered in impact craters
3. **KERWAN:** Large crater named for the Hopi spirit of sprouting corn

EASTERN HEMISPHERE

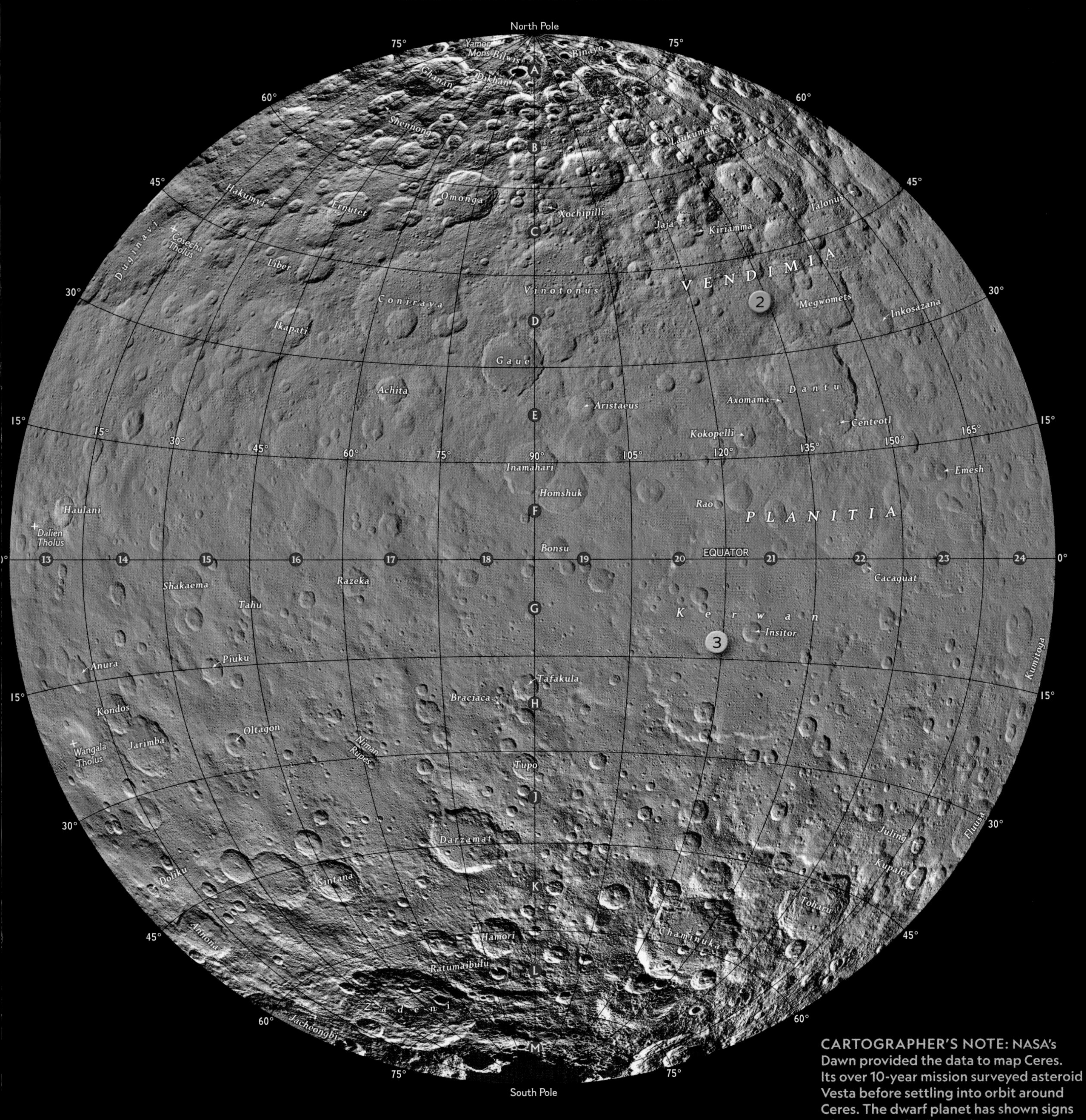

CARTOGRAPHER'S NOTE: NASA's Dawn provided the data to map Ceres. Its over 10-year mission surveyed asteroid Vesta before settling into orbit around Ceres. The dwarf planet has shown signs of recent geologic activity.

OUTER PLANETS

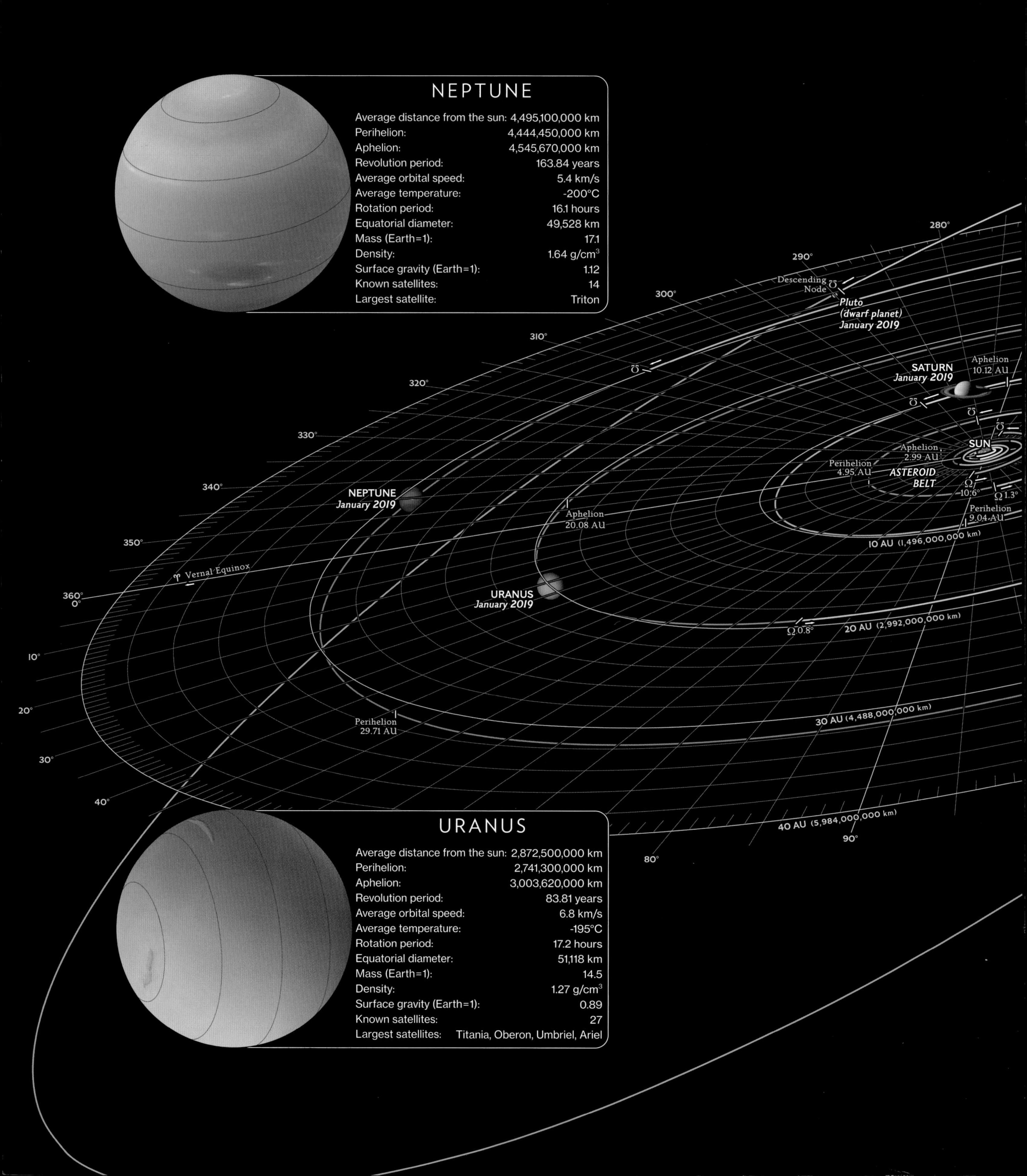

NEPTUNE

Average distance from the sun:	4,495,100,000 km
Perihelion:	4,444,450,000 km
Aphelion:	4,545,670,000 km
Revolution period:	163.84 years
Average orbital speed:	5.4 km/s
Average temperature:	-200°C
Rotation period:	16.1 hours
Equatorial diameter:	49,528 km
Mass (Earth=1):	17.1
Density:	1.64 g/cm^3
Surface gravity (Earth=1):	1.12
Known satellites:	14
Largest satellite:	Triton

URANUS

Average distance from the sun:	2,872,500,000 km
Perihelion:	2,741,300,000 km
Aphelion:	3,003,620,000 km
Revolution period:	83.81 years
Average orbital speed:	6.8 km/s
Average temperature:	-195°C
Rotation period:	17.2 hours
Equatorial diameter:	51,118 km
Mass (Earth=1):	14.5
Density:	1.27 g/cm^3
Surface gravity (Earth=1):	0.89
Known satellites:	27
Largest satellites:	Titania, Oberon, Umbriel, Ariel

Among the outer planets, Jupiter and Saturn are well studied. The Galileo probe introduced observers to each of the Jovian moons and revealed the largest planet's character. The Cassini-Huygens mission, which ended in 2017, relayed data on Saturn and its satellites, with special attention given to Titan. The other outer planets, visited only by the Voyager spacecraft, have many questions left unanswered.

CARTOGRAPHER'S NOTE: Vast distances and long years characterize the four outer planets and dwarf planet Pluto. Mighty Jupiter, largest of the giants and first of the outer planets, lies at more than 5 AU; at its most distant extent, Pluto orbits more than 48 AU from the sun. Jupiter's great mass keeps the asteroids contained and attracts visiting comets. Unlike the inner planets, the gas giants are rich in natural satellites, with more than 160 moons among them.

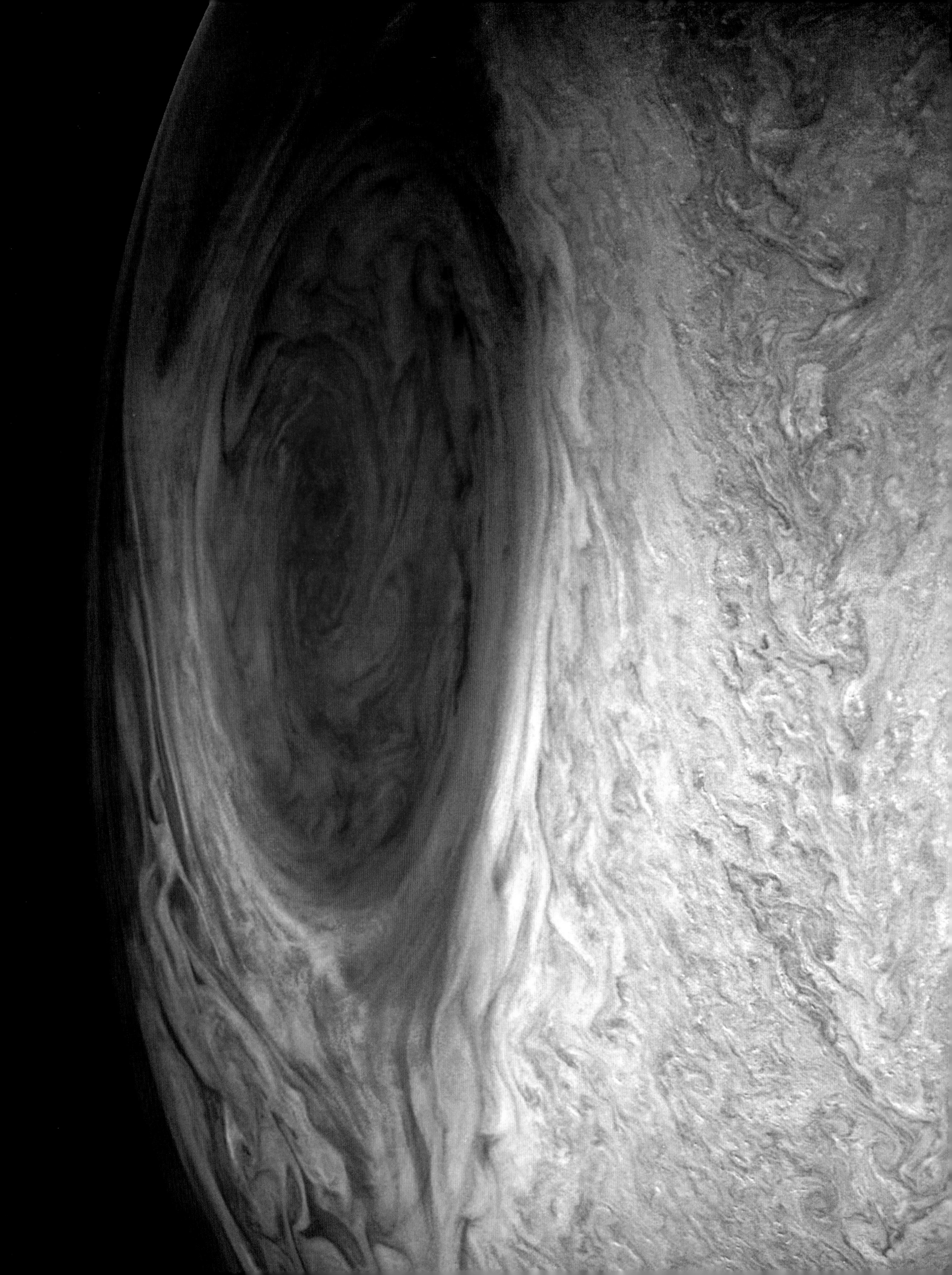

The fifth planet from the sun, Jupiter is also the largest, packing greater than two and a half times the mass of all the other planets combined. As seen through a telescope, it is a beautiful thing, with alternating stripes of different colors, surrounded by a swarm of small moons. • Jupiter, the first of the gas giants, exhibits some unusual traits that appear in all the outer planets. In the first place, Jupiter doesn't really have a "surface." Dropping down into Jupiter's atmosphere would be like sinking into a milkshake, with the density increasing as the surroundings changed from gas to liquid to slush. We wouldn't encounter anything we would call a solid surface until, perhaps, we got near the very center.

JUPITER

KING OF THE GAS GIANTS

DISCOVERER: **UNKNOWN**
DISCOVERY DATE: **PREHISTORIC**
NAMED FOR: **KING OF THE ROMAN GODS**

MASS: **317.82 × EARTH'S**
VOLUME: **1,321.34 × EARTH'S**
MEAN RADIUS: **69,911 KM (43,441 MI)**
EFFECTIVE TEMPERATURE: **−148°C (−234°F)**
LENGTH OF DAY: **9.92 HOURS**
LENGTH OF YEAR: **11.86 EARTH YEARS**
NUMBER OF MOONS: **65 (51 NAMED)**
PLANETARY RING SYSTEM: **YES**

This photo of the Great Red Spot was created with data from NASA's JUNO spacecraft. (Inset) Jupiter with shadow of Europa.

I JUPITER

Unlike the inner planets, Jupiter does not have a rocky surface with craters, mountains, or valleys. Its complex, stormy atmosphere changes over time.

Usually some shade of reddish brown, warmer belts are composed of regions where low-pressure gases are beginning to sink through the atmosphere. Where belts and zones collide, great areas of turbulence ensue, giving rise to the many storm systems that speed around Jupiter's atmosphere.

Jupiter's magnetosphere—the magnetic field that surrounds the planet—while similar to Earth's, is about 20,000 times more intense. Its pull captures charged particles from solar wind, forming vast and intense bands of radiation. The effect is similar to Earth's Van Allen belts but immensely more powerful. Any unprotected spacecraft would quickly receive a destructive dose.

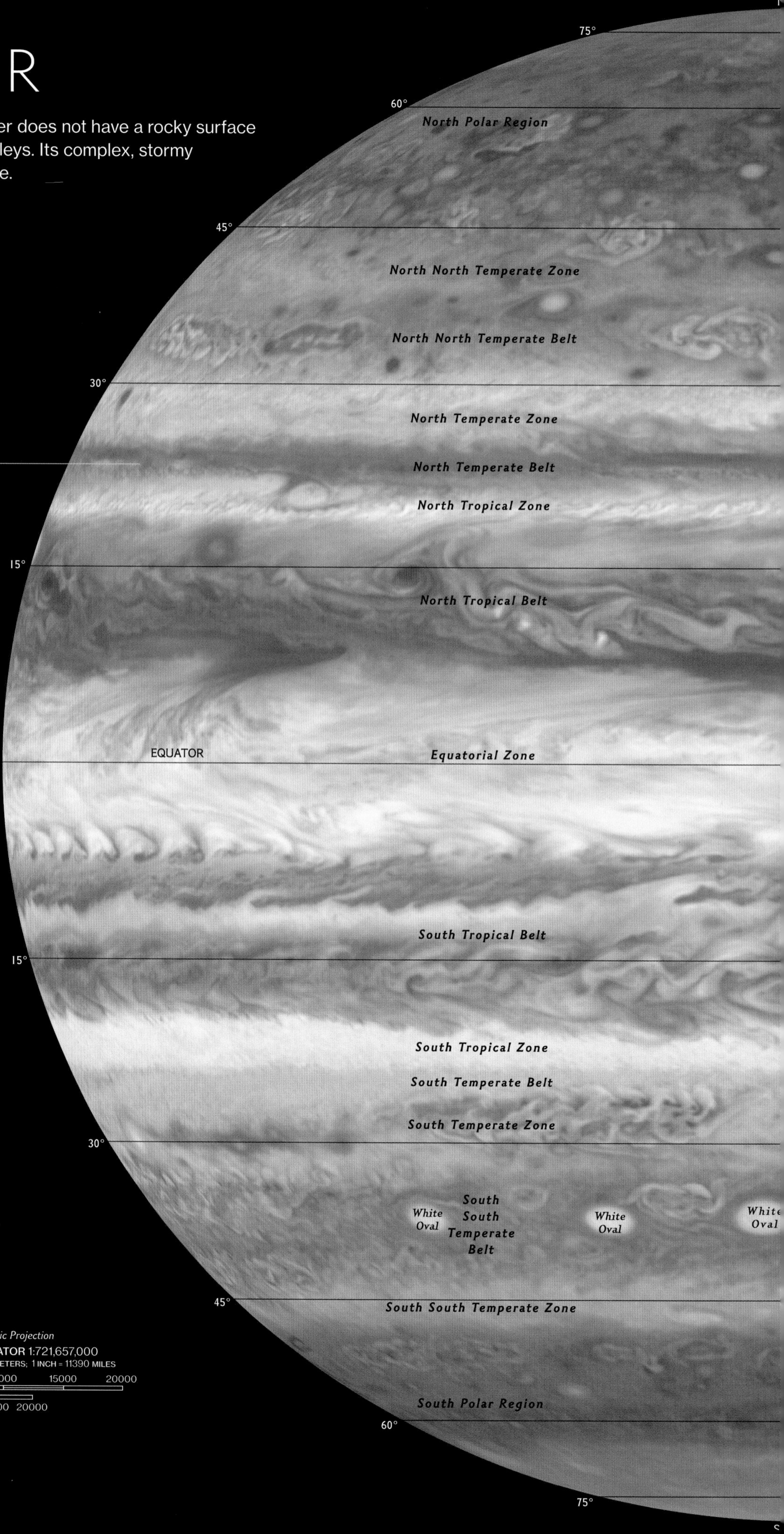

Orthographic Projection
SCALE at the EQUATOR 1:721,657,000
1 CENTIMETER = 7216 KILOMETERS; 1 INCH = 11390 MILES
STATUTE MILES 0 5000 10000 15000 20000
KILOMETERS 0 5000 10000 15000 20000

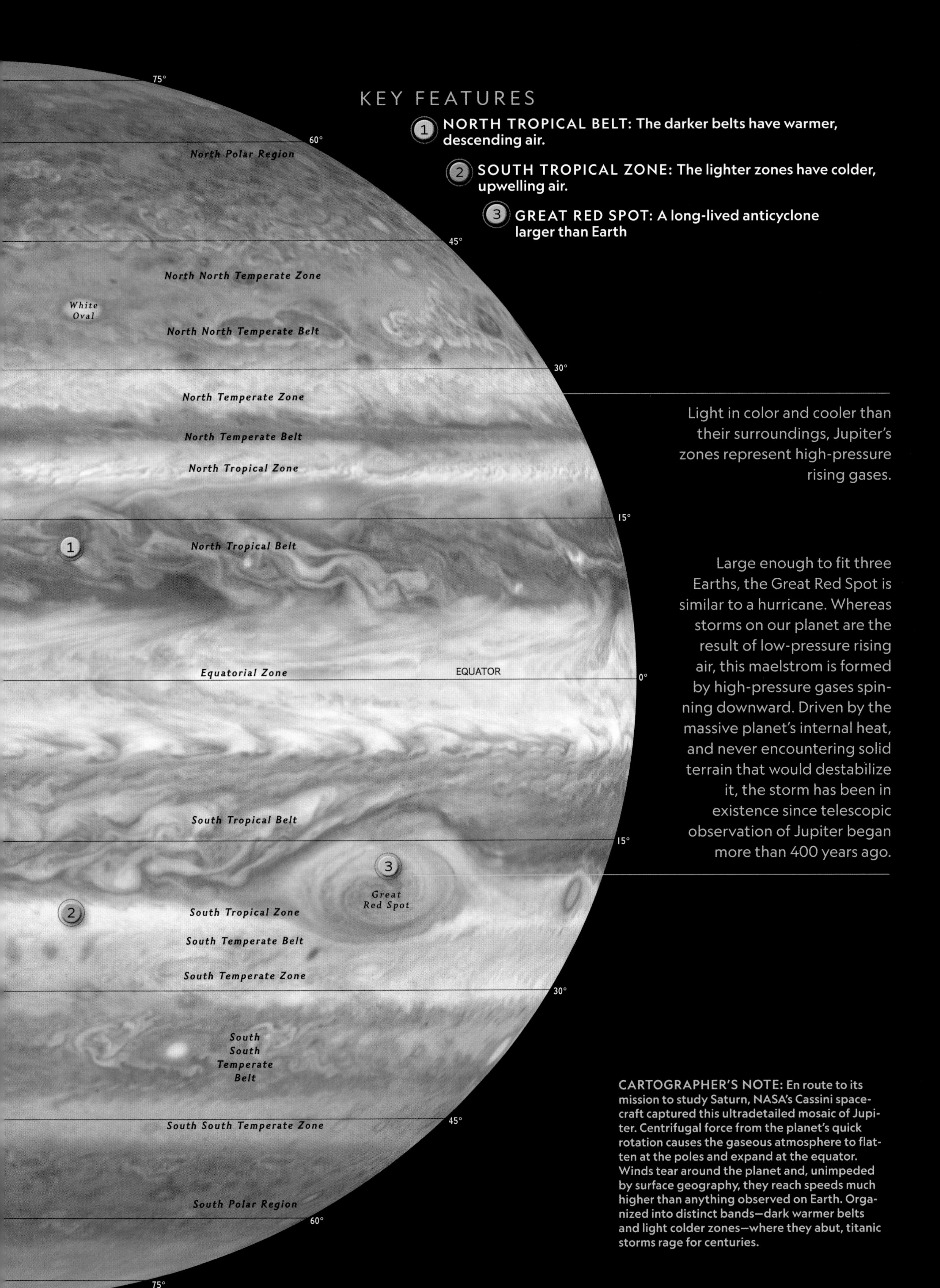

KEY FEATURES

1. **NORTH TROPICAL BELT: The darker belts have warmer, descending air.**
2. **SOUTH TROPICAL ZONE: The lighter zones have colder, upwelling air.**
3. **GREAT RED SPOT: A long-lived anticyclone larger than Earth**

Light in color and cooler than their surroundings, Jupiter's zones represent high-pressure rising gases.

Large enough to fit three Earths, the Great Red Spot is similar to a hurricane. Whereas storms on our planet are the result of low-pressure rising air, this maelstrom is formed by high-pressure gases spinning downward. Driven by the massive planet's internal heat, and never encountering solid terrain that would destabilize it, the storm has been in existence since telescopic observation of Jupiter began more than 400 years ago.

CARTOGRAPHER'S NOTE: En route to its mission to study Saturn, NASA's Cassini spacecraft captured this ultradetailed mosaic of Jupiter. Centrifugal force from the planet's quick rotation causes the gaseous atmosphere to flatten at the poles and expand at the equator. Winds tear around the planet and, unimpeded by surface geography, they reach speeds much higher than anything observed on Earth. Organized into distinct bands—dark warmer belts and light colder zones—where they abut, titanic storms rage for centuries.

The second thing to know about gas giants is that at the enormous pressures that exist inside a planet this size, matter can be forced into some pretty unusual forms, which means that the interior structures of the gas giant planets aren't like anything we've seen so far. Neither are the temperatures: These range from minus 148°C (–234°F) at the top of Jupiter's cloud layer to around 24,000°C (43,000°F)—hotter than the surface of the sun—in the core!

THE METAL OCEAN

Gravitational measurements suggest that Jupiter may have a small rocky core, possibly with as much as 20 to 40 times the mass of Earth. Surrounding this core is a layer of a strange material called metallic hydrogen. Again, we are used to thinking of hydrogen as a gas, or perhaps, at very low temperatures, as an ordinary liquid. Pressures in the interior of Jupiter are so high, however, that the atoms are forced into a state that, while still a liquid, has the properties of a metal. (If you have trouble thinking of a liquid as a metal, picture mercury.) Metallic hydrogen, an exotic rarity on Earth, actually constitutes a significant fraction of the mass of Jupiter. The layer

A cutaway view of our current idea of the structure of Jupiter. Jupiter is unusual because its rocky core is surrounded by a layer of metallic hydrogen, a material not found naturally on Earth. It exists on Jupiter because of the extremely high pressures in the interior.

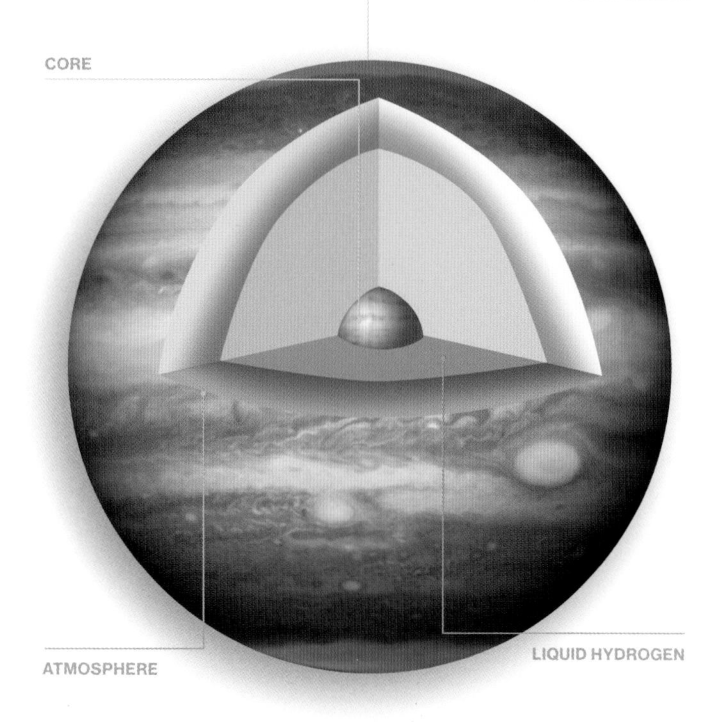

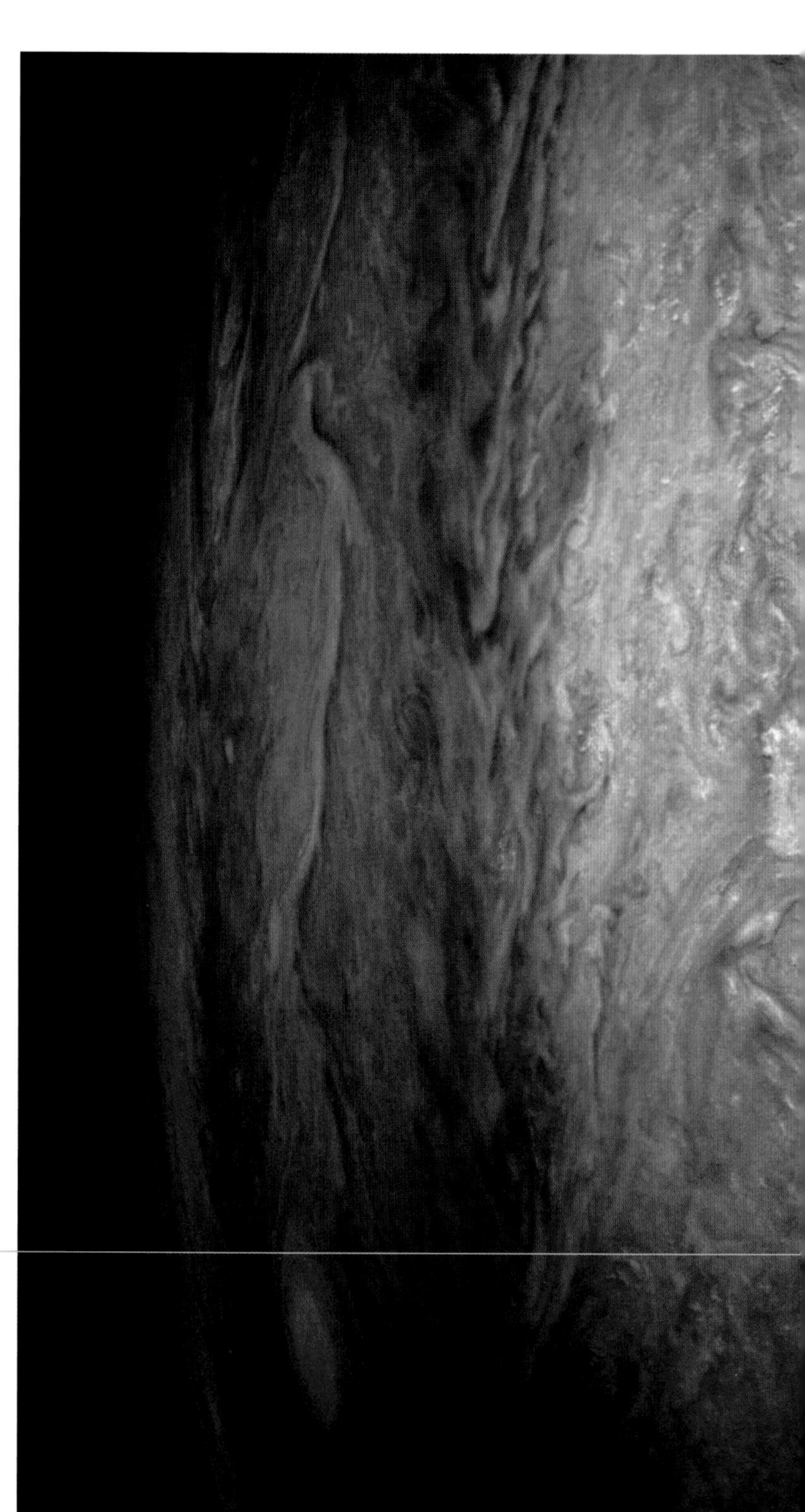

This photo of the edge of the Great Red Spot, based on JUNO data, illustrates the turbulence of the Jovian atmosphere.

of metallic hydrogen is covered by a layer of ordinary liquid hydrogen: hydrogen atoms forced together by the tremendous pressures in the planetary interior. There probably isn't a sharp transition between the two states of hydrogen—certainly nothing we would call a surface.

BELTS AND ZONES

The upper atmosphere, of course, is what we actually see when we look at Jupiter. This outermost layer of the planet is composed almost entirely of hydrogen and helium (75 percent to 24 percent by weight). The colorful bands result from a rather complex cloud structure in the top 50 kilometers (35 mi). There are two main layers of clouds. Upwelling materials from farther down in Jupiter's atmosphere mix with the lower layer and change color when they encounter the ultraviolet rays of the sun. These are the dark bands, or "belts," that we see. The lighter bands, or "zones," are clouds of crystallized ammonia carried upward, hiding the darker lower layer from view. The deepest layers, seen only occasionally, appear blue.

We are used to the fact that on Earth the prevailing winds blow in different directions at different latitudes—the tropical trade winds blowing from east to west, for example, whereas winds in higher latitudes go from west to east. The same thing ha ppens on Jupiter, but its fast rotation (its "day" is about 10 hours long) and greater size produce many more counterrotating bands than on Earth. This together with the complicated cloud dynamics discussed above is what gives rise to the colored stripes we see when we look at the planet.

As befits its role as king of the planets, Jupiter has a strong magnetic field—20,000 times as strong as Earth's—most likely produced by motions in its metallic core. This strong field has the effect of diverting charged particles streaming outward from the sun away from the planet, creating what scientists call a bow shock. The largest four moons of Jupiter have their orbits inside this protected zone.

Perhaps the most striking feature on Jupiter is the Great Red Spot, a storm in the southern hemisphere. This spot may have been seen as early as 1665, and it was sketched by astronomers in 1831. It is so big that the entire Earth could be dropped into it with no trouble. Some theorists think it may actually be a permanent feature of the planet. Interestingly enough, at the end of the 20th century, astronomers observed what may be the start of a similar but smaller storm on Jupiter, which they nicknamed Red Spot Junior.

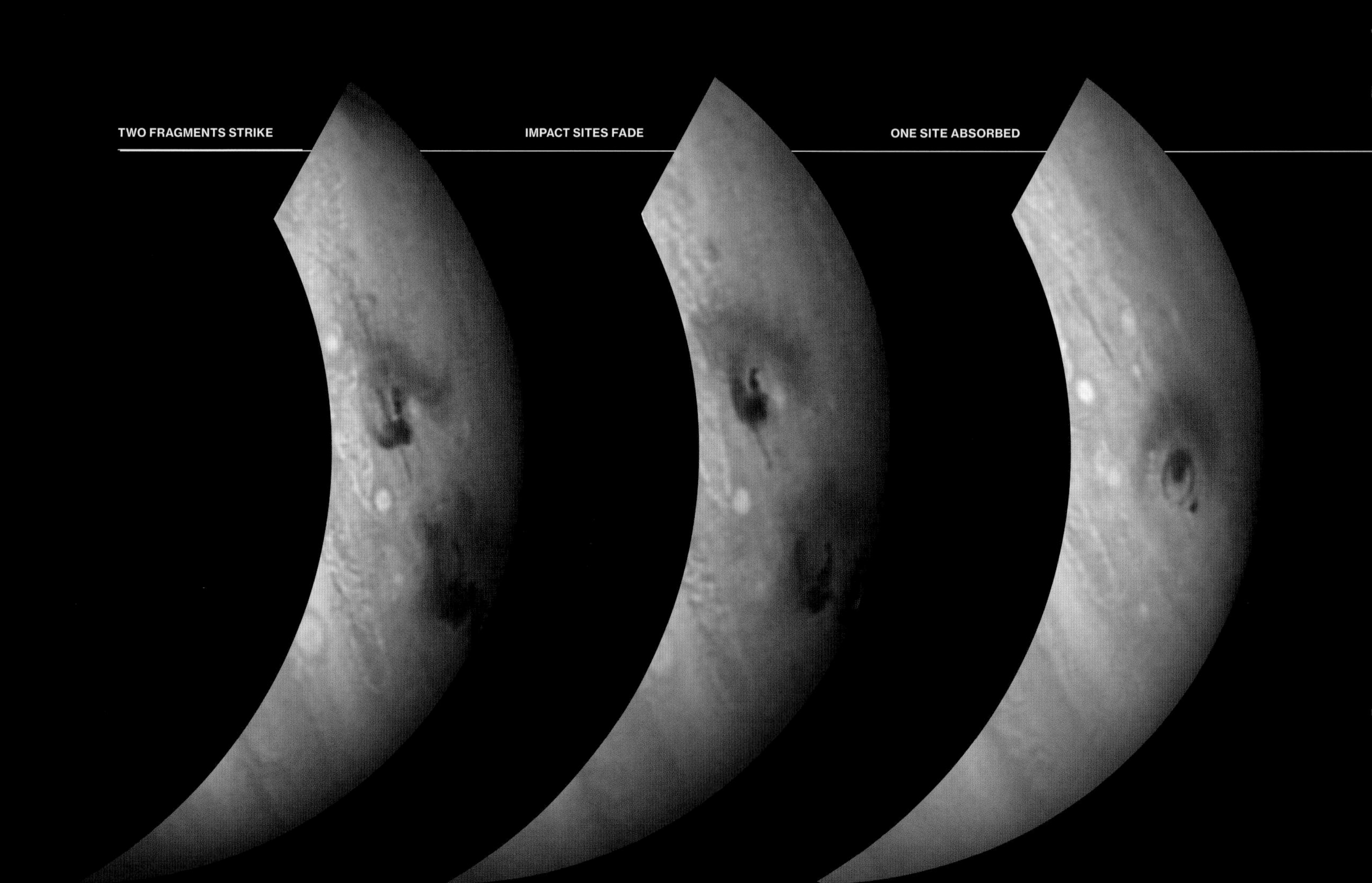

VISITING JUPITER

Jupiter is bright in the night sky. Galileo was the first to observe the planet with a telescope in 1610 and the first to document the fact that the planet has moons. Many spacecraft have flown by Jupiter on their way to other destinations. The Voyager spacecraft in 1979 discovered that Jupiter, like all of the Jovian planets, has a system of rings. In Jupiter's case, there are three rings that appear to be made out of dust ejected from nearby moons.

The main space probe associated with Jupiter, however, was the Galileo mission, launched in 1989 and inserted into orbit around the planet in 1995. Later in 1995 it dropped a probe into the Jovian atmosphere. The probe descended and returned data for almost an hour before it was crushed by the pressure 153 kilometers (95 mi) down. For seven years, the orbiting Galileo spacecraft gathered a trove of data on Jupiter and its moons. In 2003 the spacecraft met its end when it was deliberately steered into the Jovian atmosphere to eliminate any possibility that it would contaminate the moon Europa—a place where, as we shall see, scientists believe that we might find life.

In August 2011 NASA launched its most recent Jovian mission, named Juno. Juno went into orbit around Jupiter in July 2016. During the course of its 20-month mission, it will orbit the giant planet 37 times. At the end of that time it will be plunged into the Jovian atmosphere (an operation that is called "deorbiting" in NASA speak) to prevent contamination of the moons.

JUPITER TAKES A HIT

BOTH FRAGMENTS ABSORBED BY ATMOSPHERE

One of the most spectacular events in astronomical history took place during July 1994. A comet named Shoemaker-Levy (comets are customarily named after the people who discover them) collided with Jupiter, giving observers on Earth a ringside seat to a cosmic show.

The comet was first seen in orbit around Jupiter in 1993. Calculations indicated it had been captured by the planet's gravitational field and broken into fragments a year earlier. As it became clear that the fragments were going to collide with Jupiter, all of the astronomical instruments available to astronomers – land-based and in space – were turned toward the giant planet. The hope was that the impacts would churn up the Jovian atmosphere and give scientists a look at what lay between the outer cloud layers.

For six days after July 16, sky-watchers observed no fewer than 21 impacts. The Galileo spacecraft (see above) was in a position to see the actual collisions, which took place on the far side of the planet, and the Hubble Space Telescope (see sidebar page 303) took spectacular pictures. The initial blows produced fireballs and churned up areas of the Jovian atmosphere as wide as the radius of Earth. Elements such as sulfur were seen in the aftermath of the collisions but, against expectations, very little water. Scientists are still poring over the data taken during the collisions, refining their theories about the structure of the largest planet in the solar system.

Jupiter has 63 moons and counting. Most of them are rocky, irregular lumps, some only a few miles across. The consensus is that these moons are captured asteroids. In fact, only eight of the Jovian moons fit the standard image of what a moon should be—spherical bodies in equatorial orbits around their planet—and of these, four are small bodies with orbits close to Jupiter. These four inner moons may be the source of the dust that makes up Jupiter's rings. From our point of view, the most important moons of Jupiter are the remaining four—the so-called Galilean moons.

JUPITER'S MOONS

WATER WORLDS, VOLCANO WORLDS, AND ROCKS

DISCOVERER: **FIRST FOUR DISCOVERED BY GALILEO GALILEI**
DISCOVERY DATE: **JANUARY 1610**
NAMED FOR: **JUPITER'S LOVERS AND DESCENDANTS**

LARGEST MOONS, BY RADIUS:
GANYMEDE: **2,631 KM (1,635 MI)**
CALLISTO: **2,410 KM (1,498 MI)**
IO: **1,822 KM (1,132 MI)**
EUROPA: **1,561 KM (970 MI)**
HIMALIA: **85 KM (53 MI)**
AMALTHEA: **83 KM (52 MI)**
THEBE: **49 KM (30 MI)**

The volcanic surface of Jupiter's moon Io. (Inset, left to right) The moons Io, Europa, Ganymede, and Callisto.

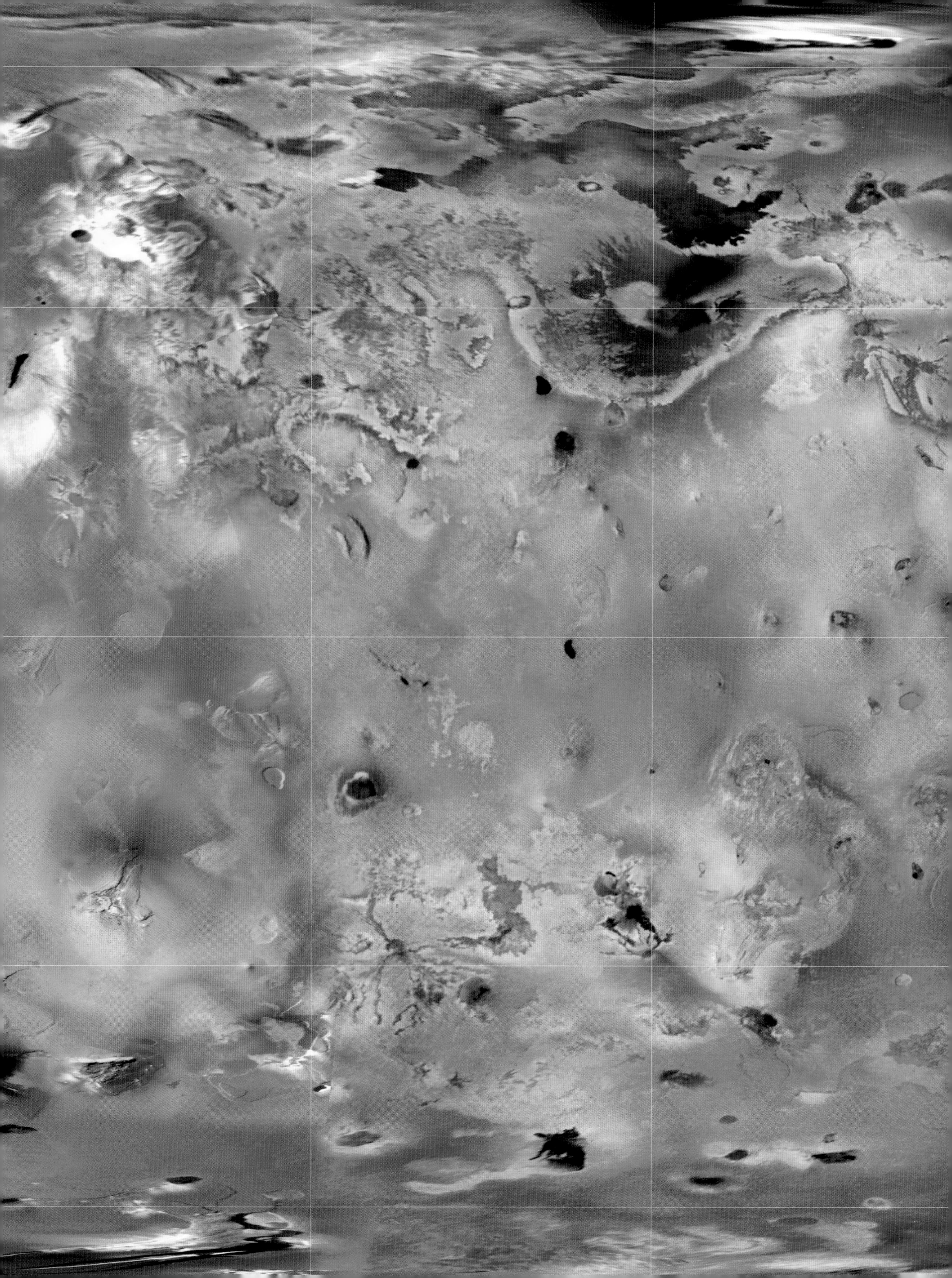

I JUPITER'S MOONS IO

Stretched and pulled between Jupiter's massive gravity and the other Galilean moons, tidal heating makes Io the most volcanic body in the solar system.

WESTERN HEMISPHERE

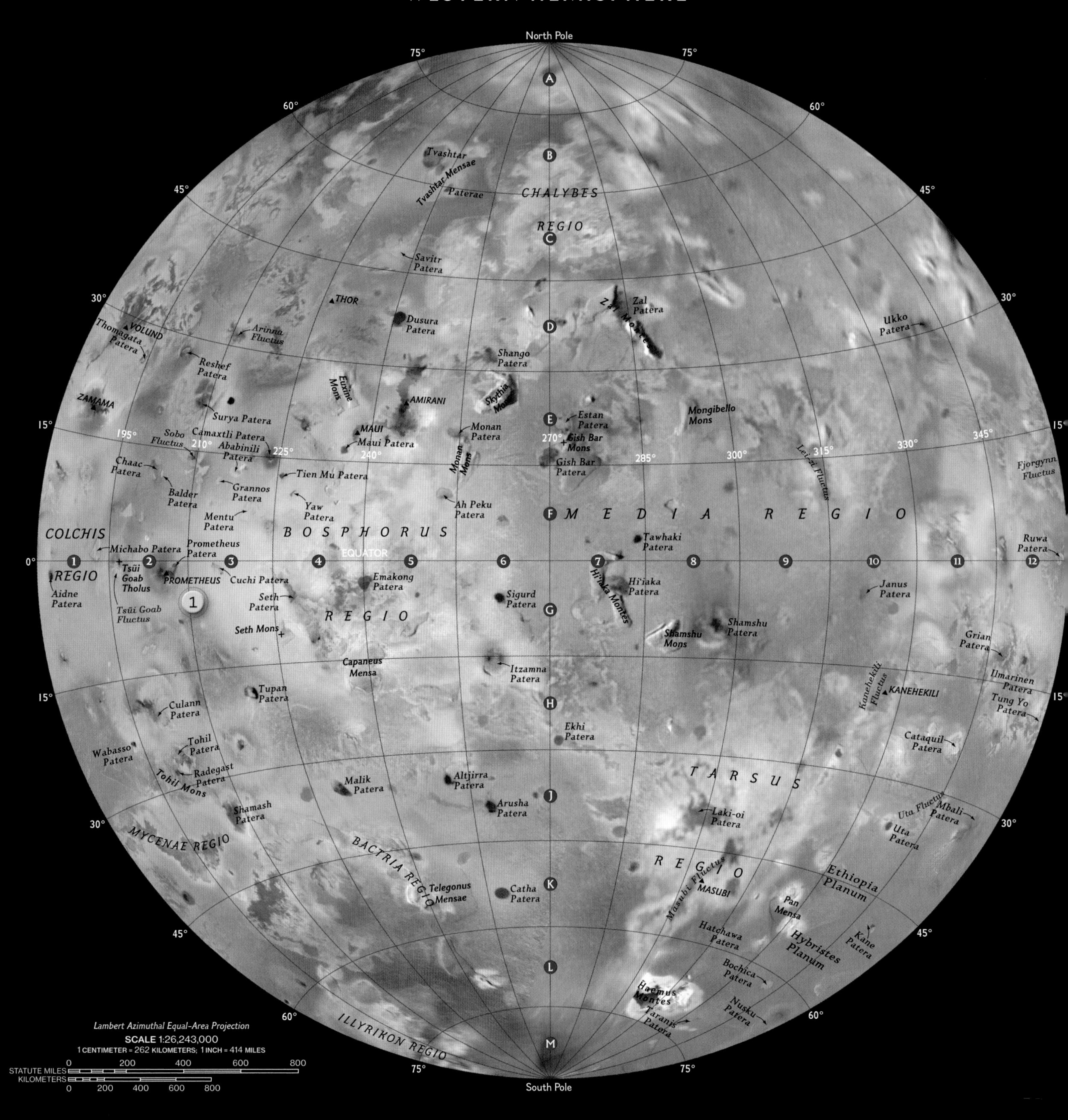

KEY FEATURES

1. **PROMETHEUS:** Large, active volcano
2. **LOKI PATERA:** Volcanic depression
3. **MAFUIKE PATERA:** Orange coloration on Io caused by sulfur compounds

EASTERN HEMISPHERE

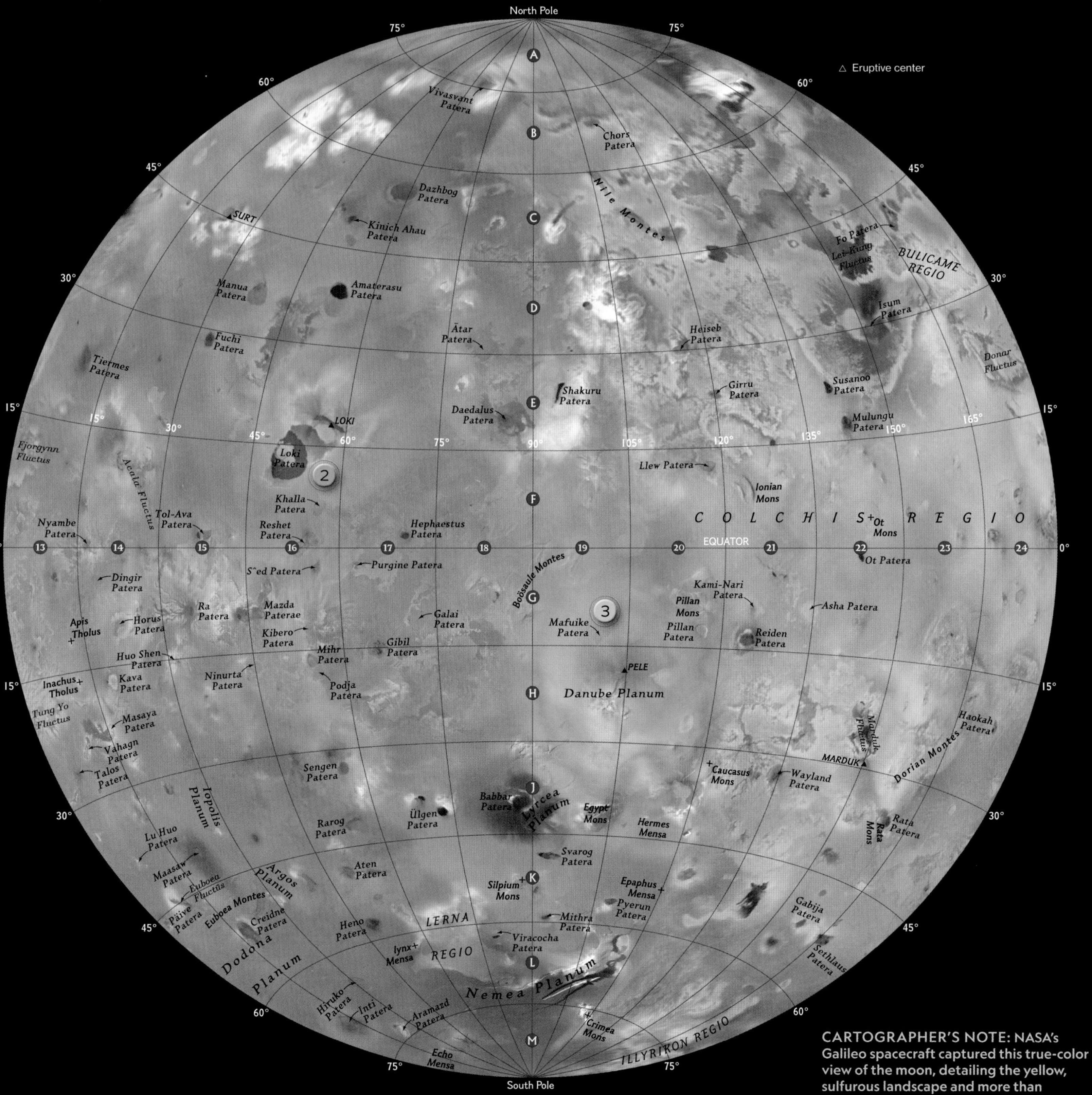

CARTOGRAPHER'S NOTE: NASA's Galileo spacecraft captured this true-color view of the moon, detailing the yellow, sulfurous landscape and more than 400 active volcanoes, named for ancient

JUPITER'S MOONS EUROPA

Scientists believe that the gravitational pull of Jupiter and other moons produces enough heat to allow liquid water to exist below Europa's icy surface.

WESTERN HEMISPHERE

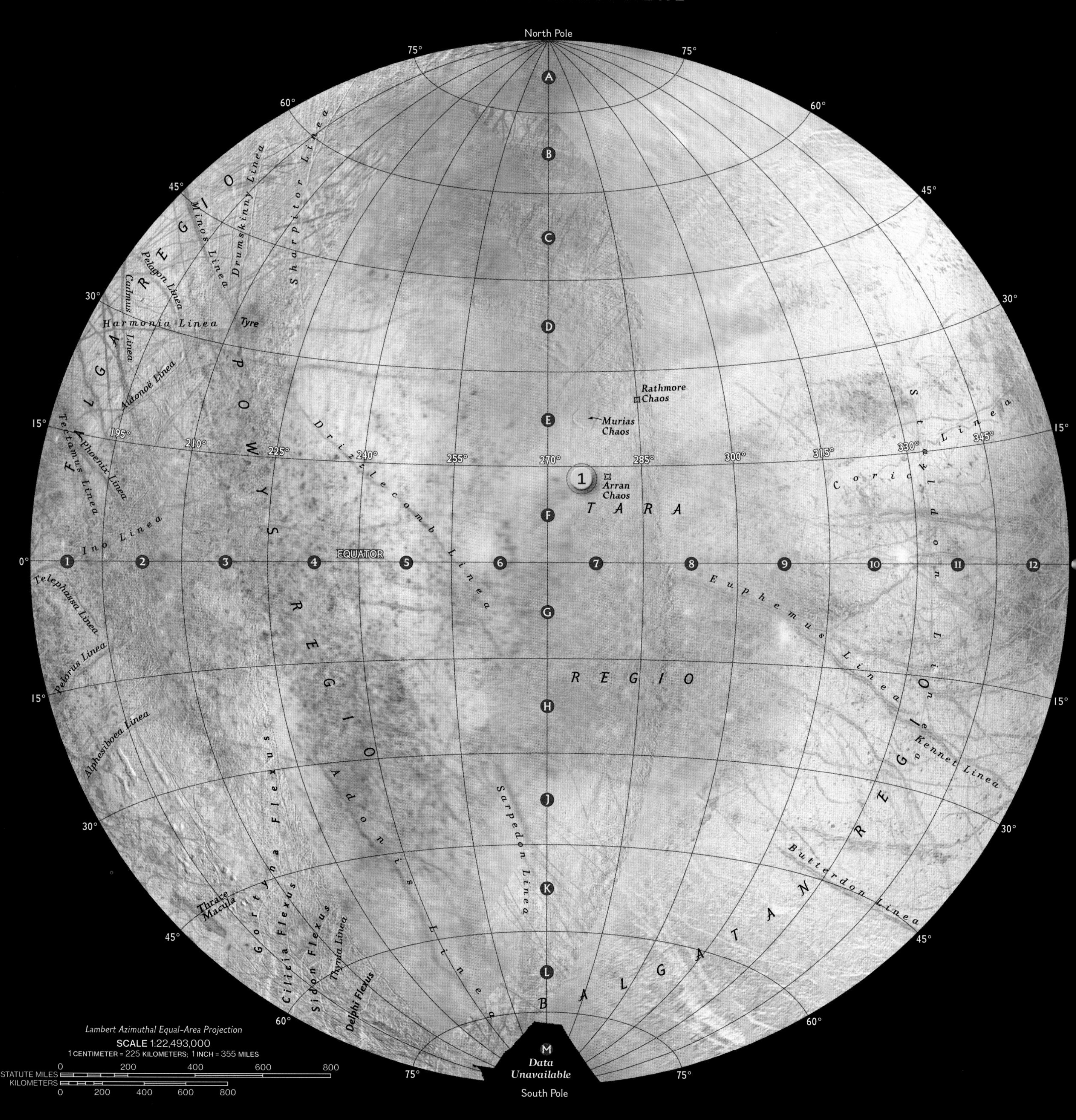

KEY FEATURES

1. **ARRAN CHAOS:** A jumbled region that may sit over a liquid lake
2. **ANDROGEOS LINEA:** One of many cracks in Europa's ice
3. **PWYLL:** A young impact crater surrounded by white water ice

EASTERN HEMISPHERE

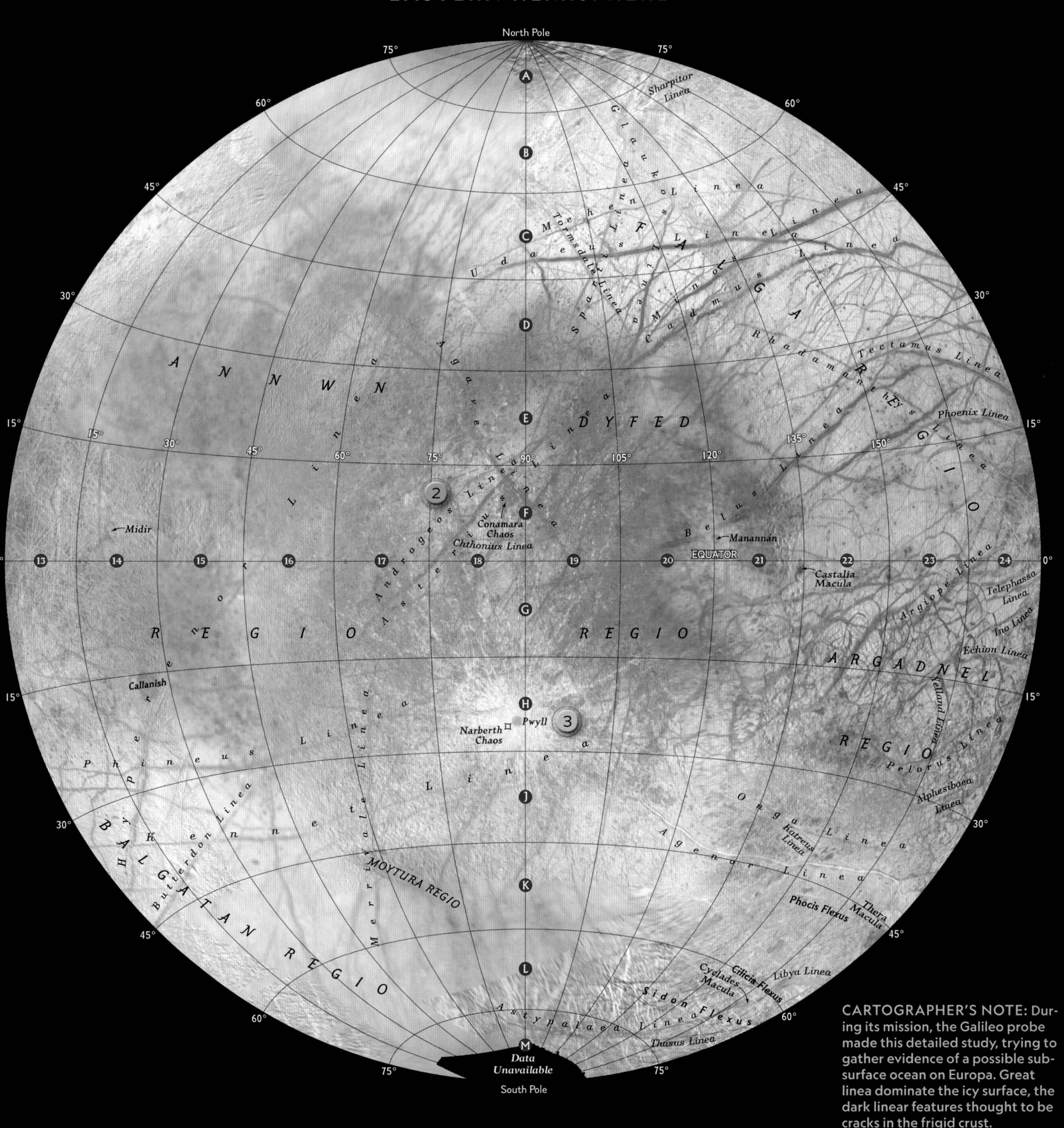

CARTOGRAPHER'S NOTE: During its mission, the Galileo probe made this detailed study, trying to gather evidence of a possible subsurface ocean on Europa. Great linea dominate the icy surface, the dark linear features thought to be cracks in the frigid crust.

JUPITER'S MOONS GANYMEDE

The third of Jupiter's moons, huge Ganymede is also the largest in the solar system, dwarfing the planet Mercury.

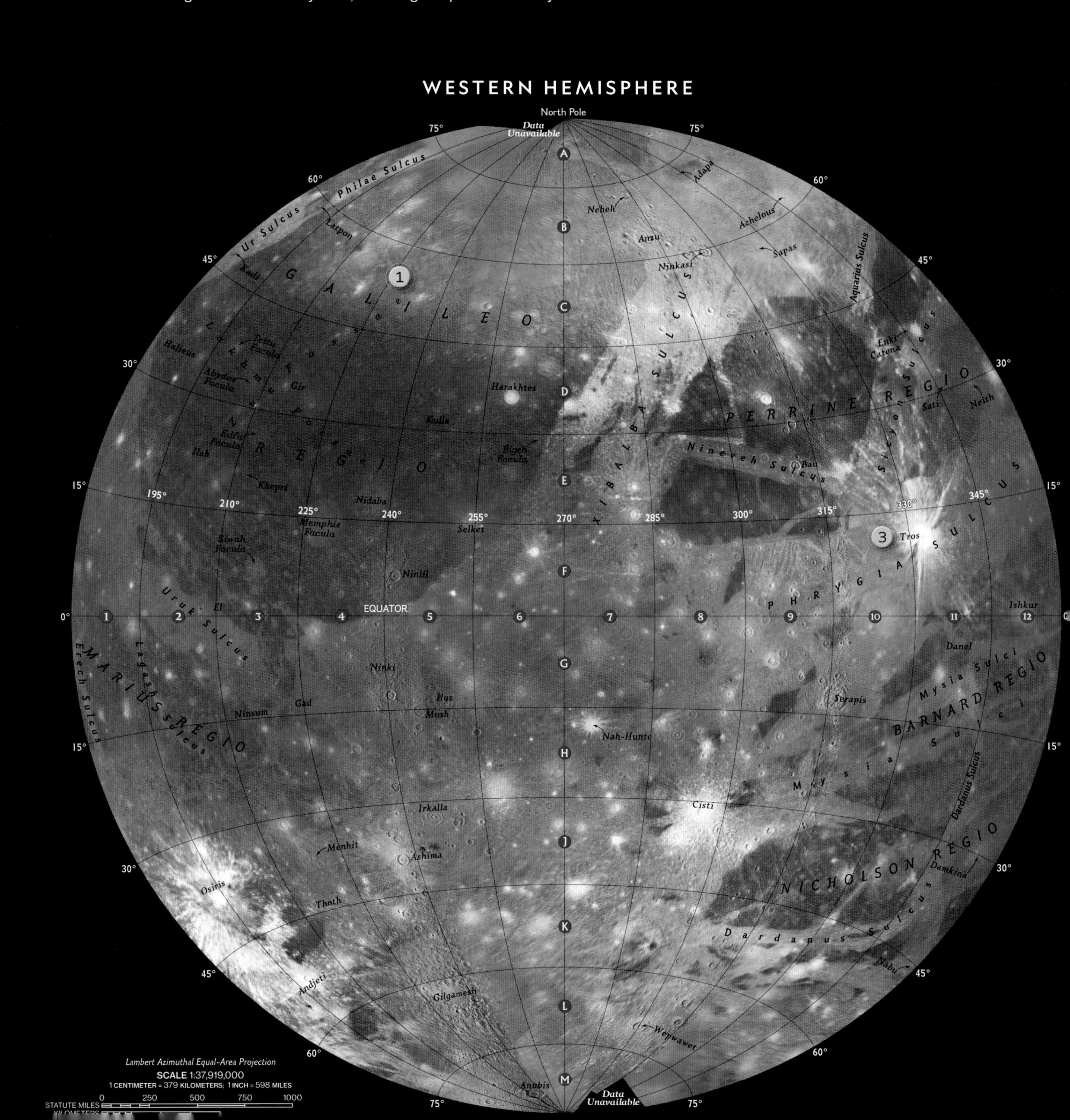

KEY FEATURES

1. **GALILEO REGIO:** Smooth, icy regions cover 40 percent of Ganymede's surface.
2. **NUN SULCI:** Grooved regions, or sulci, cover 60 percent of Ganymede's surface.
3. **TROS:** A crater; craters on Ganymede are shallow, probably because of the soft, icy surface.

EASTERN HEMISPHERE

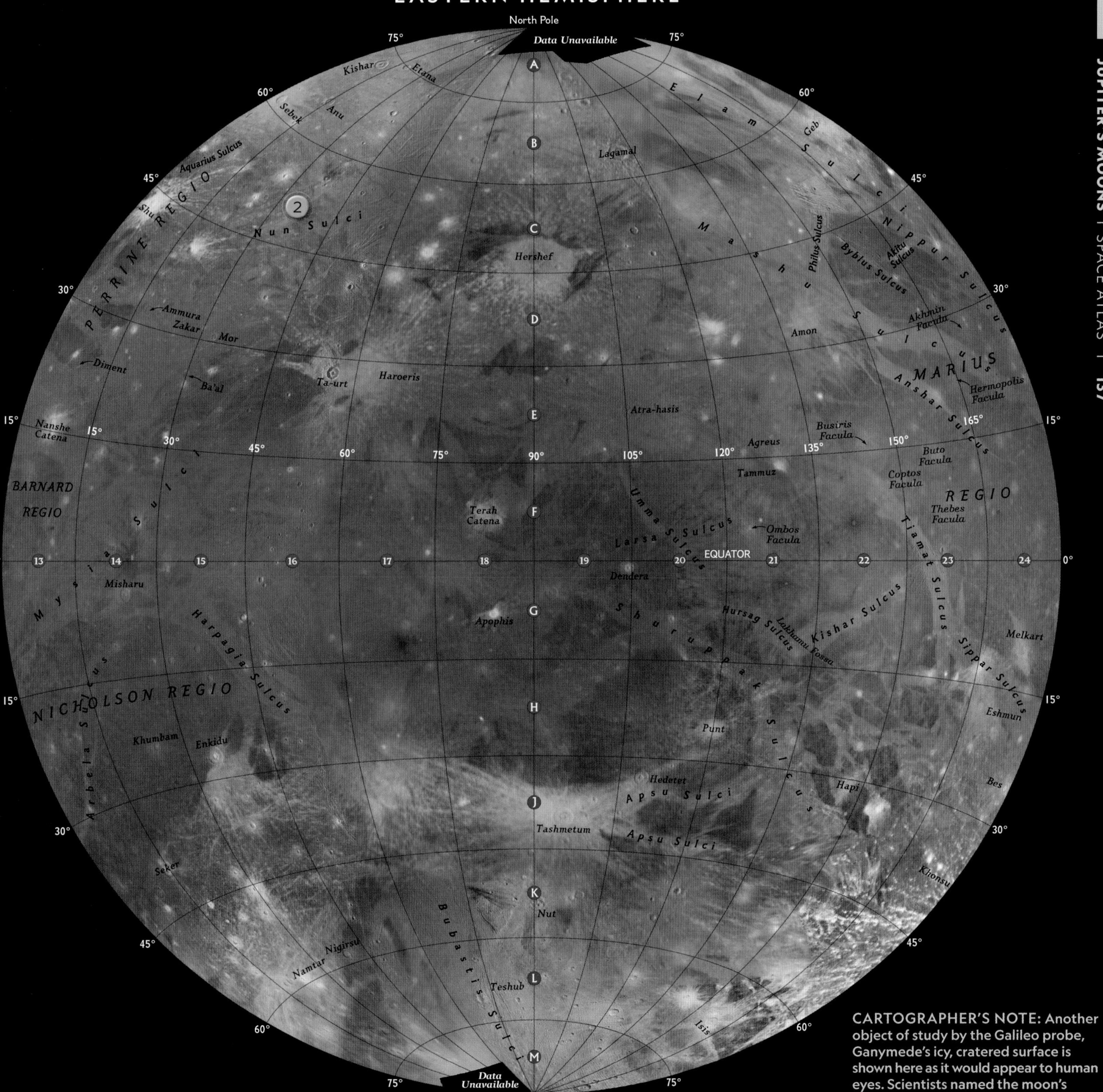

CARTOGRAPHER'S NOTE: Another object of study by the Galileo probe, Ganymede's icy, cratered surface is shown here as it would appear to human eyes. Scientists named the moon's topography from mythologies of ancient peoples who lived in the Fertile Crescent from Mesopotamia to the Levant.

JUPITER'S MOONS CALLISTO

Callisto is the third largest, and most heavily cratered, satellite in the solar system.

WESTERN HEMISPHERE

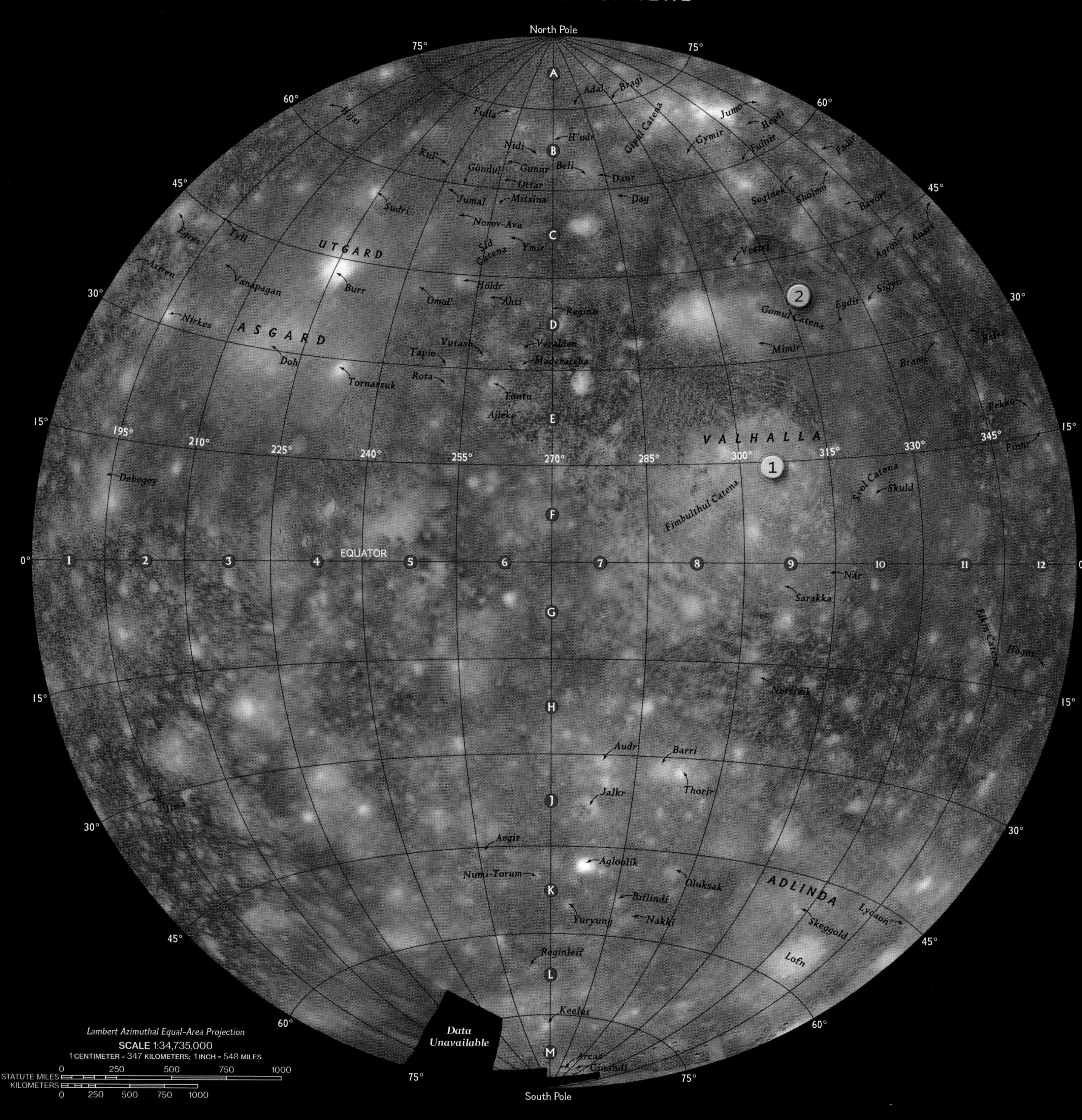

KEY FEATURES

1. **VALHALLA BASIN:** Huge, multiringed impact basin
2. **GOMUL CATENA:** Line of craters, possibly created by the same object
3. **BRAN:** Bright rayed crater

EASTERN HEMISPHERE

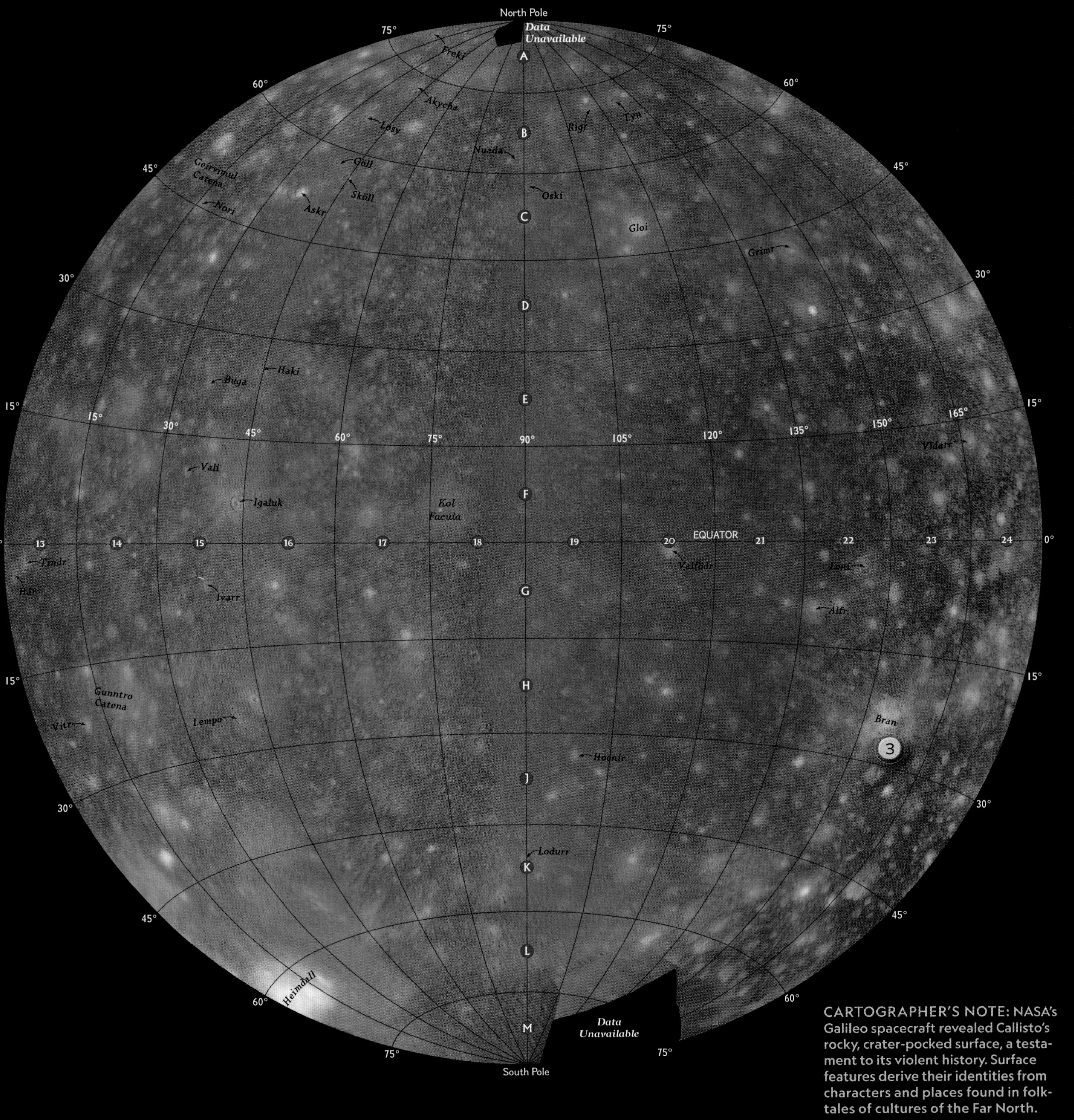

CARTOGRAPHER'S NOTE: NASA's Galileo spacecraft revealed Callisto's rocky, crater-pocked surface, a testament to its violent history. Surface features derive their identities from characters and places found in folktales of cultures of the Far North.

Jupiter's big moons are called the Galilean moons because the first telescopic sighting of these moons occurred in 1610 when Galileo turned his instrument to the Jovian system. Their discovery played a major role in scientific history: Their existence was a strong argument against the then prevailing Aristotelian view of the universe. Aristotle taught that every material object has an innate property that makes it try to move toward the center of the universe (which, in that system, corresponded to the center of Earth). When a 17th-century philosopher saw an apple fall, then he (and it was always a "he") would ascribe the fall to the apple's innate desire to be at the center of the universe. But here were the Galilean moons, perfectly content to be in orbit around Jupiter, far from that center. Maybe Earth wasn't as important in the grand scheme of things as the ancients had thought!

A cutaway view of Jupiter's moon Io. The moon has an iron-nickel core and a rocky mantle that extends to the surface. Constant flexing in Jupiter's gravitational field makes Io the most volcanically active body in the solar system.

In any case, today we see these moons as examples of the diversity of worlds that exist in our solar system. Each is unique, each has its own story to tell. Indeed, one of them (Ganymede) is bigger than Mercury, and it would be classed as a separate planet if it circled the sun instead of Jupiter.

By convention all of the Jovian moons are named for lovers or children of the god Jupiter. (If you know your mythology, you'll know that Jupiter's activities supply us with many more than 65 possible names in these categories.) For our purposes, though, we'll look at only two of the most interesting moons—Io and Europa.

IO

Io, the innermost of the Galilean moons, is the one that looks like a pizza. It is slightly larger than Earth's moon and the fourth largest moon in the solar system. The mottled appearance of its surface is the result of more than 400 active volcanoes on its surface ejecting various sulfur compounds—compounds that give it its orange and yellow colors. These volcanoes make Io the most geologically active body known.

The existence of volcanoes came as something of a shock to scientists during the flybys of the two Voyager spacecraft in 1979. On Earth volcanoes result from heat rising from deep in the mantle, heat generated in part by radioactive decay. Io is simply too small to generate heat in this way. Scientists quickly realized, however, that there was another way to heat the moon besides radioactivity. Because of the gravitational pull of the other moons, the orbit of Io is not a perfect circle. Instead, the distance of the moon from Jupiter is always changing, so that the

A bluish plume marks a volcanic eruption on the surface of Io. The gas and particles in the plume shoot 100 kilometers (60 mi) above the moon's surface.

IO HAS MORE THAN

400

ACTIVE VOLCANOES

EUROPA HAS AN
OCEAN
UNDER AN ICY SHELL

gravitational force on the moon changes as well. This means that Io is constantly being flexed and distorted, and just as a piece of metal that is bent back and forth will heat up, the moon warms. We see the result of this so-called tidal heating in Io's extensive volcanoes.

EUROPA

The most important thing about Europa, the thing that makes it one of the focal points for future space exploration, is the fact that it has an ocean of liquid water under the thick layer of ice that composes its surface. Like Io, Europa has a slightly eccentric orbit and is affected by tidal heating. Calculations quickly showed that there is enough heat generated by this effect to keep a subsurface layer of water from freezing, despite the fact that the temperature can be as low as minus 220°C (–370°F) at the surface. Because water conducts electricity, the movement of Europa through Jupiter's magnetic field produces changes in that field and in the swarm of particles around the giant planet—effects that were quickly detected by the Galileo orbiter. When you factor in that some large craters on Europa seem to contain smooth areas of recently frozen ice (presumably from upwelling liquid) and include the Galileo mission's measurements of the moon's gravitational field, you can understand why scientists now believe that, like Earth, Europa contains large amounts of liquid water. The current theory is that beneath a layer of ice some 10s of kilometers thick is an ocean containing more than twice as much water as in the oceans of our own planet. Since the initial discovery of a subsurface ocean on Europa, subsurface oceans have been found on many of the moons of Jupiter, including Ganymede and Callisto, as well as several moons of Saturn.

Cutaway view of the interior of Europa. The smallest of the four Galilean moons, it has a metallic core and a rocky mantle. These are overlain with water (probably in liquid form) and an outer shell of ice.

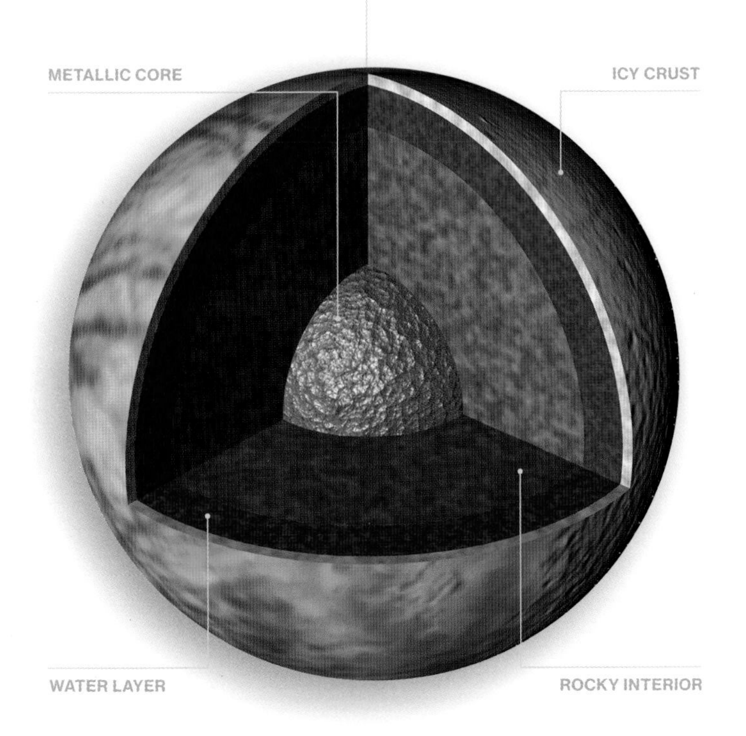

A false-color image of the surface of Jupiter's moon Europa demonstrates that the ice on the moon's surface is contaminated by material (shown in red) brought up from the interior. Icy plains are shown in blue.

Which brings us, of course, to the question of life. The European Space Agency is working on a mission called JUICE (Jupiter Icy Moons Explorer). The craft will visit all of the Galilean moons except Io and is scheduled for launch in 2022 and arrival at Jupiter in 2030. The mission may include a Ganymede lander built by the Russian Space Research Institute. NASA, in the meantime, is developing a project called the Europa Multiple Flyby Mission, which will be launched in the 2020s and, as the name implies, make many flybys of the moon.

Because of this, there are a number of proposals for sending spacecraft to Europa. Scheduled for launch in 2020, a joint NASA–European Space Agency mission known as Europa Jupiter Systems Mission seems the furthest along.

he sixth planet from the sun and the farthest visible to the naked eye, Saturn is a gas giant similar in structure to Jupiter. It is believed to have a rocky core, perhaps 10 to 20 times the size of Earth, surrounded by yer of metallic hydrogen. This, in turn, is enclosed by liquid hydrogen mixed ı helium, above which we find the gaseous atmosphere. In appearance, ırn is a little bland compared with Jupiter, although it has the same (though ter) banded cloud structure. The clouds appear to consist of water ice, nonia compounds, and ammonia crystals found in successive tiers 10s of meters thick. It is these clouds that we see when we look at the planet.

SATURN & ITS MOONS

GOLDEN GLOBE

DISCOVERER: **UNKNOWN**
DISCOVERY DATE: **PREHISTORIC**
NAMED FOR: **ROMAN GOD OF AGRICULTURE**

MASS: **95.16 × EARTH'S**
VOLUME: **763.59 × EARTH'S**
MEAN RADIUS: **58,232 KM (36,184 MI)**
EFFECTIVE TEMPERATURE: **−178°C −288°F)**
LENGTH OF DAY: **10.66 HOURS**
LENGTH OF YEAR: **29.48 EARTH YEARS**
NUMBER OF MOONS: **62 (53 NAMED)**
PLANETARY RING SYSTEM: **YES**

One of the last views of Saturn taken by the Cassini spacecraft before it was deliberately crashed into the planet's atmosphere to prevent possible contamination of the moons by terrestrial microbes. (Inset) Saturn

I SATURN

This gas giant is the most distant planet that can be seen by the naked eye.

Saturn's upper atmosphere is separated into lightly colored zones and darker belts. The dynamic winds in each are driven by the planet's quick rotation.

Given energy by Saturn's quick rotational velocity, winds rush around the planet at speeds unimaginable on Earth. Scientists have clocked the fastest around the equator at more than 1,800 kilometers per hour (1,120 mi per hour).

Weaker than Jupiter's massive field, Saturn's magnetic field is still many times stronger than Earth's. Unusual in the solar system, the magnetic field and the planet's rotation share nearly the same axis; something scientists cannot yet explain.

KEY FEATURES

1. **NORTH NORTH TEMPERATE ZONE:** Zones represent upwelling air.
2. **NORTH NORTH TEMPERATE BELT:** Belts represent descending air.
3. **EQUATOR:** Rapidly rotating Saturn bulges at the equator.

Spinning at more than 35,500 kilometers per hour (22,000 mi per hour) causes the planet to bulge around its middle. Saturn's diameter at the equator is nearly 12,000 kilometers greater than at the poles.

CARTOGRAPHER'S NOTE: Observed by NASA's Cassini mission, Saturn's dull but clear banding is visible. Like larger Jupiter, Saturn's atmosphere is divided into darker belts and lighter zones.

SATURN'S MOONS MIMAS

Heavily cratered, small Mimas is the innermost of Saturn's moons.

WESTERN HEMISPHERE

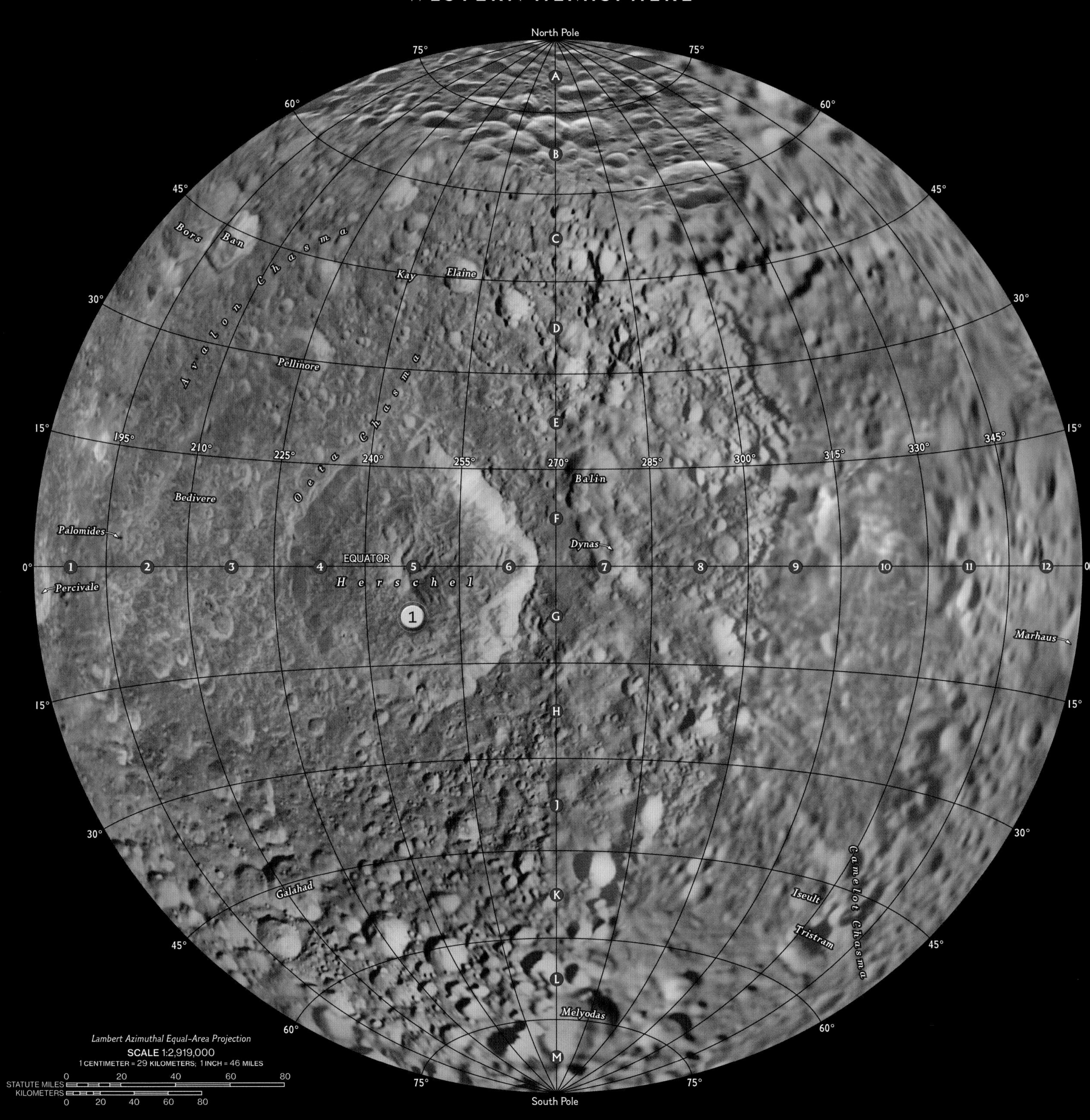

KEY FEATURES

1. **HERSCHEL:** Giant impact crater
2. **OSSA CHASMA:** One of Mimas's many deep depressions, known as chasmata
3. **TINTAGIL CATENA:** Crater chain

EASTERN HEMISPHERE

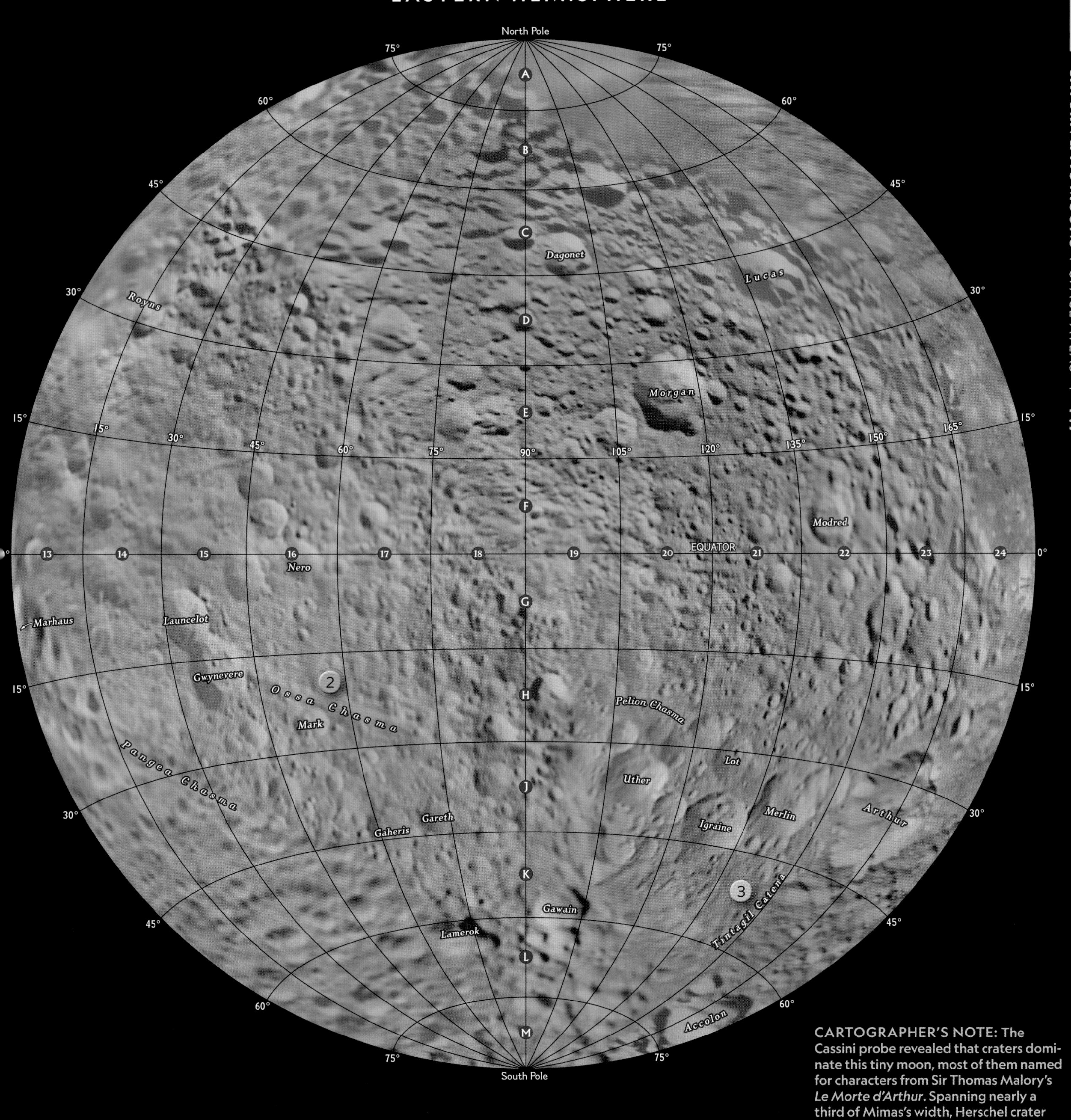

CARTOGRAPHER'S NOTE: The Cassini probe revealed that craters dominate this tiny moon, most of them named for characters from Sir Thomas Malory's *Le Morte d'Arthur*. Spanning nearly a third of Mimas's width, Herschel crater is named for the moon's discoverer.

SATURN'S MOONS ENCELADUS

Like Jupiter's Europa, icy Enceladus may have a body of subsurface liquid water.

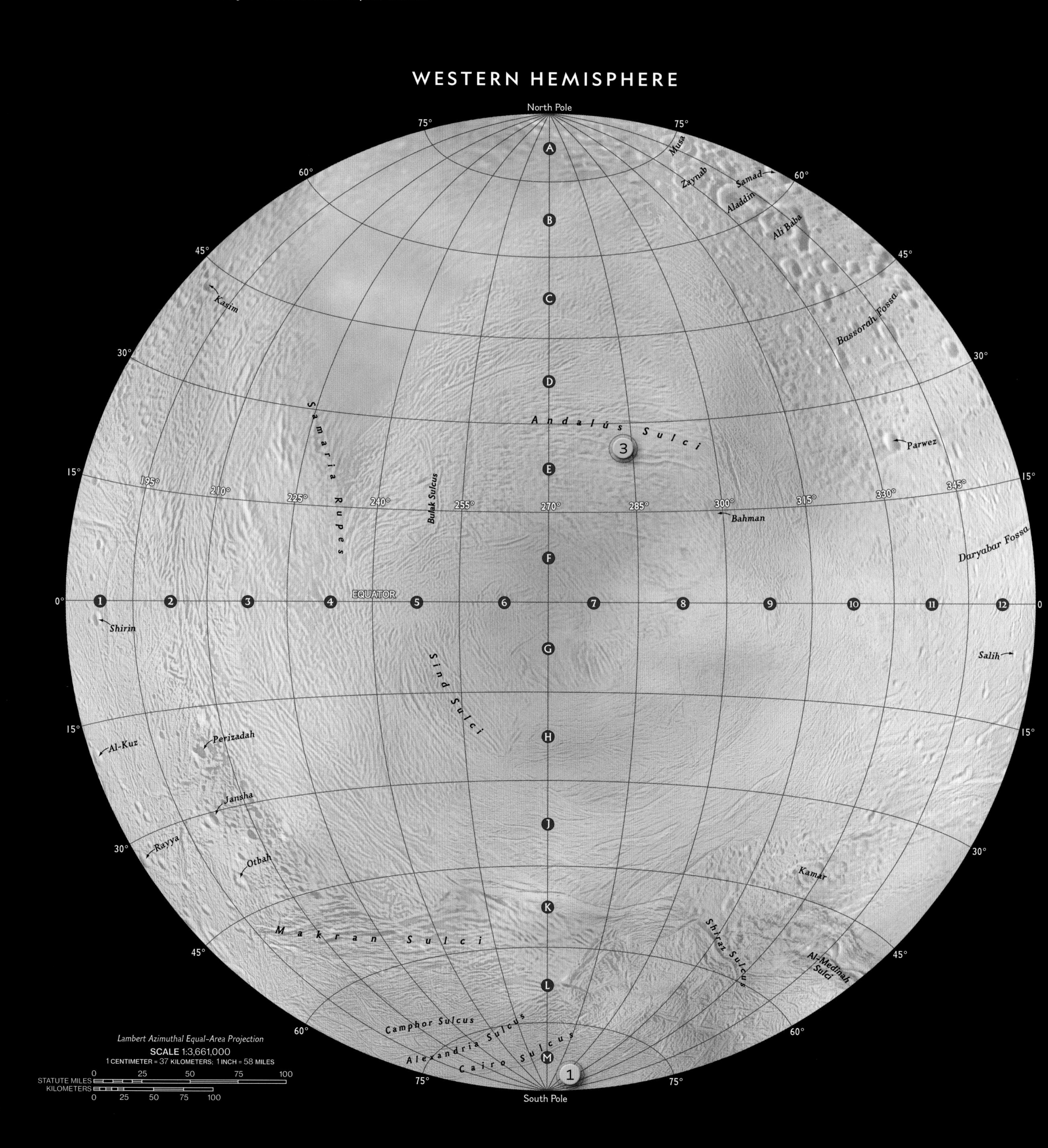

KEY FEATURES

1. **SOUTH POLE:** A plume of water vapor has been observed escaping from this pole.
2. **DIYAR PLANITIA:** Large, smooth, icy areas make Enceladus highly reflective.
3. **ANDALÚS SULCI:** Grooved sulci indicate a shifting surface.

EASTERN HEMISPHERE

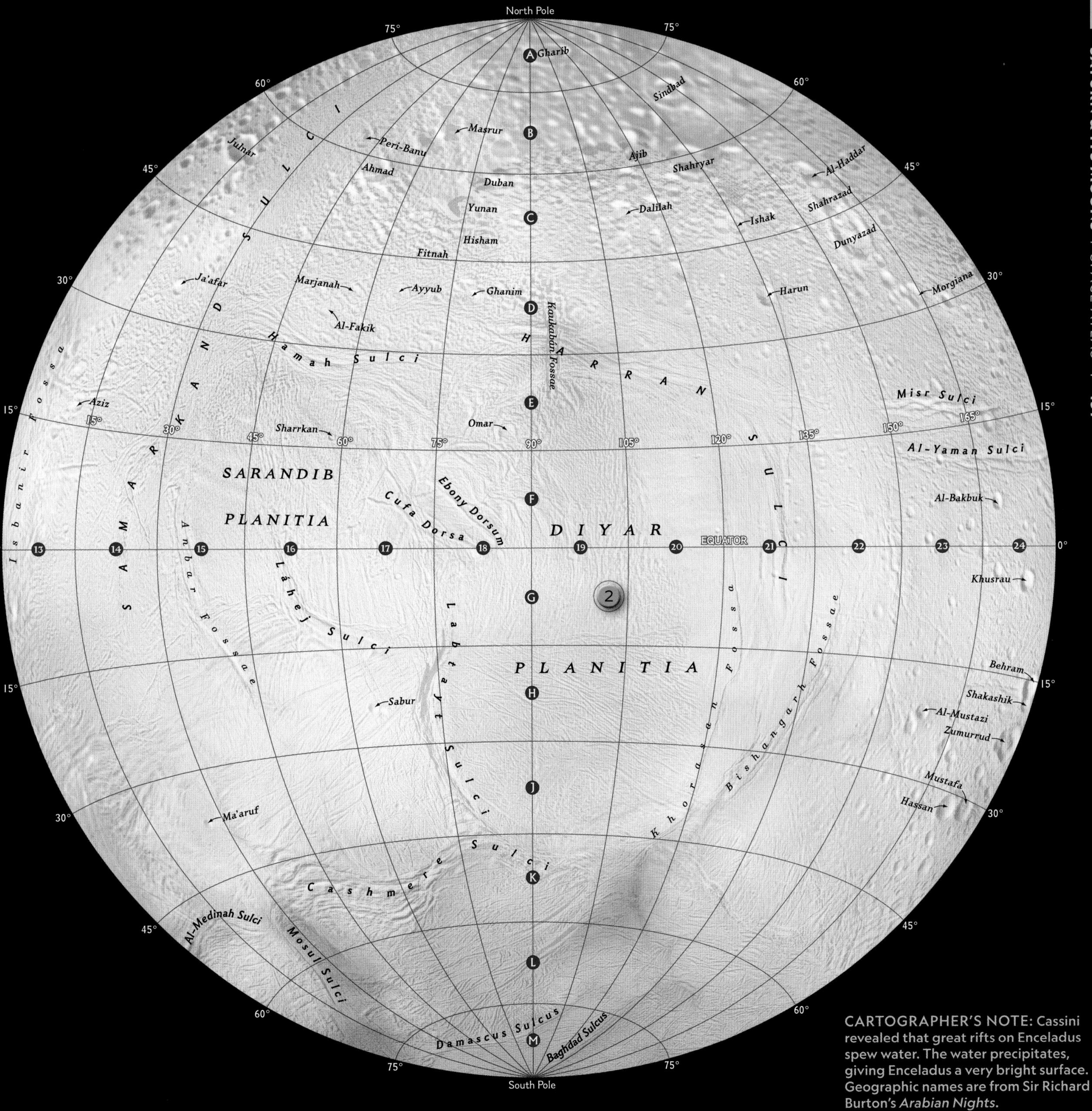

CARTOGRAPHER'S NOTE: Cassini revealed that great rifts on Enceladus spew water. The water precipitates, giving Enceladus a very bright surface. Geographic names are from Sir Richard Burton's *Arabian Nights*.

SATURN'S MOONS TETHYS

Little Tethys, heavily scarred, is made primarily of water ice.

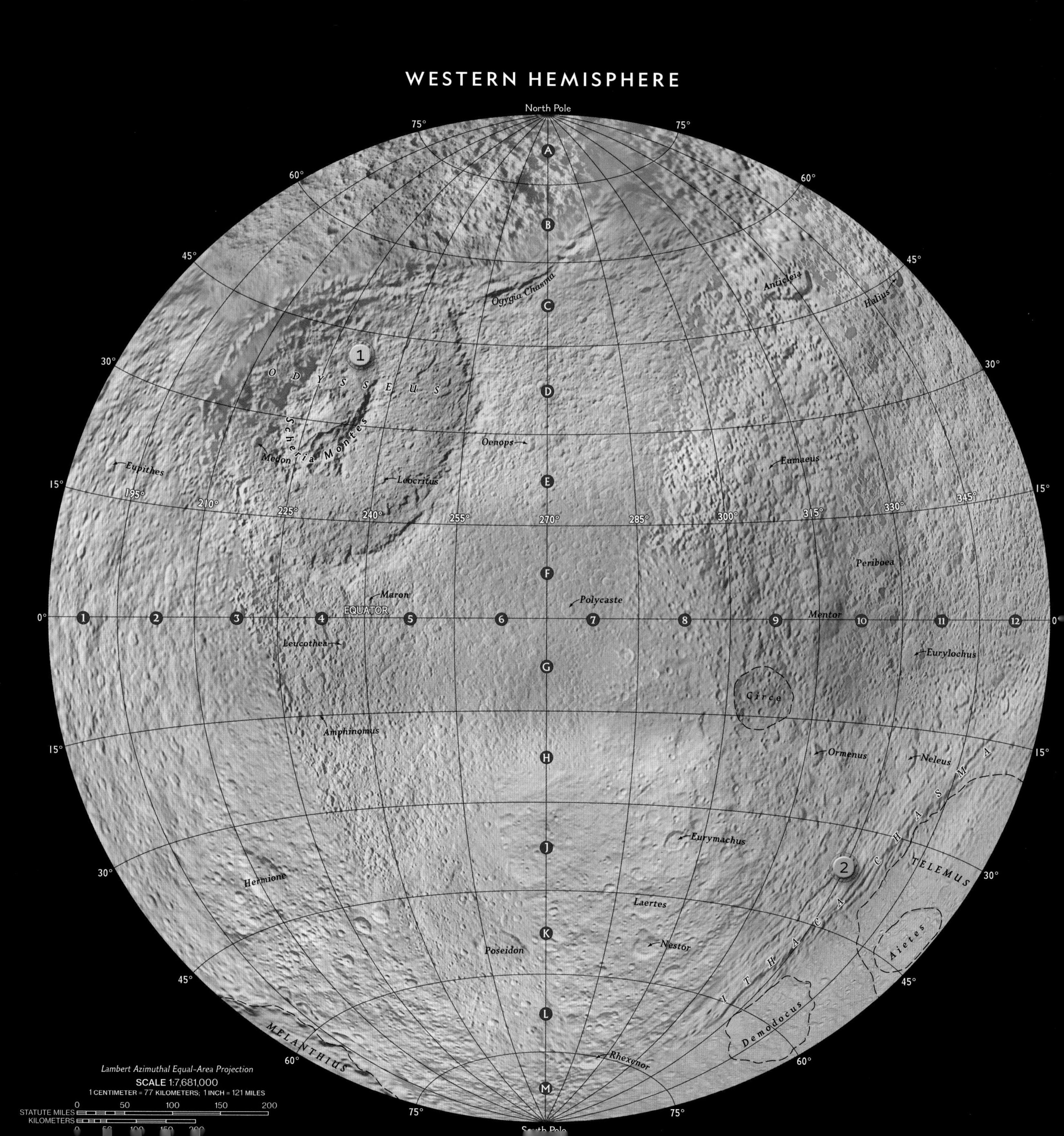

KEY FEATURES

1. **ODYSSEUS:** Huge impact basin almost two-fifths as large as Tethys
2. **ITHACA CHASMA:** Enormous, deep valley, its origins unknown
3. **PENELOPE:** Large-impact crater

EASTERN HEMISPHERE

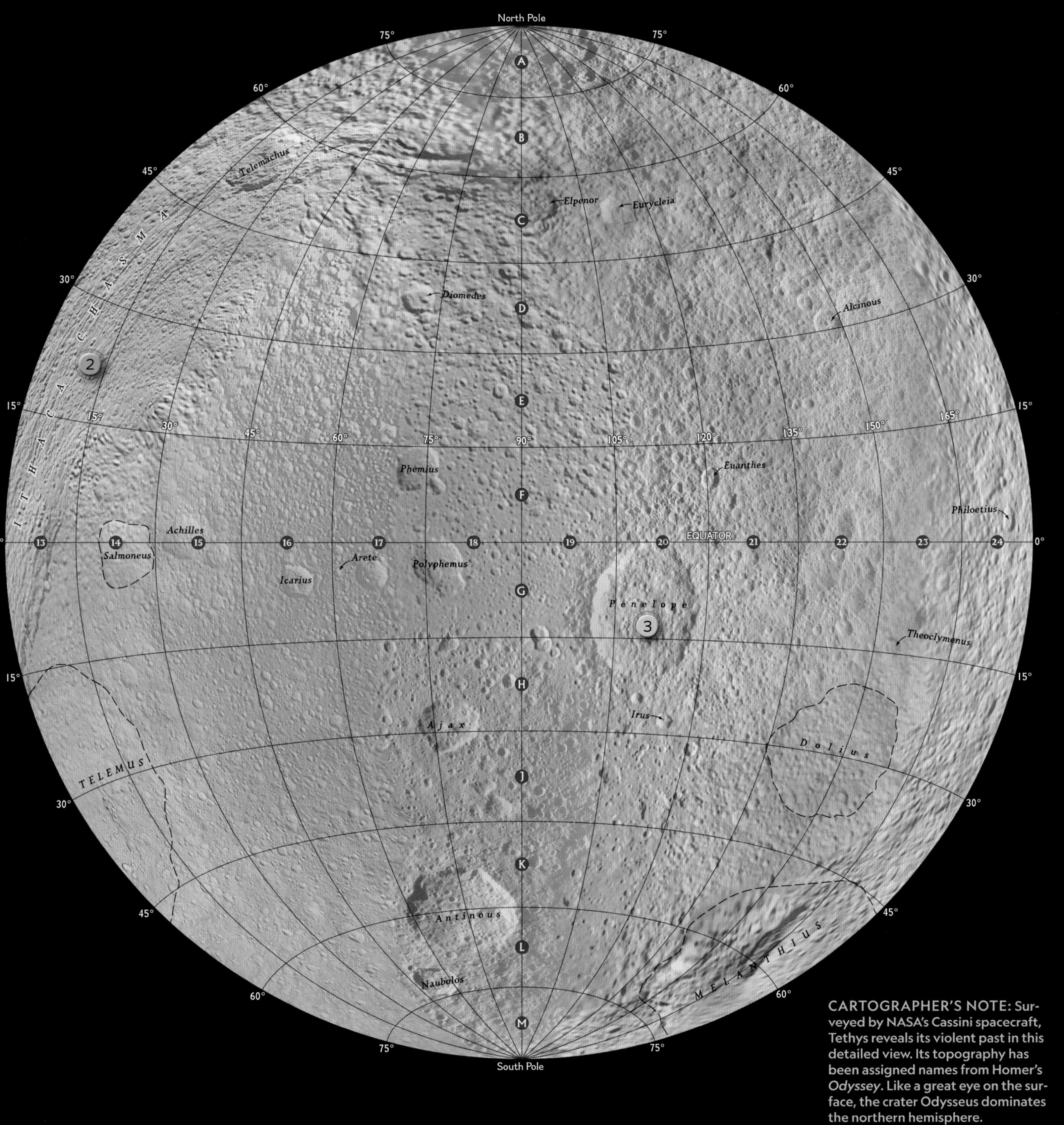

CARTOGRAPHER'S NOTE: Surveyed by NASA's Cassini spacecraft, Tethys reveals its violent past in this detailed view. Its topography has been assigned names from Homer's *Odyssey*. Like a great eye on the surface, the crater Odysseus dominates the northern hemisphere.

SATURN'S MOONS DIONE

Fractured and cratered, little Dione orbits Saturn every 2.7 Earth days.

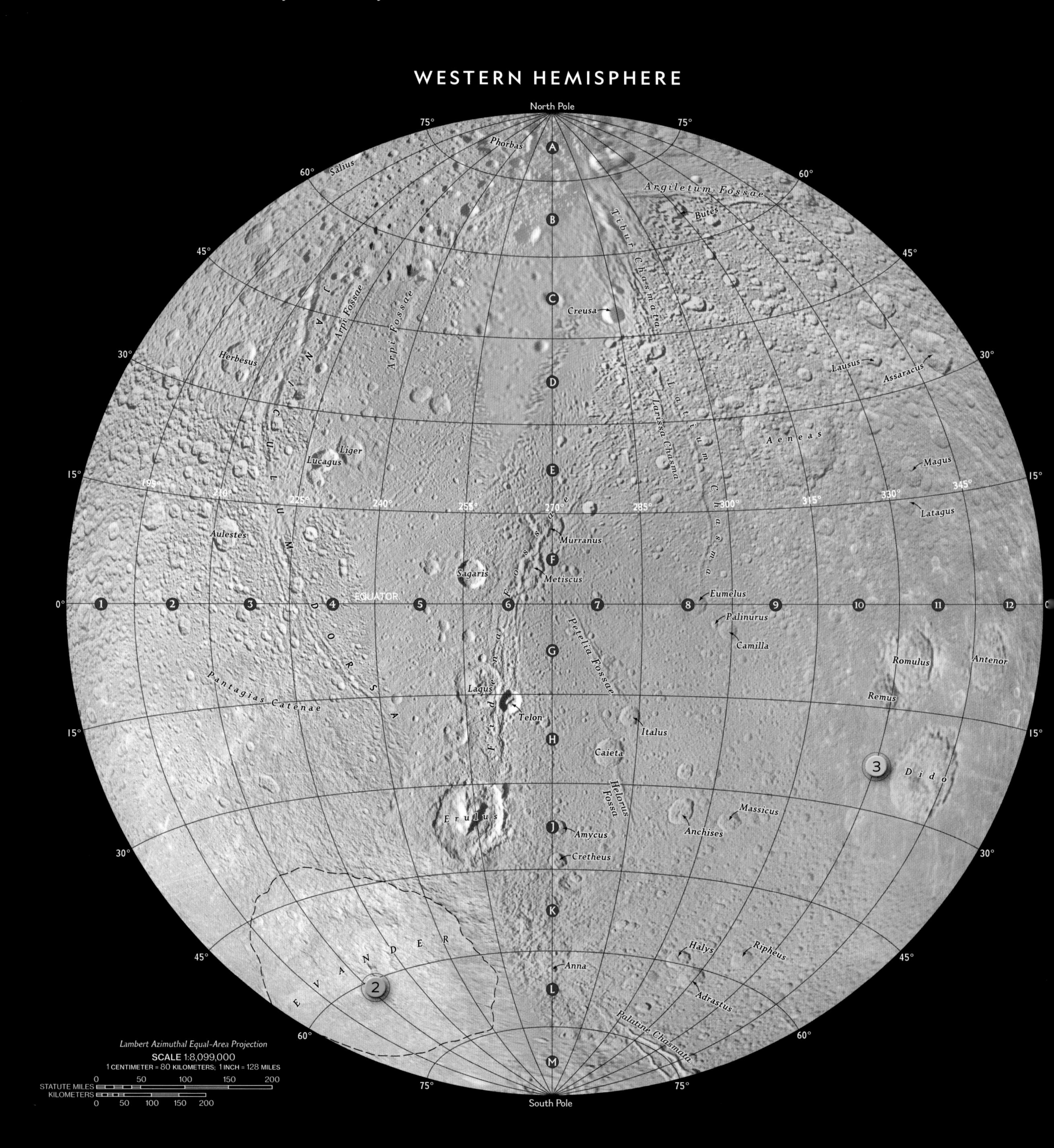

KEY FEATURES

1. **PALATINE CHASMATA:** A system of long, steep-sided depressions
2. **EVANDER:** Largest crater on Dione
3. **DIDO:** Prominent crater matched by crater Aeneas, above it

EASTERN HEMISPHERE

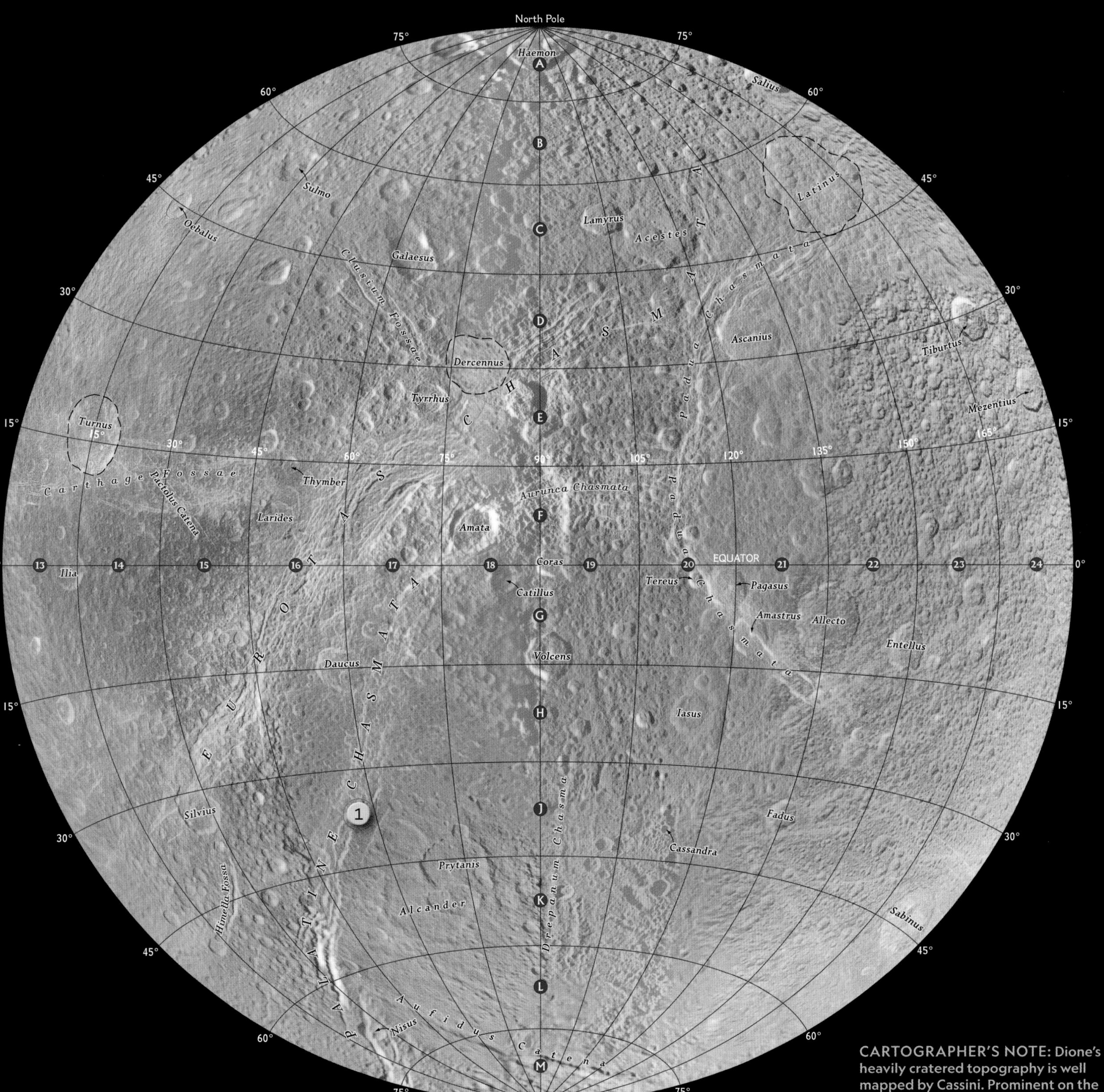

CARTOGRAPHER'S NOTE: Dione's heavily cratered topography is well mapped by Cassini. Prominent on the eastern hemisphere are great icy cliffs stretching across the landscape. Names from Virgil's *Aeneid* identify features.

I SATURN'S MOONS RHEA

Saturn's second largest moon, icy Rhea is scarred by craters and ice cliffs.

WESTERN HEMISPHERE

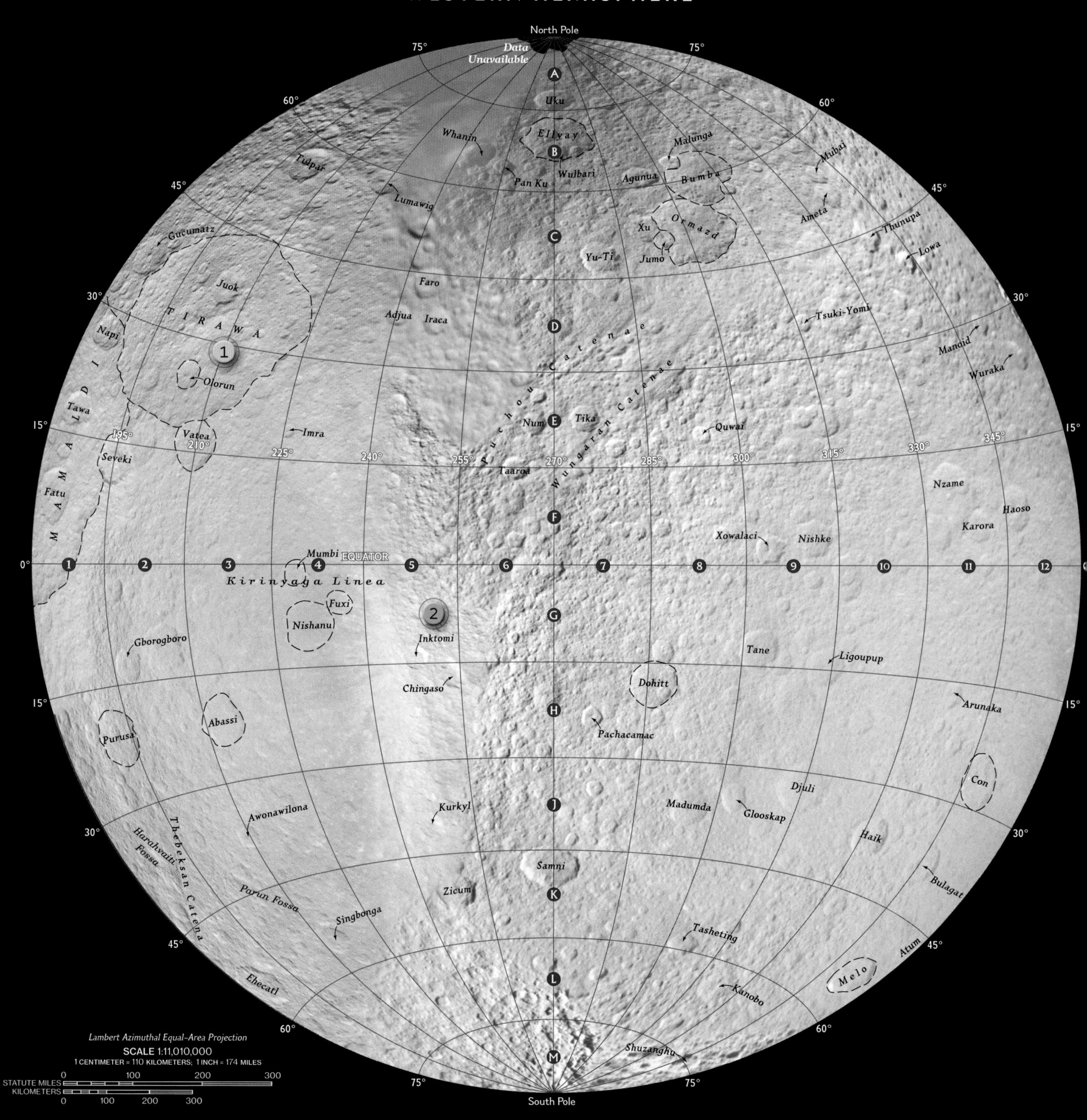

KEY FEATURES

1. **TIRAWA:** Large, well-defined impact basin
2. **INKTOMI:** Young impact basin with well-defined rays
3. **YAMSI CHASMATA:** Long ice depressions

EASTERN HEMISPHERE

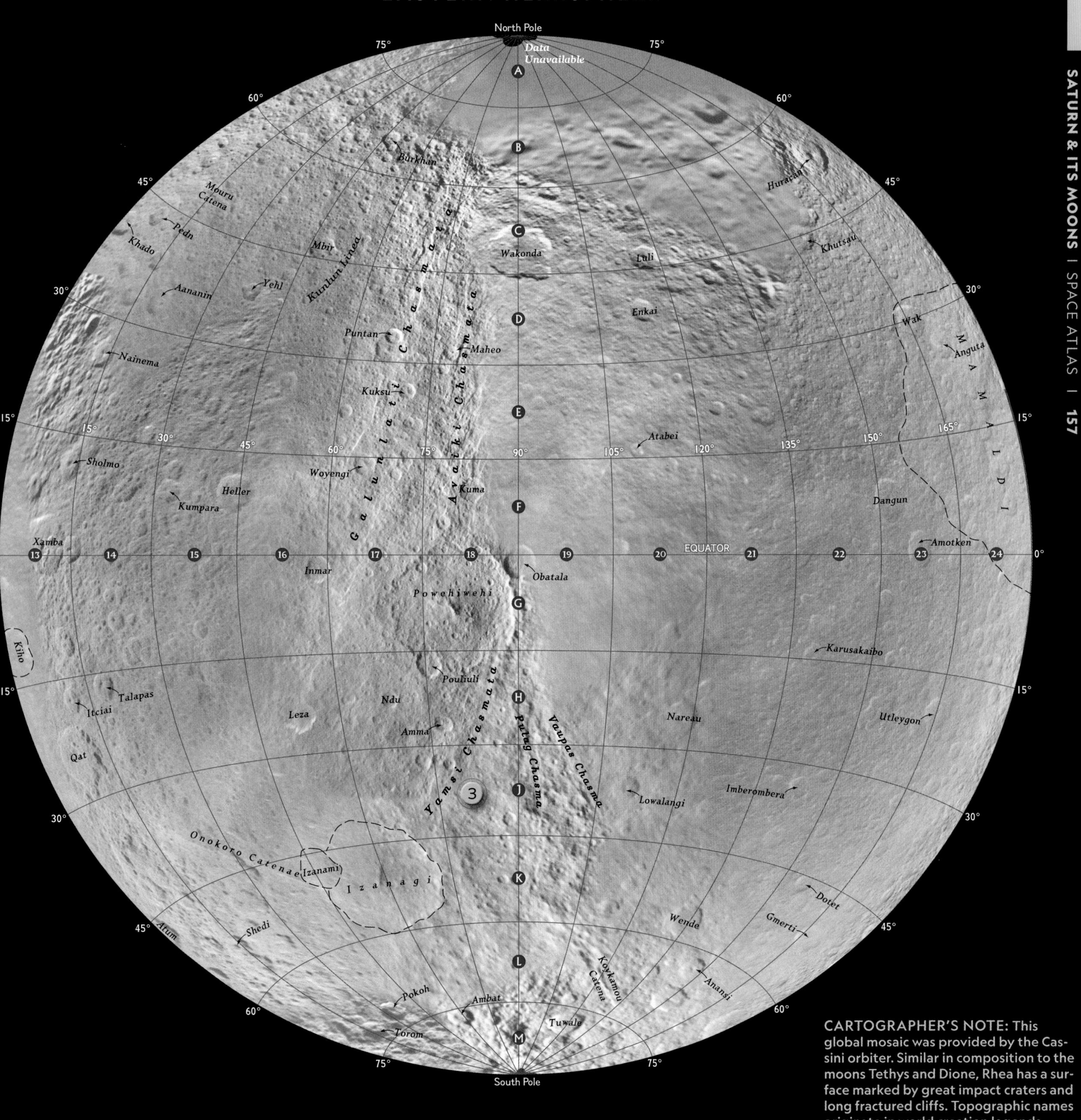

CARTOGRAPHER'S NOTE: This global mosaic was provided by the Cassini orbiter. Similar in composition to the moons Tethys and Dione, Rhea has a surface marked by great impact craters and long fractured cliffs. Topographic names originate in world creation legends.

I SATURN'S MOONS TITAN

Largest of Saturn's satellites and the only moon in the solar system with its own atmosphere

WESTERN HEMISPHERE

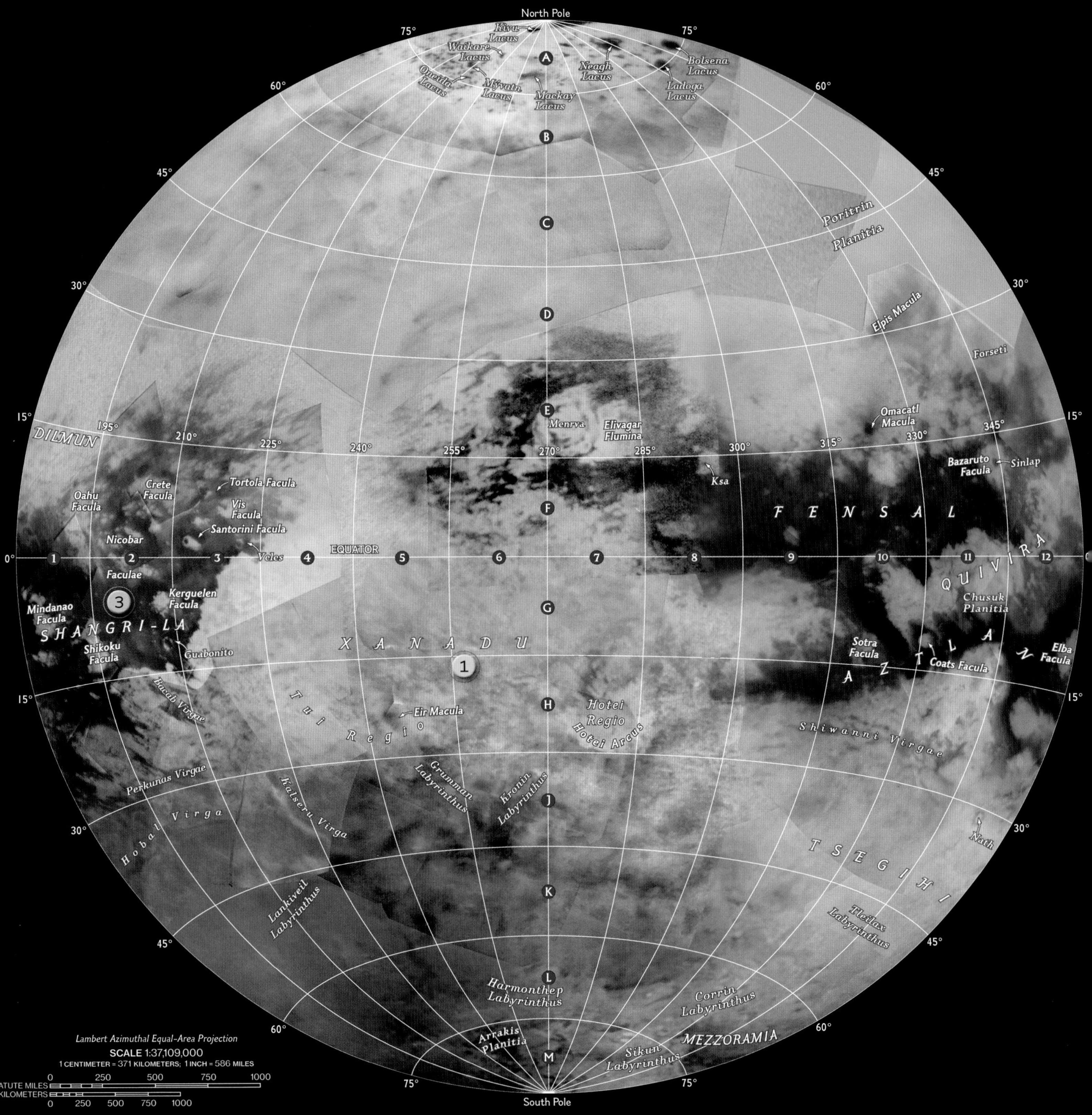

Lambert Azimuthal Equal-Area Projection

SCALE 1:37,109,000

1 CENTIMETER = 371 KILOMETERS; 1 INCH = 586 MILES

STATUTE MILES 0 250 500 750 1000

KILOMETERS 0 250 500 750 1000

KEY FEATURES

1. **XANADU:** Area of hills and valleys
2. **ONTARIO LACUS:** Shallow lake of liquid hydrocarbons
3. **SHANGRI-LA:** Landing place of Huygens probe

EASTERN HEMISPHERE

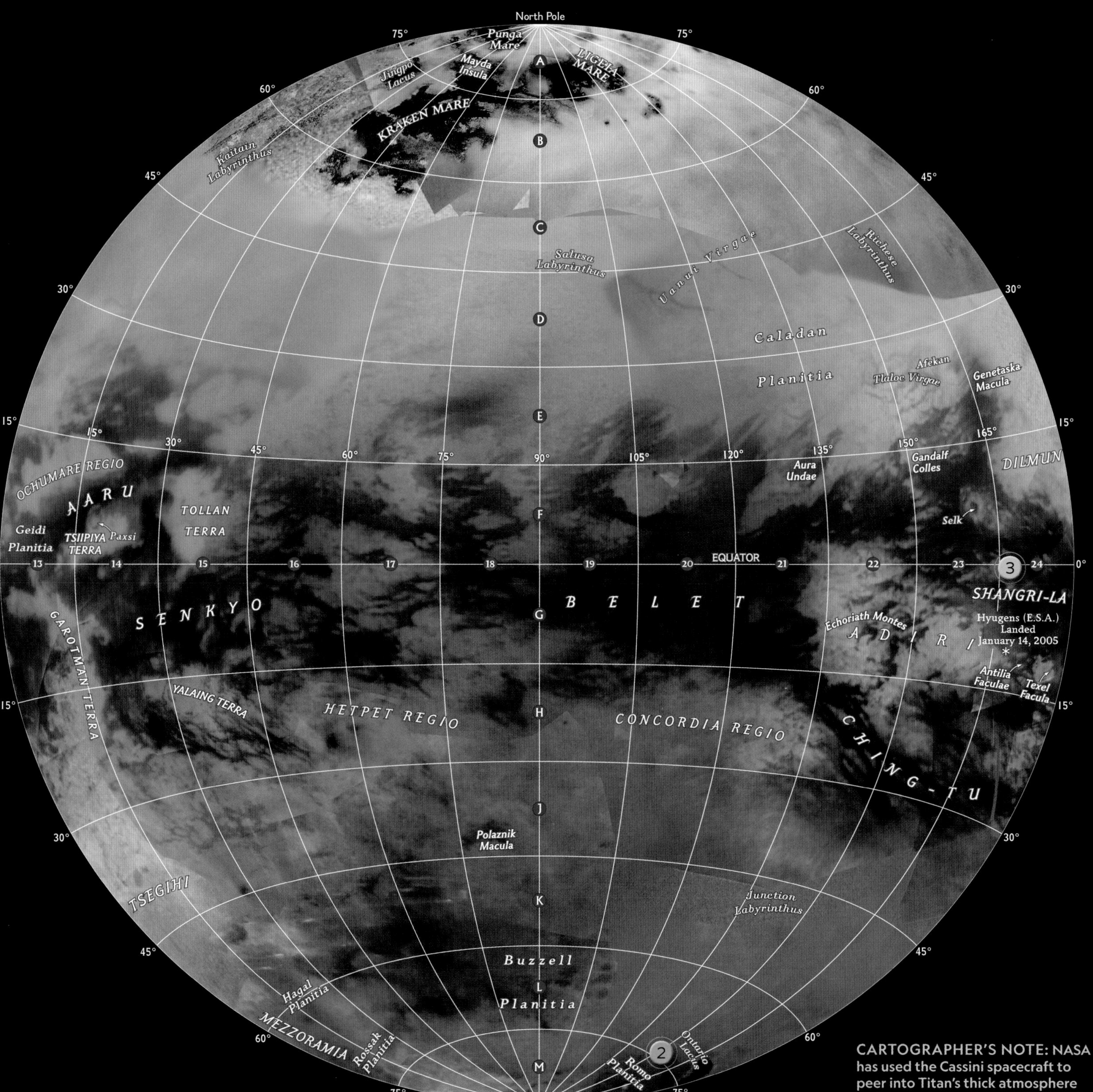

CARTOGRAPHER'S NOTE: NASA has used the Cassini spacecraft to peer into Titan's thick atmosphere and study the landscape of this unique moon. Names for the largest surface features—the terrae—refer to sacred or enchanted lands from mythology.

SATURN'S MOONS IAPETUS

Iapetus's dramatically two-toned surface may result from dark materials left behind by evaporation on its warmer, leading hemisphere.

WESTERN HEMISPHERE

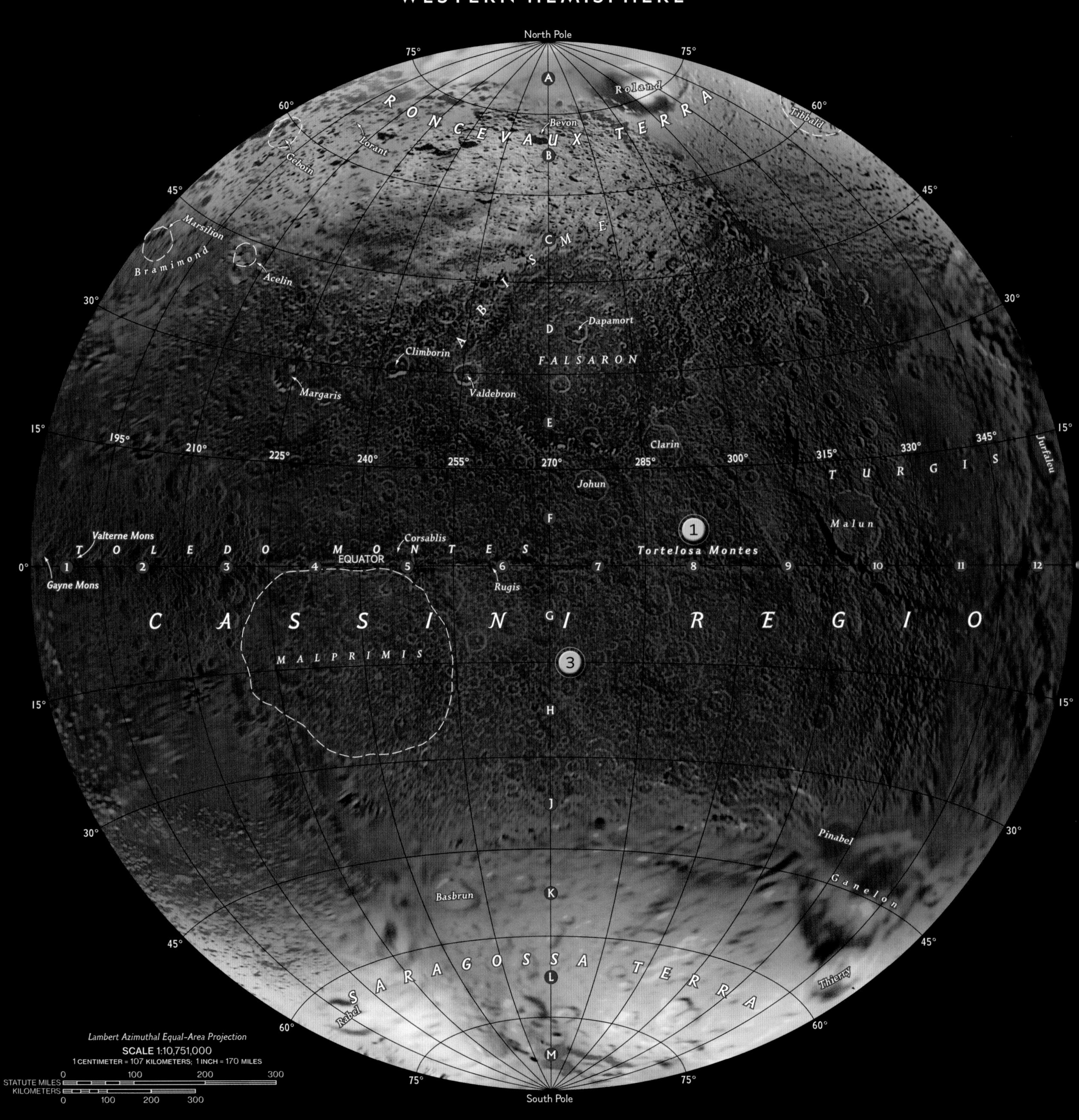

KEY FEATURES

1. **TORTELOSA MONTES:** 10-kilometer-high (6 mi) mountains
2. **ENGELIER:** Impact basin 500 kilometers (300 mi) wide
3. **CASSINI REGIO:** Darker region

EASTERN HEMISPHERE

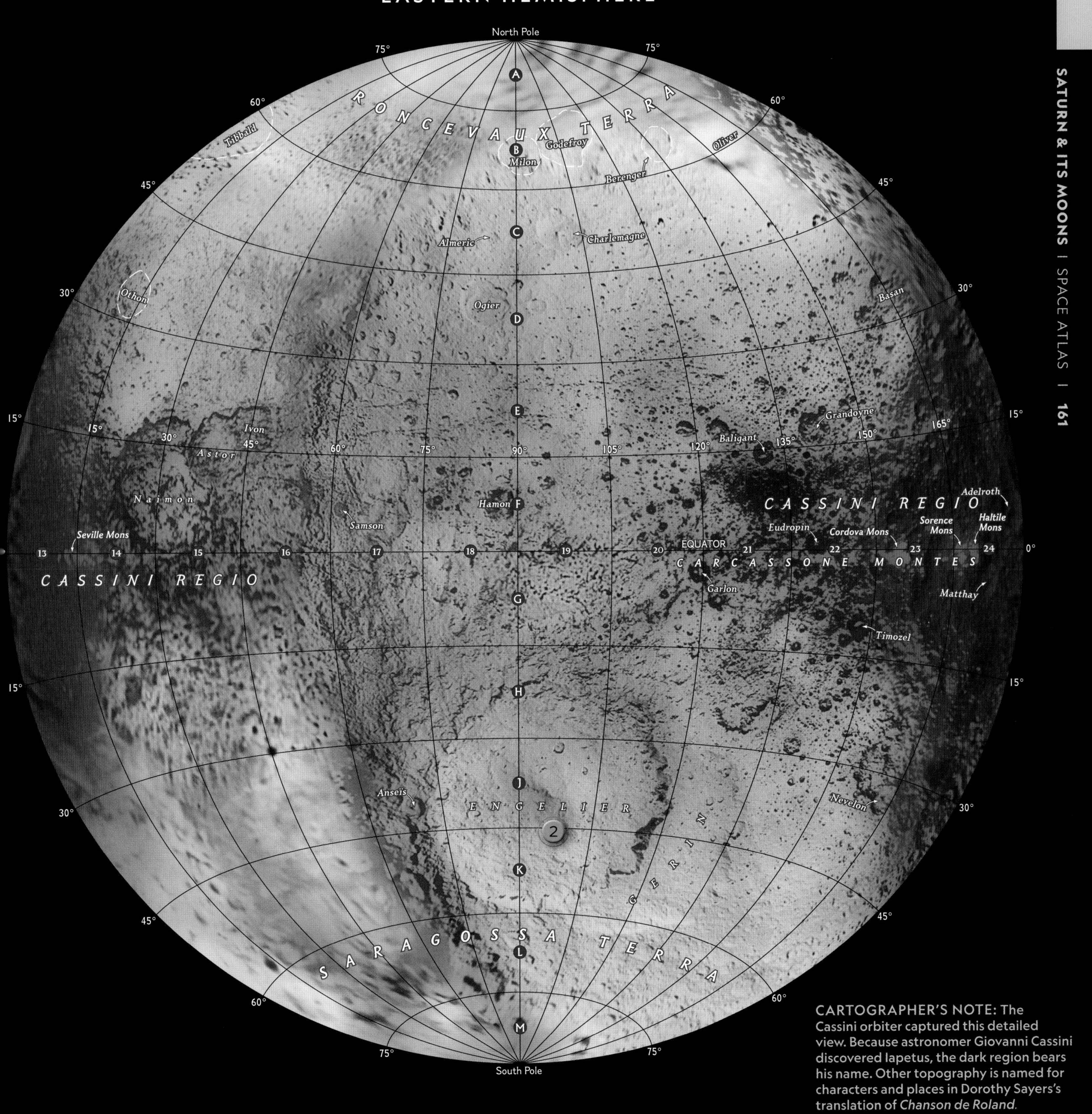

CARTOGRAPHER'S NOTE: The Cassini orbiter captured this detailed view. Because astronomer Giovanni Cassini discovered Iapetus, the dark region bears his name. Other topography is named for characters and places in Dorothy Sayers's translation of *Chanson de Roland.*

Saturn's winds are a striking feature of its atmosphere, registering at up to 1,800 kilometers an hour (1,120 mph)—among the fastest winds anywhere in the solar system. The planet also sports a short-lived storm structure known as the Great White Spot. This structure shows up every Saturn year (about 30 Earth years) at about the time of the summer solstice in Saturn's northern hemisphere. Similar in appearance to Jupiter's Great Red Spot, the Great White Spot was first observed in 1876 and has been seen intermittently since then.

Like Jupiter, Saturn has a large and powerful magnetic field—and like Earth, it has shimmering auroras at both poles, caused by magnetic particles spiraling along the magnetic field lines.

MISSIONS TO SATURN

In addition to observing Saturn by telescope, scientists have sent a number of spacecraft to the Saturnian system. Between 1979 and 1982, Pioneer and Voyager spacecraft flew by the planet on their way out of the solar system, discovering new moons and new aspects of the rings in the process. Then, in 1997, the Cassini spacecraft was launched, reaching and entering its orbit around Saturn in 2004. Most of the detailed information we have about the planet, its rings, and its moons comes from this. In 2017, after 13 years of exemplary service, the Cassini spacecraft was deliberately sent into Saturn's atmosphere, where it kept sending data until it was destroyed. As with the Galileo spacecraft at Jupiter, this was done to prevent any possible contamination of Saturn's moons by terrestrial organisms.

When scientists look at Saturn, however, they seldom concentrate on the planet itself. Instead, they focus on its spectacular ring system (which we will discuss in the next section) and its moons. Like Jupiter, Saturn has a large array of satellites—62 discovered as of this writing, of which 53 have been officially named. As was the case with Jupiter, many of these moons are small—some less than a mile across—and only 13 are more than 50 kilometers (30 mi) in diameter; of these large moons, the two that have attracted the most scientific attention are Titan and Enceladus.

TITAN

Titan was discovered in 1655 by the Dutch astronomer Christiaan Huygens (1629–1695). It is larger than Mercury and the only moon in the solar system to have a significant atmosphere. It is also the one world in the system where we can, perhaps, get a glimpse of the

A cutaway view of our best current idea about the structure of Saturn. Like Jupiter, it probably has a rocky core surrounded by a layer of metallic hydrogen, which in turn is surrounded by layers of liquid hydrogen and helium.

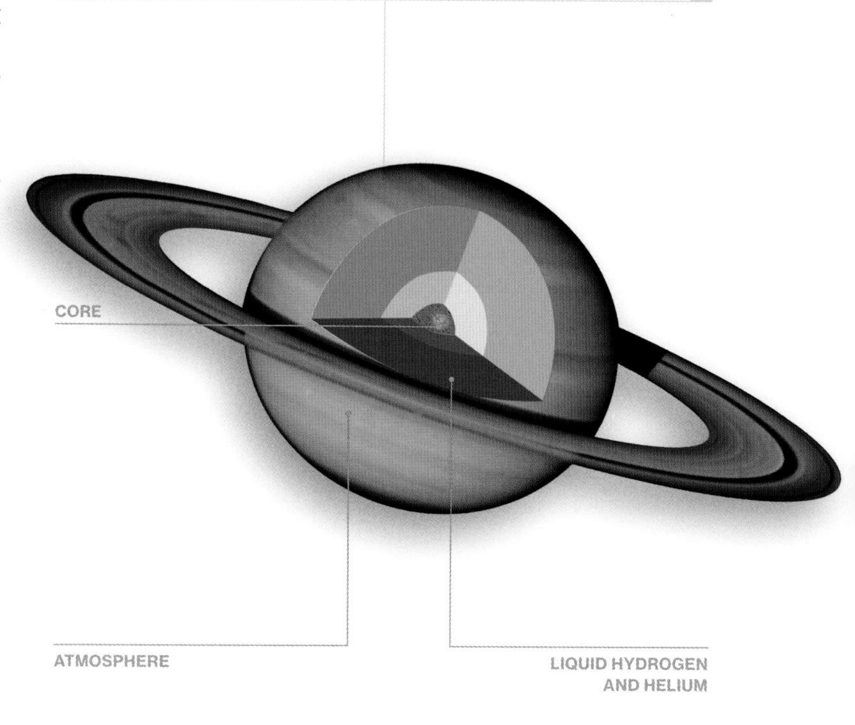

Dwarfed by the bulk of its parent planet, little Mimas (at bottom of image) circles in orbit against the bluish mass of Saturn's northern hemisphere. The dark bands at the top are shadows cast on Saturn's surface by its rings.

SATURN'S VOLUME IS

764

TIMES THAT OF EARTH

kind of chemistry that led to life on Earth billions of years ago—which is why it has become a center of scientific exploration.

Titan is thought to have a rocky core covered by thick layers of ice, with a water-ammonia ocean beneath the icy surface. Its atmosphere is mostly nitrogen and supports several cloud layers that make it impossible to see the surface from outside. Like Venus, Titan waited for visiting spacecraft before it revealed its surface.

A few months after its arrival at Saturn in July 2004, Cassini dropped a probe, appropriately named Huygens, into Titan's atmosphere. (Eventually, Cassini made more than 125 close approaches to the moon.) Descending by parachute, Huygens sent back the first images of the surface. To the surprise of the observers, the surface of Titan turned out to look almost exactly like the familiar surface of Earth! As one researcher put it, "What was alien about Titan was its eerie familiarity."

In fact, Titan's equatorial region is covered by long sand dunes as big as those in the Sahara, interspersed with rocky hills, while near the poles are large liquid lakes. One is slightly larger than Lake Ontario, after which it is named.

You can understand the similarity of Titan to Earth by noting two things: First, Titan is really cold, with surface temperatures hovering at about minus 180°C (–320°F). Second, at these temperatures, familiar materials take on unexpected forms. Water ice is as hard as granite, for example, and methane, which we know as natural gas on our own balmy planet, is a liquid. In essence, what we see on Titan are familiar processes involving unfamiliar materials.

For example, Titan's high clouds, made of hydrocarbon molecules like methane and ethane (methane's chemical cousin), interact with ultraviolet light from the sun to produce a haze that wouldn't be out of place in Los Angeles on a bad day. The hydrocarbons rain down onto the surface, producing, among other things, that "sand" in the equatorial dunes. (One researcher compared the material to a heap of coffee grounds.) The lakes are liquid methane, which condenses and falls out of Titan's sky as rain to produce other familiar landscape features just as water does on Earth.

THE FLOATING WORLD

The planets of our solar system exhibit a wide range of densities, from Earth (the densest) to Saturn (the least dense). The density of any object measures how much "stuff" is packed into it, or, technically, its mass divided by its volume. Water's density, 1 gram per cubic centimeter, is used as a standard. Iron, for example, has a density around 7.9, while Earth has a density of about 5.5. Interestingly, ice has a density of about 0.92–less than that of water. This explains why ice stays at the surface of frozen lakes and ponds – any material with a density less than 1 will float in water.

Finding the density of a planet requires that we measure both its mass and its volume. The latter is easy – if we know how far away the planet is and how big its disk appears to be, we can find its radius. From there, simple geometry gets us the volume. Finding the mass is a little more difficult, but if the planet has moons we can deduce the planetary mass by observing the orbits of the moons.

When we apply these methods to Saturn, we find that it is a planet with a volume 764 times bigger and a mass 95 times larger than Earth. This translates into a density of about 0.69 – less than the density of water and less, even, than the density of the sun. If you could find a body of water big enough, Saturn would actually float!

SATURN COULD FLOAT IN WATER.

An illustration depicts the insertion of the spacecraft Cassini into orbit around Saturn. One of the largest and most complex spacecraft ever built, Cassini was launched in 1997 and reached Saturn in 2004. In December 2004 it released the probe Huygens into the atmosphere of Saturn's moon Titan. The product of a collaboration between 16 European countries and NASA, Cassini's mission ended in 2017.

Because hydrocarbons are organic molecules, close relatives of the kinds of molecules we believe produced life on Earth, scientific attention has focused on Titan. We shall see on page 228 that the creation of organic molecules from nonliving material is the first step in the development of living things. The hope, of course, is that by studying the way this process is proceeding on Titan right now we can learn something about how it happened on Earth four billion years ago.

ENCELADUS

The sixth largest moon of Saturn, Enceladus long was known to have a surface made of water ice. In 2005, however, Cassini recorded a geyser spewing near the moon's south pole: a geyser that contained liquid water. Like Jupiter's moon Io, Enceladus appears to be heated by the flexing it undergoes in Saturn's gravitational field. Cassini had many encounters with Enceladus, including one that brought it to within 30 kilometers (20 mi) of the moon's surface. These encounters verified that the geysers were mostly water vapor, and scientists now believe that there is a subsurface ocean on the moon, one approximately as big as Lake Superior.

In 2017, NASA announced that further measurement of material in the geysers indicated that the subsurface ocean of Enceladus is in contact with hot material being brought up from the interior of the moon, a phenomenon analogous to deep ocean vents in Earth's oceans. Because life developed around the Earth's vents, scientists argue that there is a possibility that it might have developed on Enceladus (and other moons) as well.

The rings of Saturn are the most spectacular—and least substantial—objects in the solar system. Made up primarily of chunks of water ice in orbit around the planet, they create a stunning display of reflected sunlight visible from Earth through even small telescopes. • Like many objects in the solar system, Saturn's rings were first seen by Galileo when he turned his telescope toward the skies in 1610. Because his instrument was crude by modern standards, he saw the rings as two dots on either side of the main planet and mistook them for moons—at one point he described Saturn as having "ears." It wasn't until 1655 that the Dutch astronomer Christiaan Huygens, using an improved telescope, identified the ears as rings circling the planet.

SATURN'S RINGS

HOOPS OF GLITTERING ICE

DISCOVERER: **GALILEO GALILEI**

DISCOVERY DATE: **JULY 1610**

NAMED FOR: **RINGS, GAPS, AND DIVISIONS ARE NAMED ALPHABETICALLY IN ORDER OF DISCOVERY AND FOR SCIENTISTS**

LARGEST RINGS AND DISTANCE FROM PLANET'S CENTER:

D: **67,000-74,490 KM (41,630-46,290 MI)**

C: **74,490-91,980 KM (46,290-57,150 MI)**

B: **91,980-117,500 KM (57,150-73,010 MI)**

A: **122,050-136,770 KM (75,840-84,980 MI)**

G: **166,000-174,000 KM (103,150-108,120 MI)**

E: **180,000-480,000 KM (111,850-300,000 MI)**

False-color image of Saturn's rings. (Inset) Saturn's A and F rings stretch in front of Titan and tiny Epimetheus.

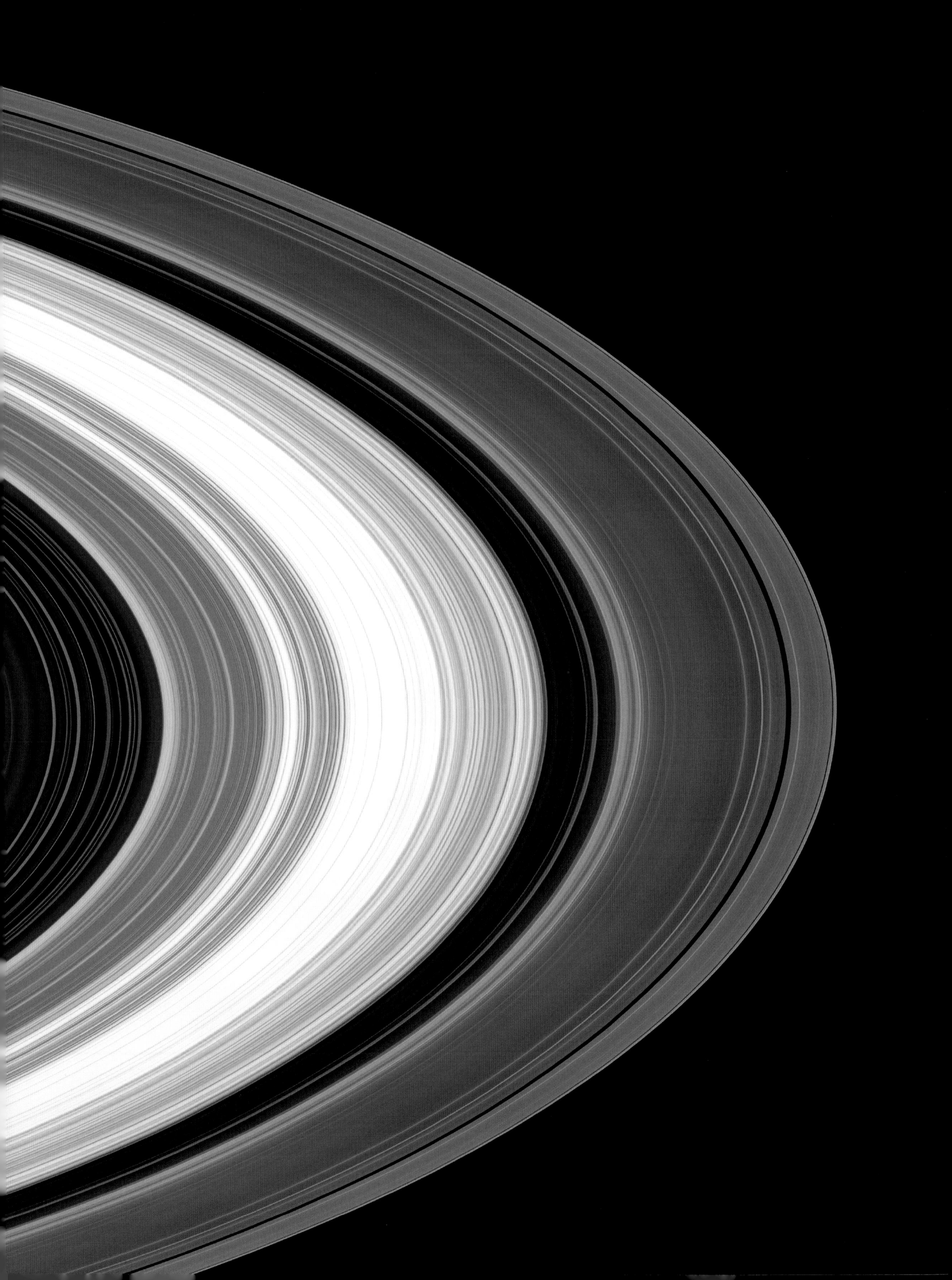

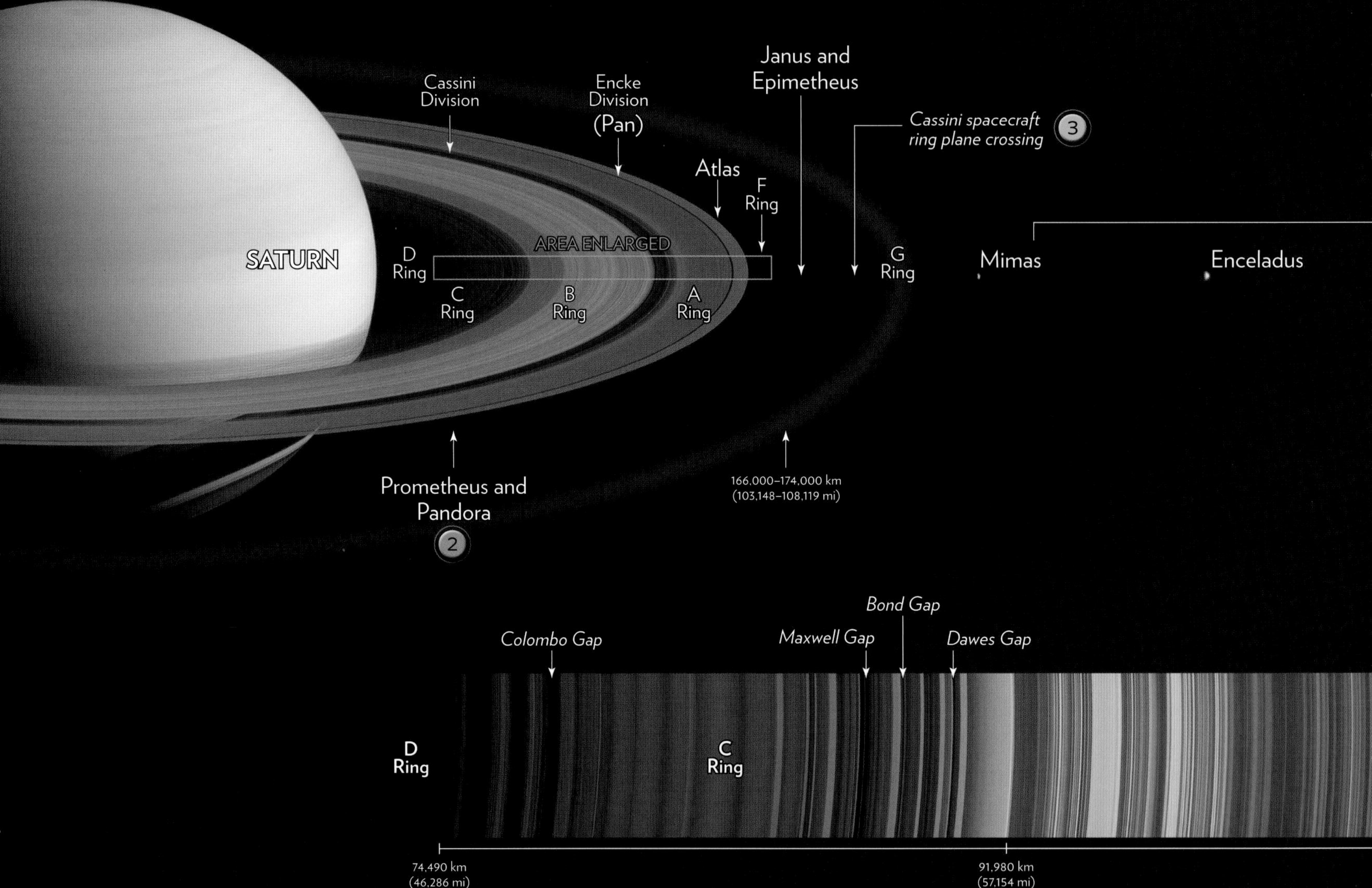

Cassini Division
Encke Division (Pan)
Janus and Epimetheus
Cassini spacecraft ring plane crossing
3
Atlas
F Ring
AREA ENLARGED
SATURN
D Ring
C Ring
B Ring
A Ring
G Ring
Mimas
Enceladus
Prometheus and Pandora
2
166,000–174,000 km (103,148–108,119 mi)
Bond Gap
Colombo Gap
Maxwell Gap
Dawes Gap
D Ring
C Ring
74,490 km (46,286 mi)
91,980 km (57,154 mi)

KEY FEATURES

1. **CASSINI DIVISION:** Largest gap in the ring system
2. **PROMETHEUS AND PANDORA:** Tiny "shepherd" moons
3. **CASSINI RING PLANE CROSSING:** Gap through which the Cassini orbiter flew

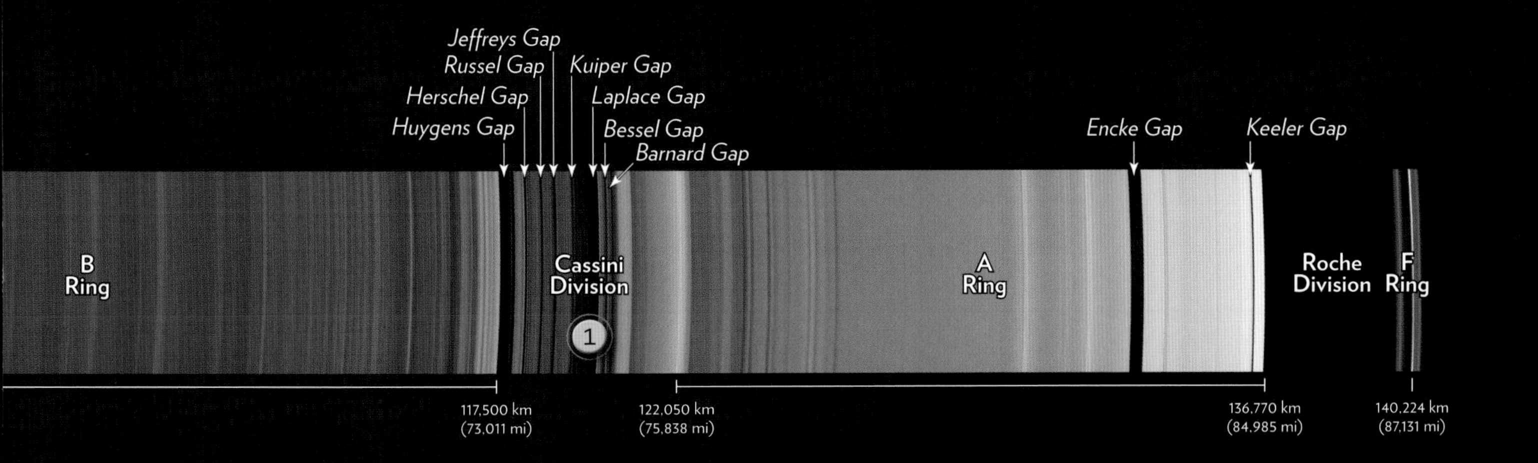

Although today we know that all the giant planets have ring systems, the rings of Saturn remain the largest and most complete set in the solar system. Almost from their discovery, they have stimulated the human imagination. For example, a book published in 1837 by English cleric Thomas Dick, with the grandiloquent title of *Celestial Scenery, or the Wonders of the Planetary System Displayed, Illustrating the Perfections of Deity and a Plurality of Worlds,* estimates the number of human beings living on the rings of Saturn at 8,141,963,826,080!

Unfortunately for the good reverend, in the 18th and 19th centuries scientific calculations established clearly that the rings of Saturn could not be solid, like a cosmic hula hoop, or liquid. In either case, the forces acting in the ring would make it unstable. Early on, then, scientists knew that the rings had to be collections of orbiting particles. In fact, we now know that the constituents of the rings are mainly pieces of water ice, ranging in size from bits a fraction of a centimeter wide to boulders several meters across. The most astonishing thing is that in spite of their optical prominence, the rings are actually quite thin. Estimates put the average thickness of the rings at about 10 meters (30 ft)—an ordinary two-story building would barely stretch from top to bottom.

Saturn's rings were named alphabetically in the order of discovery. Gaps and divisions between the rings are named for astronomers; these gaps seem to be maintained by the influence of Saturn's many moons.

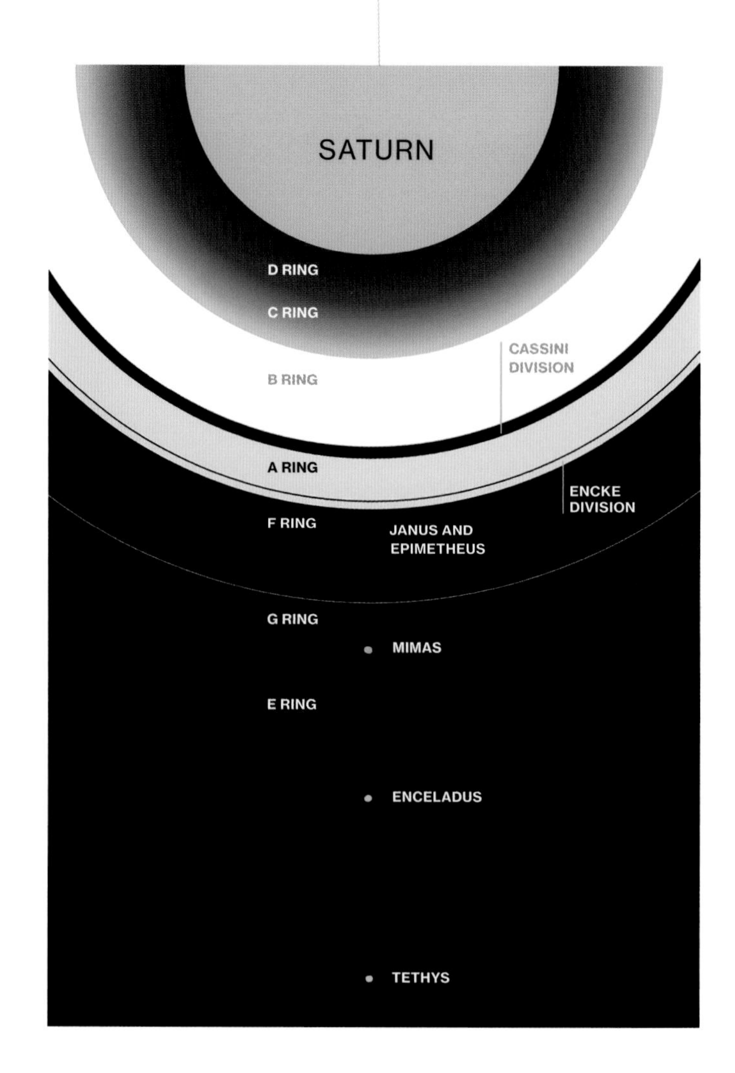

RINGS, RINGLETS, AND GAPS

At the same time that scientists were gaining a theoretical understanding of the basic dynamics of the rings, astronomers were using telescopes to discover something about their structure. In 1675 the Italian astronomer Giovanni Cassini (1625–1712) found that the rings were not a single unbroken hoop, but had dark divisions between different sections. The largest of these gaps is now called the Cassini division. In fact, as seen from Earth, there are two main dark gaps in the rings, dividing the rings into three parts. These parts were named, with an unusual lack of imagination, the A, B, and C rings. The naming of the many rings discovered since then takes the alphabet up to G, with some also named for the moons that most influence them. Although there are faint rings of dust millions of kilometers from Saturn, the primary rings extend from about 74,000 kilometers (46,000 mi) from the center of the planet out to about 137,000 kilometers (85,000 mi). (For reference, our own moon is 375,000 kilometers/250,000 mi from Earth.)

An artist's conception portrays the icy rubble that makes up one of Saturn's rings. The particles in the rings are actually quite small, ranging from pebble to boulder size. Saturn's moons play a shepherding role, since their gravity keeps these particles in tight orbits.

With the flyby of the Voyager spacecraft in 1980 and, of course, the arrival of the Cassini spacecraft in 2004, our knowledge of the rings has increased considerably. We now understand that the rings are actually complex structures, with thousands of thin gaps separating ringlets. The gaps aren't really empty—they just have less material in them—and seem to be maintained by a variety of gravitational interactions with Saturn's moons. In some cases, close-in moons simply clear out a path, whereas in other cases the process is more complex, but in the end it is the combination of the moons and their gravity that maintains the structure of the rings.

The one possible exception to this rule is the appearance of phenomena known as spokes, seen intermittently by spacecraft. These are lines running across the rings, bright or dark depending on whether light is being reflected from the spokes or transmitted through them. They are thought to be made of microscopic dust particles, perhaps created in storms on Saturn, that are suspended above the rings by static electricity.

A MOON'S REMAINS?

Finally, we can turn to the question of how such a ring system could form. The most popular theory is that the particles are debris remaining from an ancient moon. The amount of material is about the same as that found in many of Saturn's surviving moons. Different versions of the hypothesis have the moon breaking up because it got too close to Saturn or because of a collision. If the collision hypothesis is correct, then it is likely that the rings of Saturn are one more legacy of the Late Heavy Bombardment (see page 59), that great reshuffling of the solar system that took place four billion years ago.

The last two planets in the solar system are the least known and the least explored. They probably formed closer to the sun than their present orbits, and, although they are similar to each other, they differ significantly from the gas giants Jupiter and Saturn. The atmospheres of Uranus and Neptune contain a lot more of the substances astronomers call ices—dense mixtures of water, ammonia, and methane. Thus, they are often referred to as ice giants. • The ice giants are intermediate in size between the terrestrial planets and the gas giants. Neptune, for example, has 17 times the mass of Earth, but only 1/19 the mass of Jupiter.

URANUS & NEPTUNE

| ICE GIANTS |

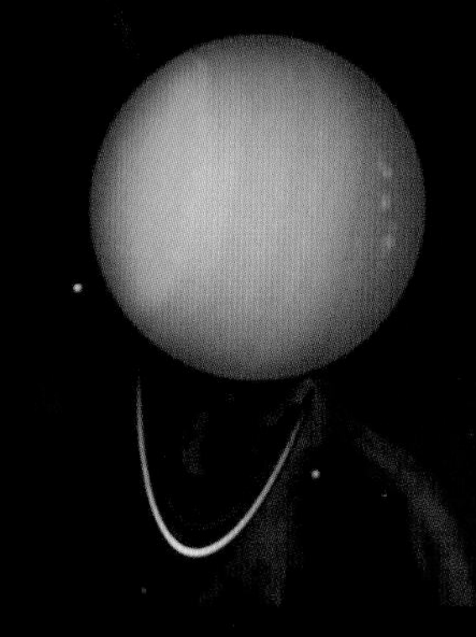

URANUS DISCOVERER: **WILLIAM HERSCHEL**
URANUS DISCOVERY DATE: **MARCH 13, 1781**
URANUS NAMED FOR: **GREEK GOD OF THE SKY**
URANUS MASS: **14.54 × EARTH'S**
URANUS VOLUME: **63.09 × EARTH'S**

NEPTUNE DISCOVERERS: **URBAIN LEVERRIER, JOHN COUCH ADAMS, AND JOHANN GALLE**
NEPTUNE DISCOVERY DATE: **SEPTEMBER 23, 1846**
NEPTUNE NAMED FOR: **ROMAN GOD OF THE SEA**
NEPTUNE MASS: **17.15 × EARTH'S**
NEPTUNE VOLUME: **57.72 × EARTH'S**
URANUS AND NEPTUNE MOONS: **27 AND 13**
PLANETARY RING SYSTEM: **YES FOR BOTH PLANETS**

Art of Neptune with rings and moon Triton. (Inset) Uranus.

URANUS

Distant and still somewhat mysterious, Uranus has not been visited by a spacecraft since 1986.

With its axis of rotation nearly parallel to the plane of its orbit, Uranus can be thought of as "rolling" through its year. As it revolves, one pole is always lit by the sun, while the other is in darkness. Taking nearly 84 Earth years to complete an orbit, each pole receives about 40 years of day followed by 40 of night.

Methane gas in the atmosphere absorbs the red spectrum of visible light. Only blue-green light is reflected, giving Uranus its characteristic color.

Unlike Earth, which has a magnetic field rotated 11 degrees from its geographic poles, Uranus's is shifted 59 degrees. Another baffling feature is that the axis of the magnetic poles does not pass through the center of the planet.

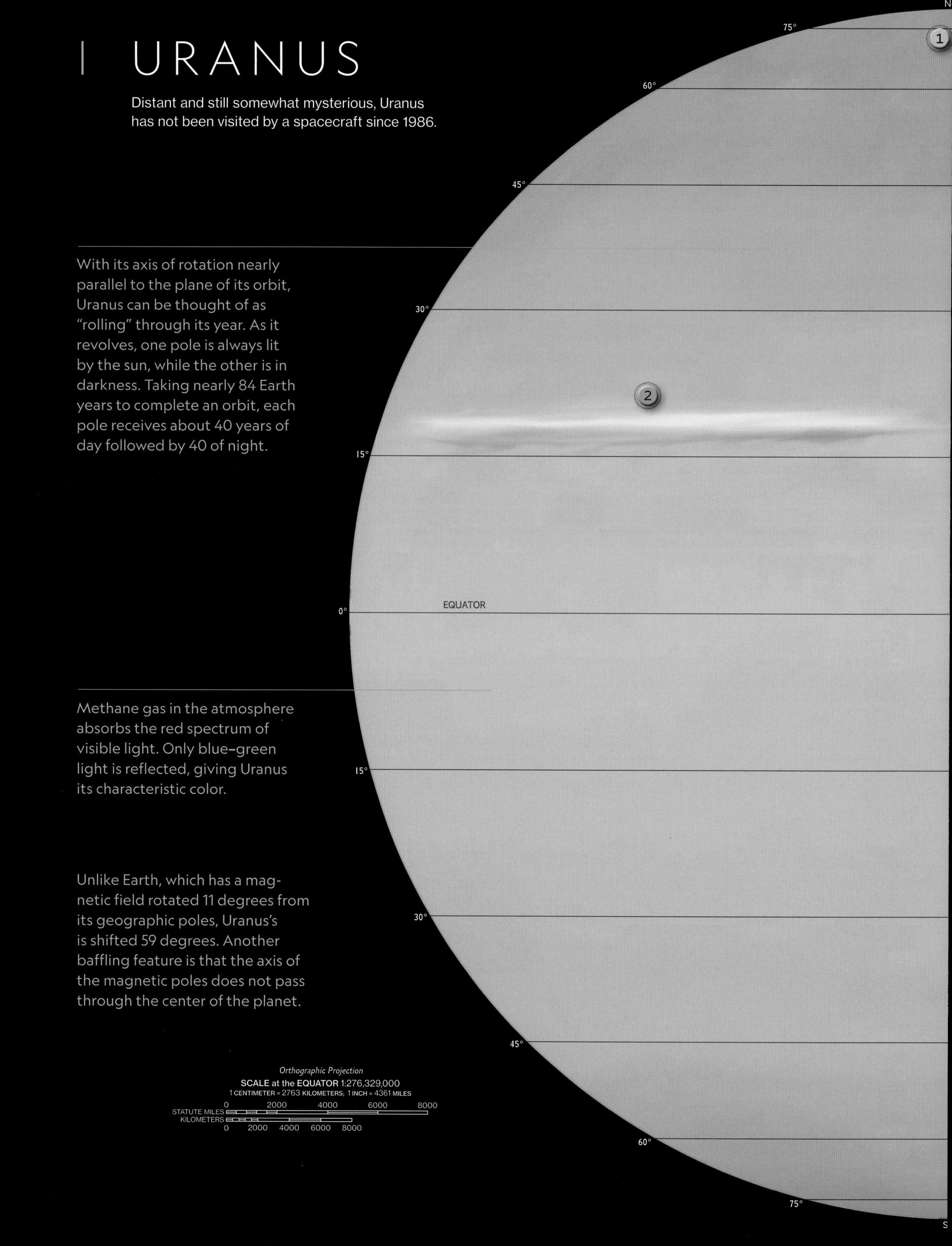

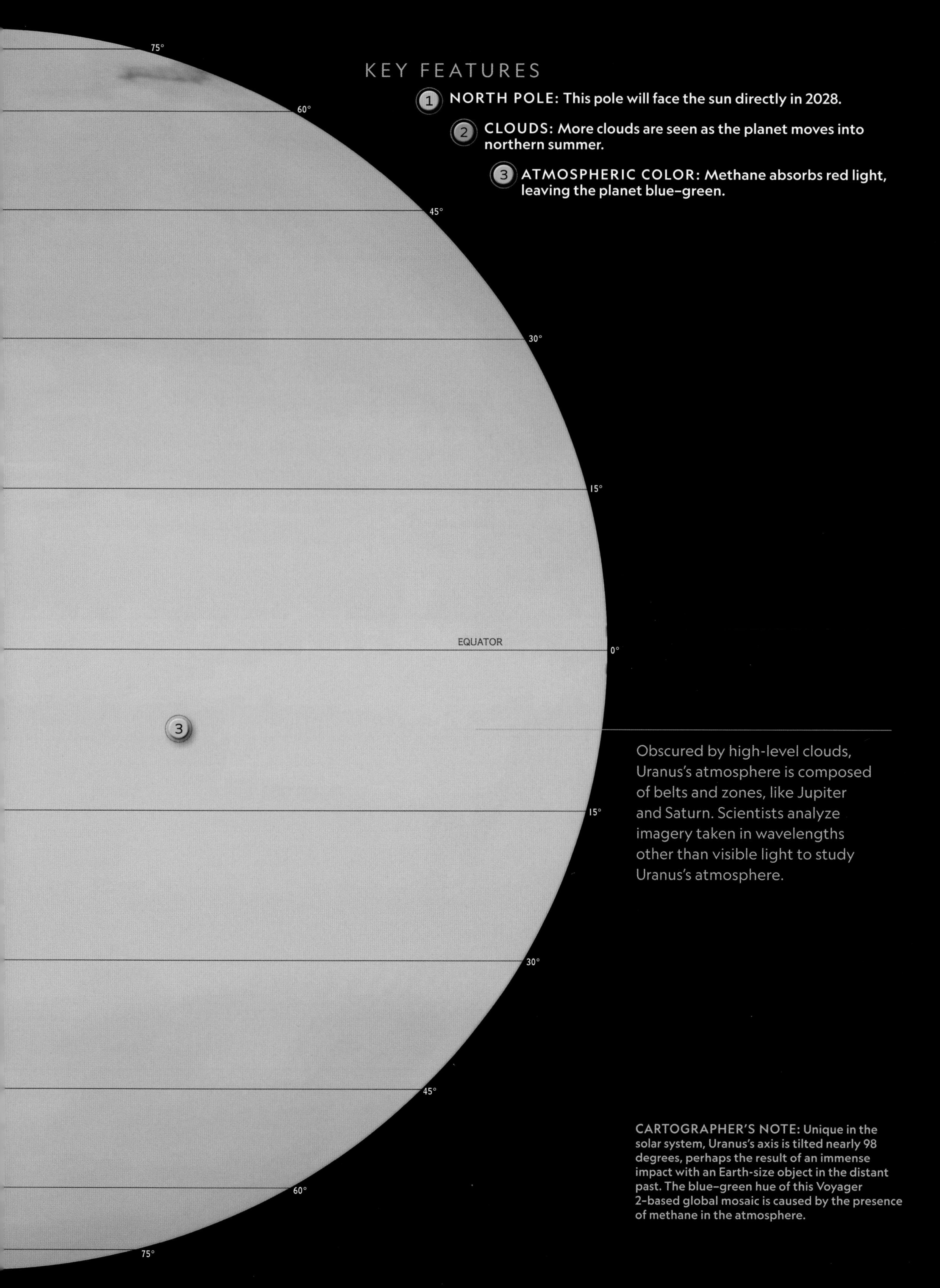

KEY FEATURES

1. **NORTH POLE:** This pole will face the sun directly in 2028.
2. **CLOUDS:** More clouds are seen as the planet moves into northern summer.
3. **ATMOSPHERIC COLOR:** Methane absorbs red light, leaving the planet blue-green.

Obscured by high-level clouds, Uranus's atmosphere is composed of belts and zones, like Jupiter and Saturn. Scientists analyze imagery taken in wavelengths other than visible light to study Uranus's atmosphere.

CARTOGRAPHER'S NOTE: Unique in the solar system, Uranus's axis is tilted nearly 98 degrees, perhaps the result of an immense impact with an Earth-size object in the distant past. The blue-green hue of this Voyager 2-based global mosaic is caused by the presence of methane in the atmosphere.

URANUS'S MOONS
MIRANDA AND ARIEL

Miranda has one of the most shattered surfaces in the solar system; Ariel may have had recent geological activity.

MIRANDA — SOUTHERN HEMISPHERE

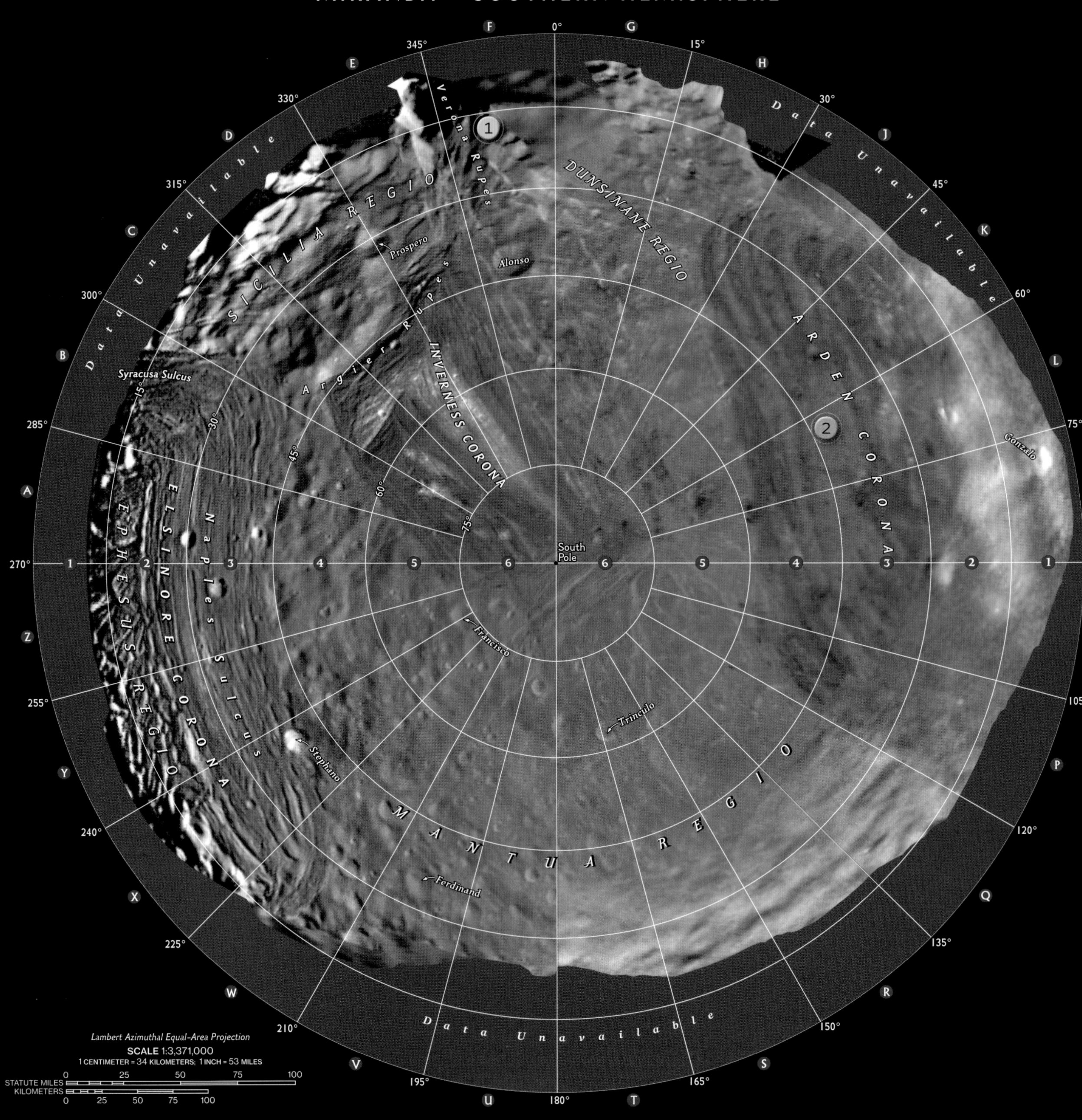

KEY FEATURES

1. **VERONA RUPES:** One of many sharp cliffs and canyons
2. **ARDEN CORONA:** Grooved landscape, possibly over warmer ice
3. **KACHINA CHASMATA:** Canyons possibly formed by faulting

ARIEL – SOUTHERN HEMISPHERE

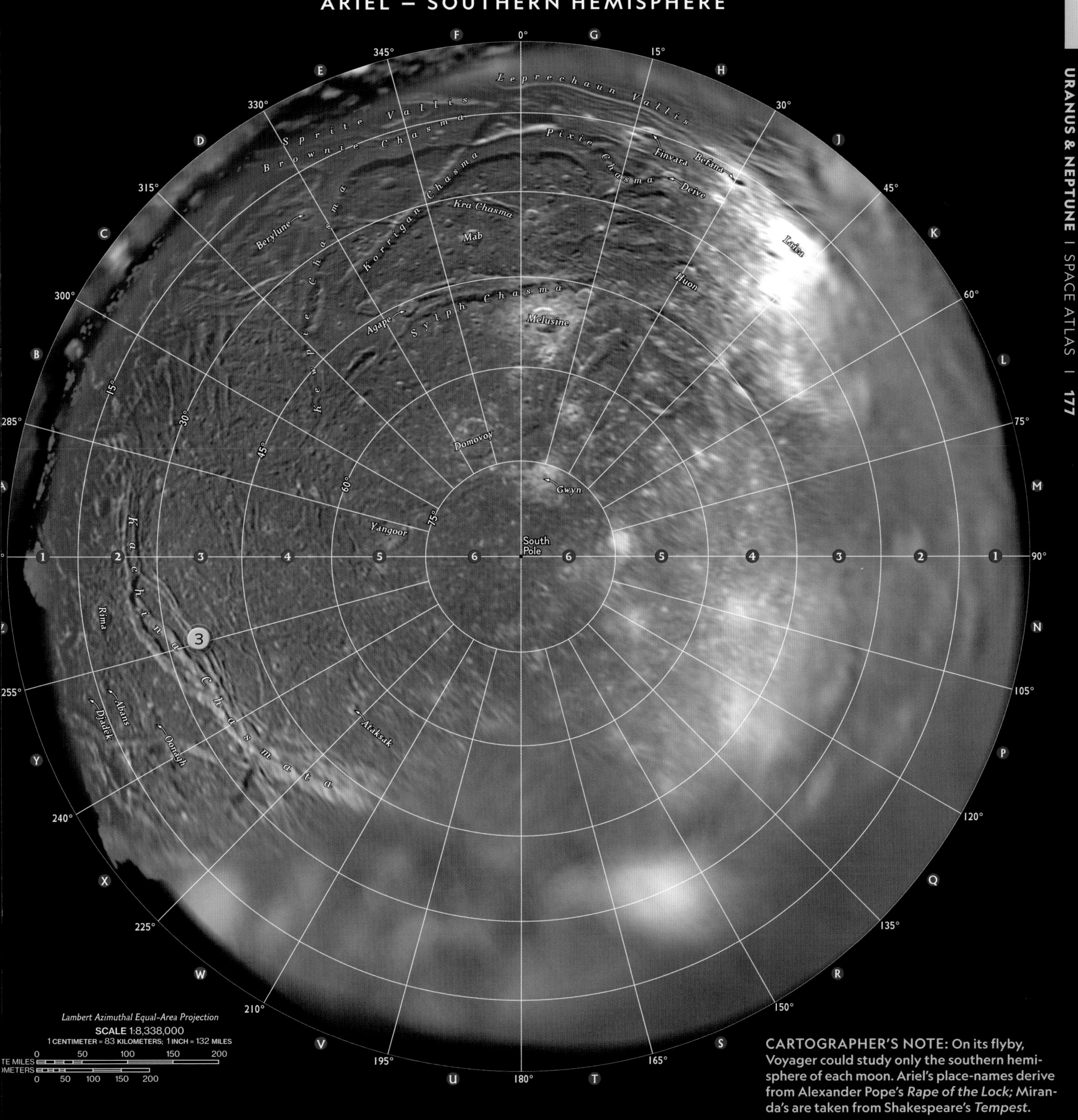

Lambert Azimuthal Equal-Area Projection
SCALE 1:8,338,000
1 CENTIMETER = 83 KILOMETERS; 1 INCH = 132 MILES

CARTOGRAPHER'S NOTE: On its flyby, Voyager could study only the southern hemisphere of each moon. Ariel's place-names derive from Alexander Pope's *Rape of the Lock;* Miranda's are taken from Shakespeare's *Tempest*.

| URANUS'S MOONS

UMBRIEL AND TITANIA

Mysteriously dark, Umbriel has an ancient surface. Titania, largest of the planet's moons, may be geologically active.

UMBRIEL — SOUTHERN HEMISPHERE

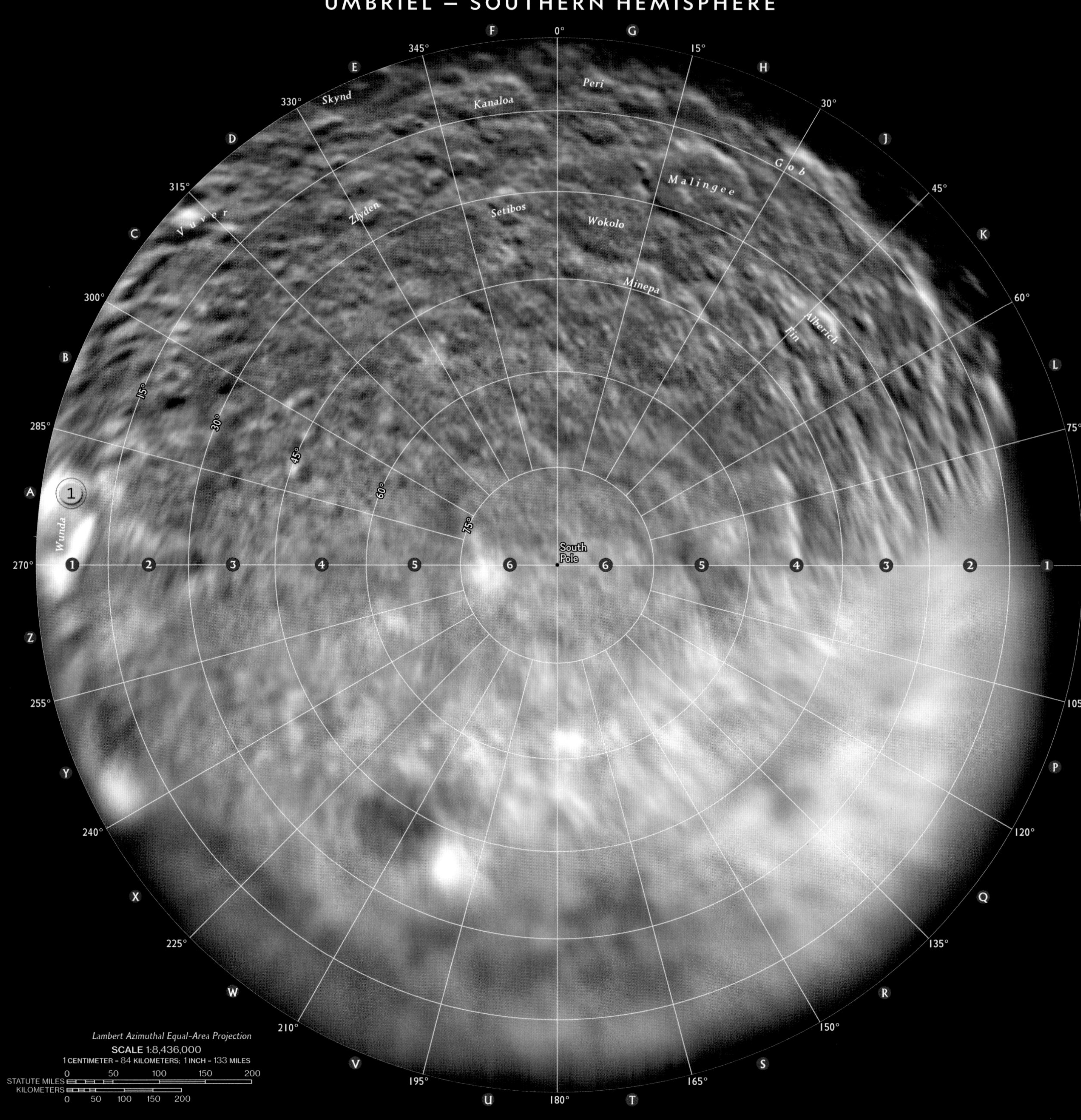

KEY FEATURES

1. **WUNDA:** Crater with oddly bright rim, possibly from frost or impact deposits
2. **MESSINA CHASMATA:** Canyon almost 1,500 kilometers (930 mi) long
3. **ROUSILLON RUPES:** Young scarp (cliff) 400 kilometers (250 mi) long

TITANIA – SOUTHERN HEMISPHERE

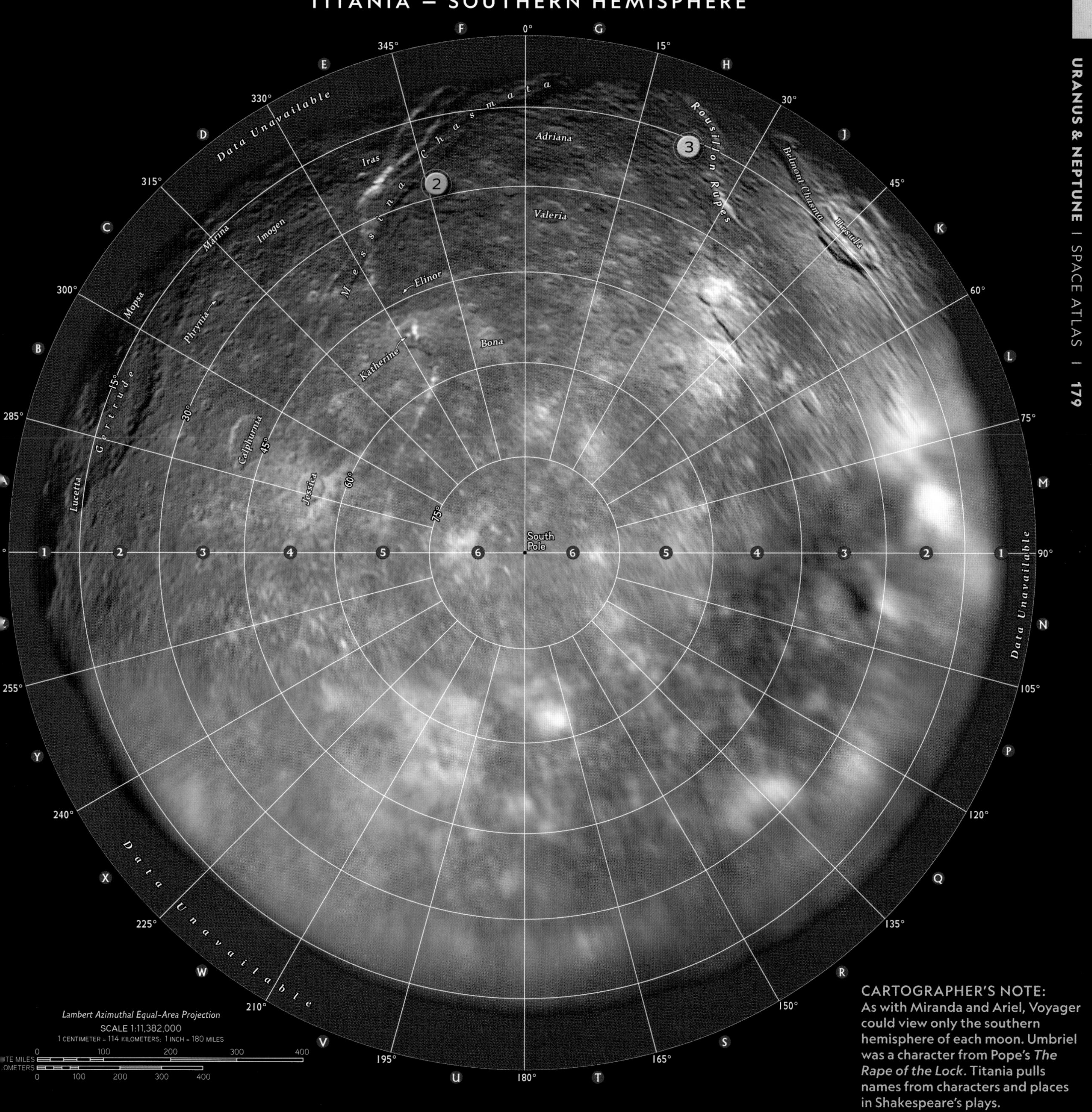

CARTOGRAPHER'S NOTE:
As with Miranda and Ariel, Voyager could view only the southern hemisphere of each moon. Umbriel was a character from Pope's *The Rape of the Lock*. Titania pulls names from characters and places in Shakespeare's plays.

NEPTUNE

Sometimes orbiting beyond Pluto, icy Neptune shines bright blue.

Like the other outer giants, Uranus's atmosphere is divided into belts and zones. Unique to this planet, those winds move easterly, counter to Neptune's rotation.

Similar to Jupiter and Saturn, Neptune radiates more heat than it receives from the sun. Scientists don't know the process by which this occurs.

Seen in many Voyager images of Neptune, some scientists think the dark feature may not be a storm like the Great Red Spot on Jupiter. Variable in size, it could be a vast hole in the atmosphere's upper cloud layers. Once the Hubble Space Telescope began observing the blue planet, the feature had disappeared.

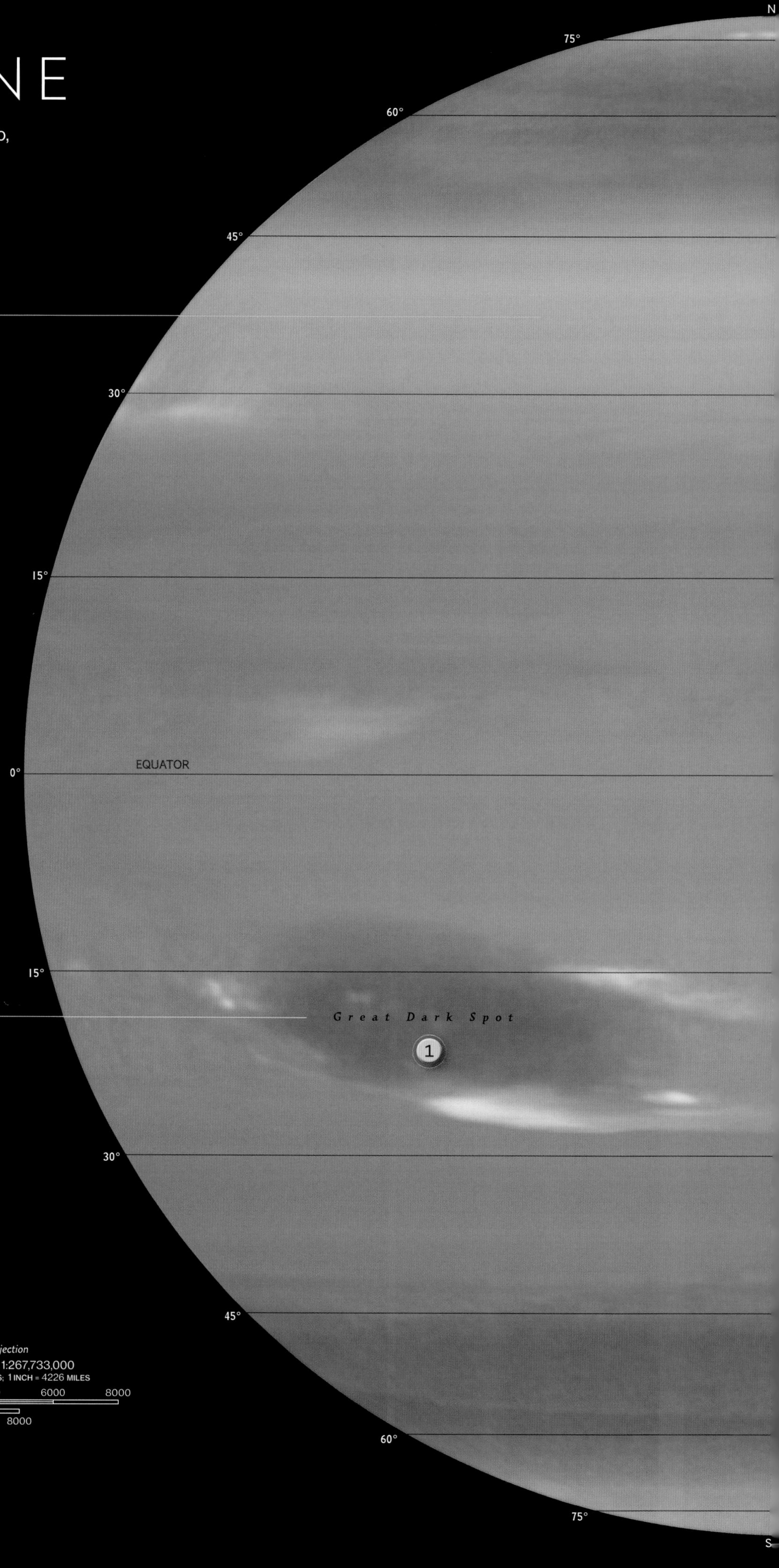

Orthographic Projection

SCALE at the EQUATOR 1:267,733,000

1 CENTIMETER = 2677 KILOMETERS; 1 INCH = 4226 MILES

STATUTE MILES 0 2000 4000 6000 8000

KILOMETERS 0 2000 4000 6000 8000

KEY FEATURES

1. **ANTICYCLONE:** Huge storms appear and disappear in the atmosphere.
2. **WINDS:** High-altitude winds blow at supersonic speeds.
3. **BANDS:** Southern cloud bands are increasing during the 40-year summer.

Voyager 2 flew by Neptune, achieving its closest approach to the planetary system on August 24, 1989. Passing just under 5,000 kilometers (3,100 miles) from the planet's cloud surface, nearly all detailed information on Neptune came from this sojourn.

CARTOGRAPHER'S NOTE: Visited only by the Voyager probe, Neptune is mapped here in a construction based on those data. Warmth from inside the planet stirs active weather in the hydrogen–helium atmosphere; winds are immensely fast.

NEPTUNE'S MOONS TRITON

Largest of Neptune's moons, at minus 240°C (−400°F) Triton is colder even than Pluto.

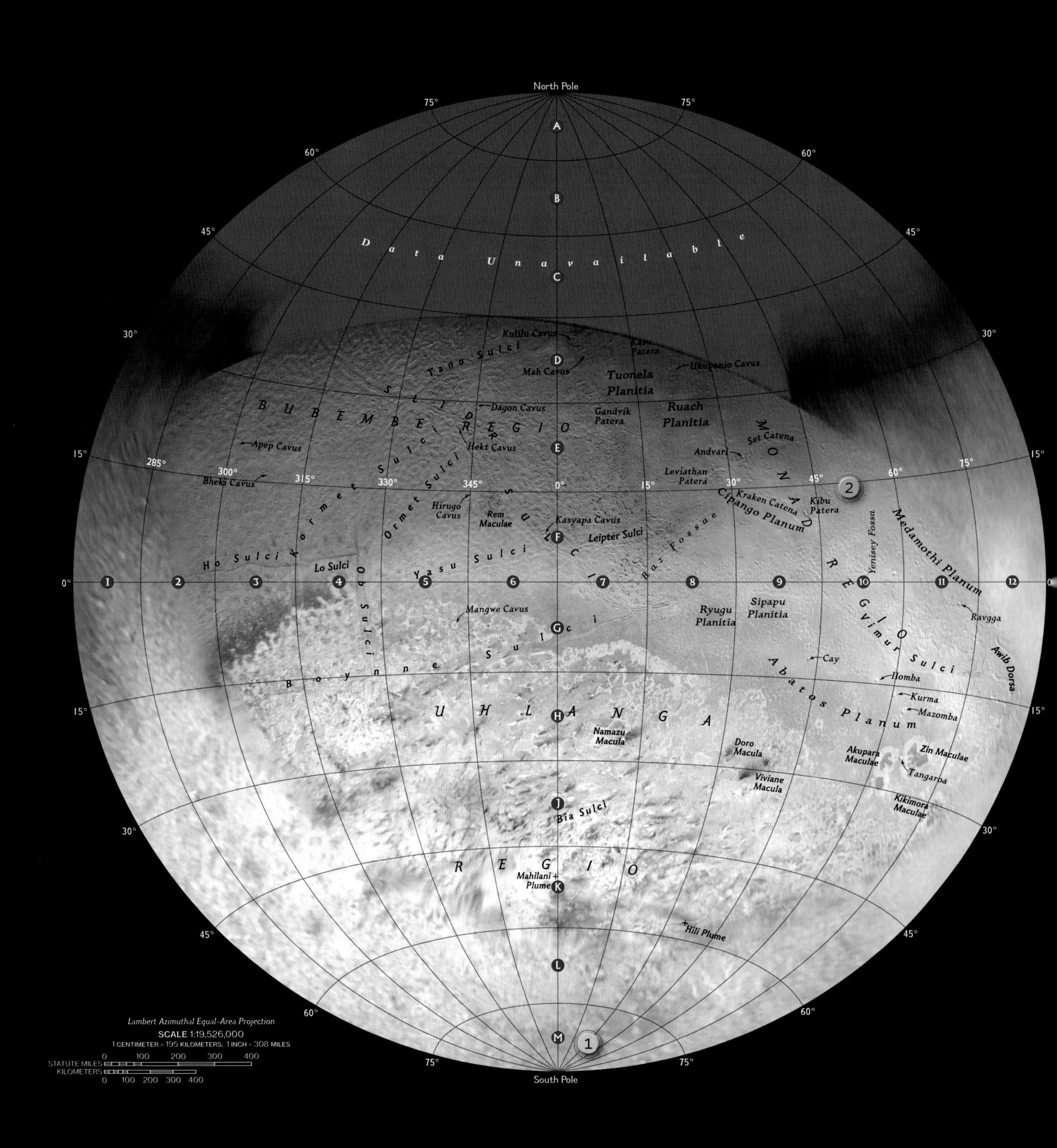

KEY FEATURES

1. **SOUTH POLE:** This pole has an ice cap of frozen nitrogen and methane.
2. **CRYOVOLCANOES:** Dark streaks may indicate deposits from ice volcanoes.
3. **CANTALOUPE TERRAIN:** An image highlights Triton's frozen, buckled surface.

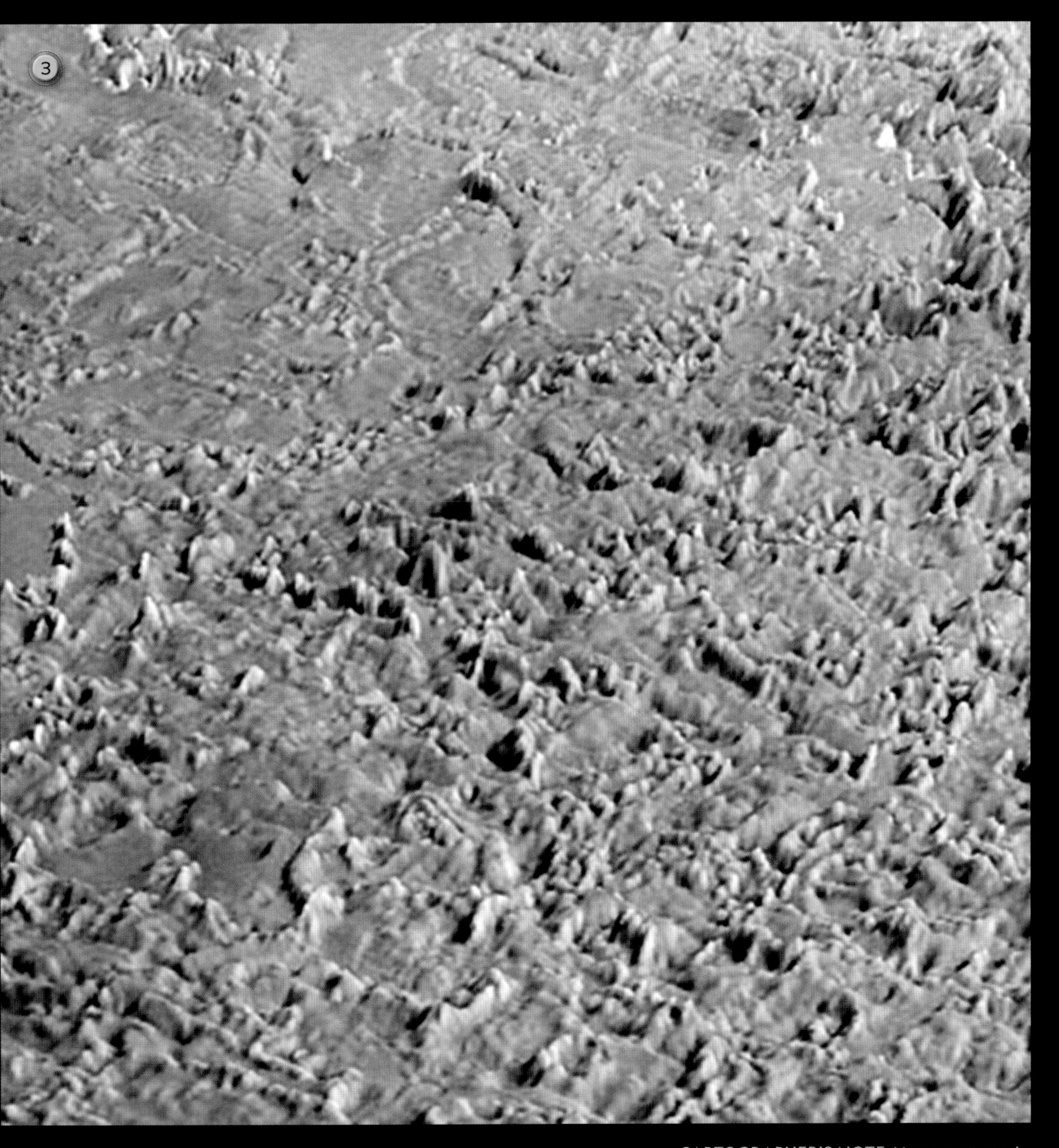

CARTOGRAPHER'S NOTE: Voyager returned images of only portions of the moon as the spacecraft streaked past on its way to the outer reaches of the solar system. Aquatic names from Earth are applied to the moon's physiographic features. Cryovolcanoes dot its surface, erupting with what some scientists think could be nitrogen.

Beneath their upper cloud layers, Uranus and Neptune are thought to have very similar structures. Each has an atmosphere consisting of hydrogen, helium, and methane gases that gets more and more dense as you descend. At some point there is a transition, without a clearly defined surface, to a hot liquid composed of the same materials. The liquid is hot because of the intense pressure inside the planets—they are much too far from the sun for solar heat to have much of an effect. (It is an idiosyncrasy of astronomical nomenclature that a mixture like this is called an ice, even though it is a dense liquid at a temperature of several thousand degrees.) At the very center of each planet is a small rocky core roughly the size of Earth. Like all the giant planets, both Uranus and Neptune have multiple moons and ring systems.

Cutaway view of Uranus. Smaller than Jupiter and Saturn, the ice giant has a rocky core, an icy mantle, and an outer envelope of gaseous hydrogen and helium. This atmosphere makes up 20 percent of the planet's radius.

PLANET SIDEWAYS

In many ways, the most interesting thing about the ice giants is the way they were discovered.

Uranus was actually detected several times before it was officially discovered—it is so faint and so slow-moving that it was mistaken for a star. The realization that this faint, slow-moving object was actually a planet was made by the amateur British astronomer William Herschel. The German-born Herschel worked as chief organist in a chapel in Bath. In his spare time he made telescopes and observed the heavens, and on March 13, 1781, he saw a strange object in his telescope. At first he thought it might be a comet, but as data accumulated on the new object, it became clear that Herschel had become the first human being in recorded history to discover a new planet. (Herschel wanted to name the planet Georgium Sidus in honor of George III. The name didn't stick, but Herschel did get a lifetime pension from the king.)

Much of our modern information about the planet was garnered when the spacecraft Voyager 2 flew by Uranus in 1986. Methane (natural gas) is the third most abundant element in the Uranian atmosphere, and it is this gas that gives the planet its aquamarine color. However, the property that catches our attention is the planet's axis of rotation: It lies in the same plane as its orbit, meaning that the planet is tilted 98 degrees, rotating on its side. The skewed rotation is probably the result of a collision shortly after the planet formed. Each pole gets 42 Earth years of sunlight followed by 42 Earth years of darkness as Uranus completes an 84-year orbit around the sun.

Uranus has 27 moons, named after such Shakespearean characters as Ariel and Miranda, and 13 narrow rings.

PLANET OF WINDS

If the discovery of Uranus depended on a bit of observational serendipity, Neptune's discovery was the result of careful calculation. As the newly discovered Uranus was tracked around its orbit, discrepancies started to appear between what was observed and what the law of gravity predicted. Two young astronomers—John Couch Adams in England and Urbain Leverrier in France—independently asked themselves whether these discrepancies could be due to the gravitational tug of another, unseen planet still farther from the sun. Following a series of complex maneuvers and communications, astronomers at the Berlin Observatory trained their telescopes on the place where the unknown planet was predicted to be. On September 23, 1846, the planet we now call Neptune was seen and recorded by Johann Galle.

The tilt of Neptune's axis of rotation is similar to that of Earth, so, unlike Uranus, it has seasons and weather. Visited by Voyager 2 in 1989 (after the craft had visited Uranus), Neptune was revealed to have large storms on its surface. Called the Great Dark Spot, the Small Dark Spot, and, incongruously, Scooter, these storms are similar in appearance to Jupiter's Great Red Spot, but they seem to last for only months rather than centuries. Neptune is home to the strongest sustained winds in the solar system—2,100 kilometers an hour (1,300 mph).

Like Uranus, Neptune has thin rings, whose arcs include some with the quintessentially French names of Liberté, Égalité, and Fraternité.

TRITON

Like all the giant planets, Neptune has a large number of moons (13 and counting), but one of them stands out from the others. Triton is a large body—larger than Pluto, for example—and is big enough to be pulled into a spherical shape by its own gravity. What is interesting about it is that, alone among the solar system's large moons, it moves around Neptune in the opposite direction from the planet's rotation (astronomers call this a retrograde orbit). This fact suggests that the moon did not form at the same time as the planet, but that it took

An illustration depicts Uranus (in the background), its thin rings, and its five largest moons—(left to right) Umbriel, Miranda, Oberon, Titania, and Ariel—shown to scale.

NEPTUNE'S GREAT DARK
STORM
WAS LARGER THAN EARTH

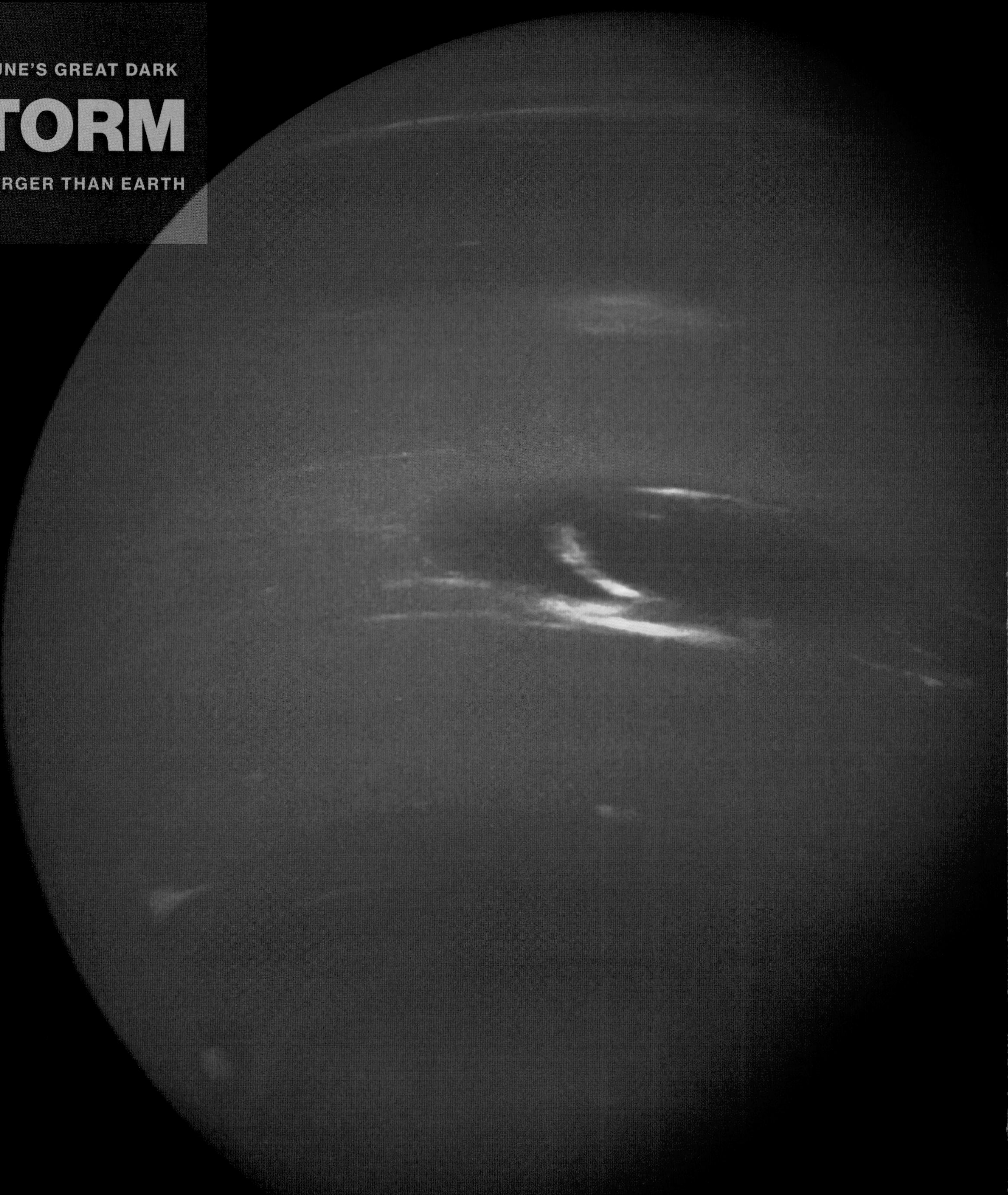

A cutaway view of Neptune shows its rocky, Earth-size core; dense water and ammonia mantle; and gaseous atmosphere, composed of hydrogen, helium, and methane. The outer atmosphere is extremely windy and cold.

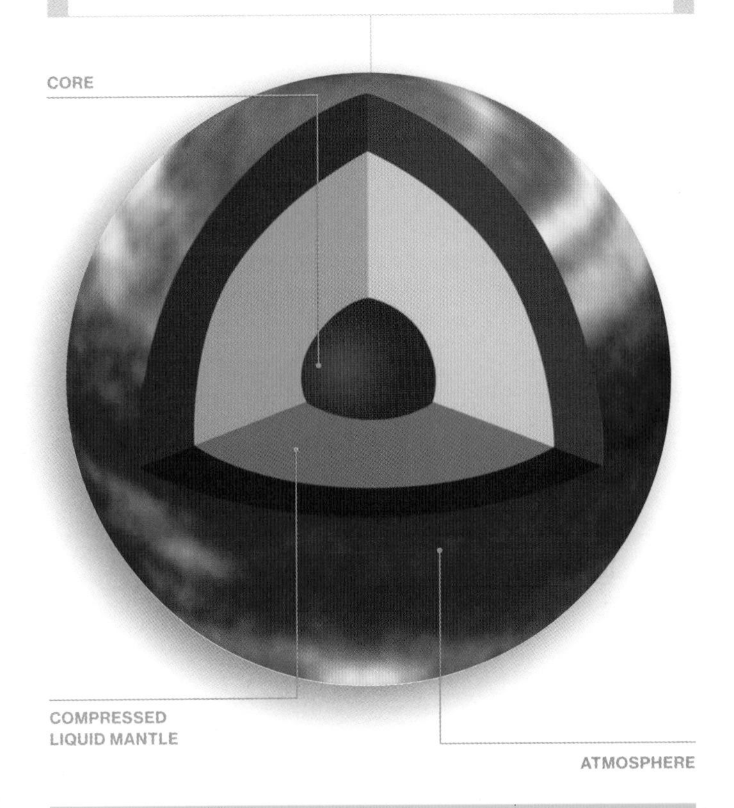

A view of the planet Neptune taken by the Voyager spacecraft shows the Great Dark Spot—a titanic storm raging on the planet's surface—and several smaller storms, including one named Scooter (triangular spot at lower left).

shape elsewhere and was captured. In fact, current thinking is that, like Pluto (which is discussed in the next section), Triton is actually an object that formed in the Kuiper belt (see pages 200–203). It is believed to have a rocky core covered by a layer of frozen nitrogen, and it is one of the coldest objects in the solar system, with a temperature only about 40 degrees above absolute zero. Geysers of frozen nitrogen may even spout from its surface. Although missions to the ice planets in the 2020s and 2030s have been proposed by a number of groups, none has been approved by either NASA or the European Space Agency. For the foreseeable future, then, these planets will retain their mysteries.

Pluto has always been something of an enigma for astronomers. It is small and rocky in a place where we expect gas giants to form. Its orbit seems a little out of whack. What is going on? • Oddly enough, we can begin answering this question by visiting a Kansas farm in the late 1920s. A young farm boy by the name of Clyde Tombaugh, barely past his 20th birthday, had just finished building a small telescope, scavenging parts from old machines to complete his construction project. Using his new instrument, he made some sketches of the surface of Mars and sent them off to Lowell Observatory in Flagstaff with a note asking the professional astronomers there for a critique.

PLUTO

LAST PLANET, FIRST PLUTOID

DISCOVERER: **CLYDE TOMBAUGH**
DISCOVERY DATE: **MARCH 13, 1930**
NAMED FOR: **ROMAN GOD OF THE UNDERWORLD**

MASS: **0.002 × EARTH'S**
VOLUME: **0.006 × EARTH'S**
MEAN RADIUS: **1,151 KM (715 MI)**
MIN./MAX. TEMPERATURE: **−233/−223°C (−387/−369°F)**
LENGTH OF DAY: **6.39 EARTH DAYS (RETROGRADE)**
LENGTH OF YEAR: **248 EARTH YEARS**
NUMBER OF MOONS: **4 (3 NAMED)**
PLANETARY RING SYSTEM: **NO**

A view of the surface of Pluto taken by the New Horizons spacecraft. (Inset) Pluto and Charon flanked by tiny Nix and Hydra.

The astronomers at Lowell were so impressed by what they saw in Tombaugh's sketches that they hired him as an assistant and brought him to Arizona, where they put him to work on a project known as the search for Planet X.

A word of explanation: It was thought at the time (incorrectly, as it turned out) that there were discrepancies in the observed orbit of Neptune that indicated the presence of an unknown planet orbiting farther out from the sun, a body astronomers dubbed Planet X. Searching for Planet X was a straightforward, though tedious, process. Astronomers would take photographs of the same section of the sky a couple of weeks apart, then look for an object that had moved just the right amount to be a planet. Boring work—just the thing to foist off on the new hire.

So as Earth swept around its orbit, Tombaugh alternated between photographing the sky (during the dark of the moon) and examining his plates. On February 18, 1930, six months into the project,

CLYDE TOMBAUGH

"I'd better look at my watch – this could be a historic moment."

I had the opportunity to interview Clyde Tombaugh before his death in 1997. At the time he was an emeritus professor at New Mexico State University in Las Cruces. He was given to tooling around campus in a yellow pickup truck, wearing a baseball cap. This is probably why the students I talked to referred to him as a "cool dude."

Tombaugh told me the story of his job offer from the Lowell Observatory ("Hell, it beat pitching hay") and his reaction to the discovery of Pluto. ("It made my day." Then, more seriously, he thought: "I'd better look at my watch – this could be a historic moment.")

What intrigued me most was his story about what happened after the discovery. Realizing that he wanted a career in astronomy, he enrolled as a student at the University of Kansas. How did professors deal with the fact that they had a famous scientist in their classes? "I got along well with the professors," Tombaugh recalled, "but they wouldn't let me take introductory astronomy. They cheated me out of five hours of credit!"

After dinner the skies were clear, and Tombaugh offered to take me on a tour of his backyard telescopes. It was a magical experience – in his fur hat he was like a wizard from The Lord of the Rings, introducing me to a beautiful new world. We saw craters on the moon, the rings of Saturn, the moons of Jupiter. Pointing to the telescope, Tombaugh commented, "This shaft came from a 1910 Buick, this stand's from a cream separator." With a jolt, I realized that this was the very telescope that had started his career and had eventually led to the discovery of Pluto.

"Aren't you going to give that to the Smithsonian?" I asked. He laughed. "They want it, but they can't have it. I ain't

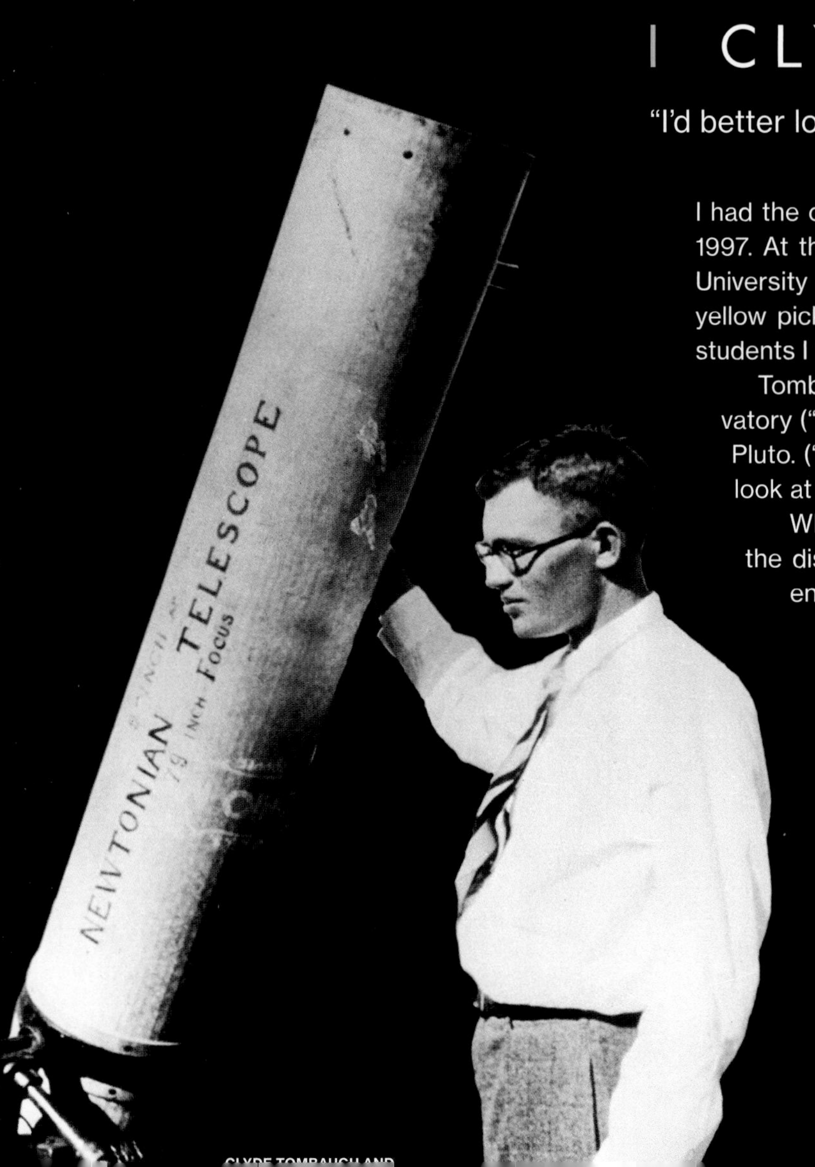

CLYDE TOMBAUGH AND

his hard work paid off. The body we now call Pluto showed up, moving just the right amount between photos to be a new planet. After taking three weeks to verify their results—an astonishingly long interval by today's standards—scientists at Lowell announced that a new planet had been discovered, and Tombaugh (see sidebar opposite) joined the select circle of those who had discovered new worlds. The distant, frigid new planet, fittingly enough, was named for the Greek god of the underworld—a name suggested by 11-year-old British schoolgirl Venetia Burney and passed on to the Lowell Observatory astronomers through a relative connected with Oxford University.

A cutaway shows the probable structure of Pluto. Observations from the Hubble Space Telescope suggest that it is about 50 to 70 percent rocky core, with the remainder ice. At Pluto's temperature, even substances like nitrogen and methane can be found as ices.

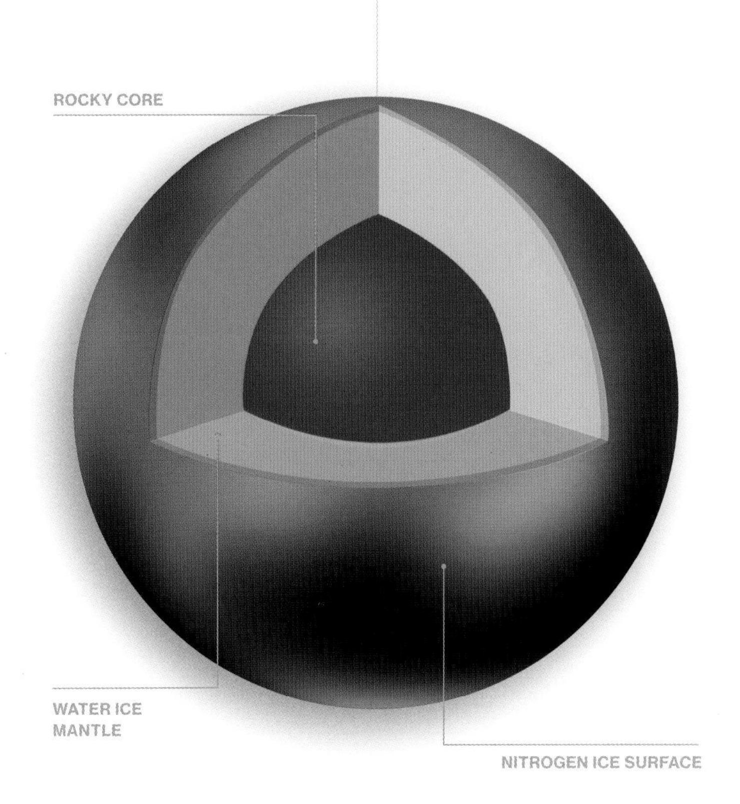

PROBLEMS WITH PLUTO

It wasn't long, though, before problems began to develop with Pluto. For one thing, the plane of its orbit was tilted with respect to the orbits of the other planets. For another, the orbit was somewhat unusual—from 1977 to 1999, for example, Pluto was actually closer to the sun than Neptune. Finally, in 1978 detailed observations revealed that Pluto had a large moon, eventually named Charon after the mythological figure who ferried souls to the underworld. (In addition to Charon, Pluto is now known to have five moons, including Charon; the rest are named Styx, Nix, Kerebos, and Hydra). With the exception of Charon, the moons are quite small; none has a diameter greater than 55 kilometers (37 mi). It is thought that Charon was formed in a collision—much like the one that created the Earth's moon—and that the other moons are simply leftover bits of debris. The discovery of Charon allowed astronomers to calculate Pluto's mass, which turned out to be smaller than that of our own moon. In fact, if you weigh 100 pounds on Earth, you would weigh only 8 pounds on Pluto. Instead of the gas giant that our hypotheses led us to expect out there, we had found a small, rocky, icy world. Because of all these issues, throughout the latter part of the 20th century Pluto was something of an odd man out—it was there, but nobody really wanted to talk about it.

But mystery continued to follow mystery. Theorists calculated (by means described in the next section) that the solar system didn't end with Pluto, but extended outward in a ring of rocky debris called the Kuiper belt. This is discussed in more detail in the next section, but for the moment we'll simply note that in 2005 astronomer Mike Brown at the Palomar Observatory discovered

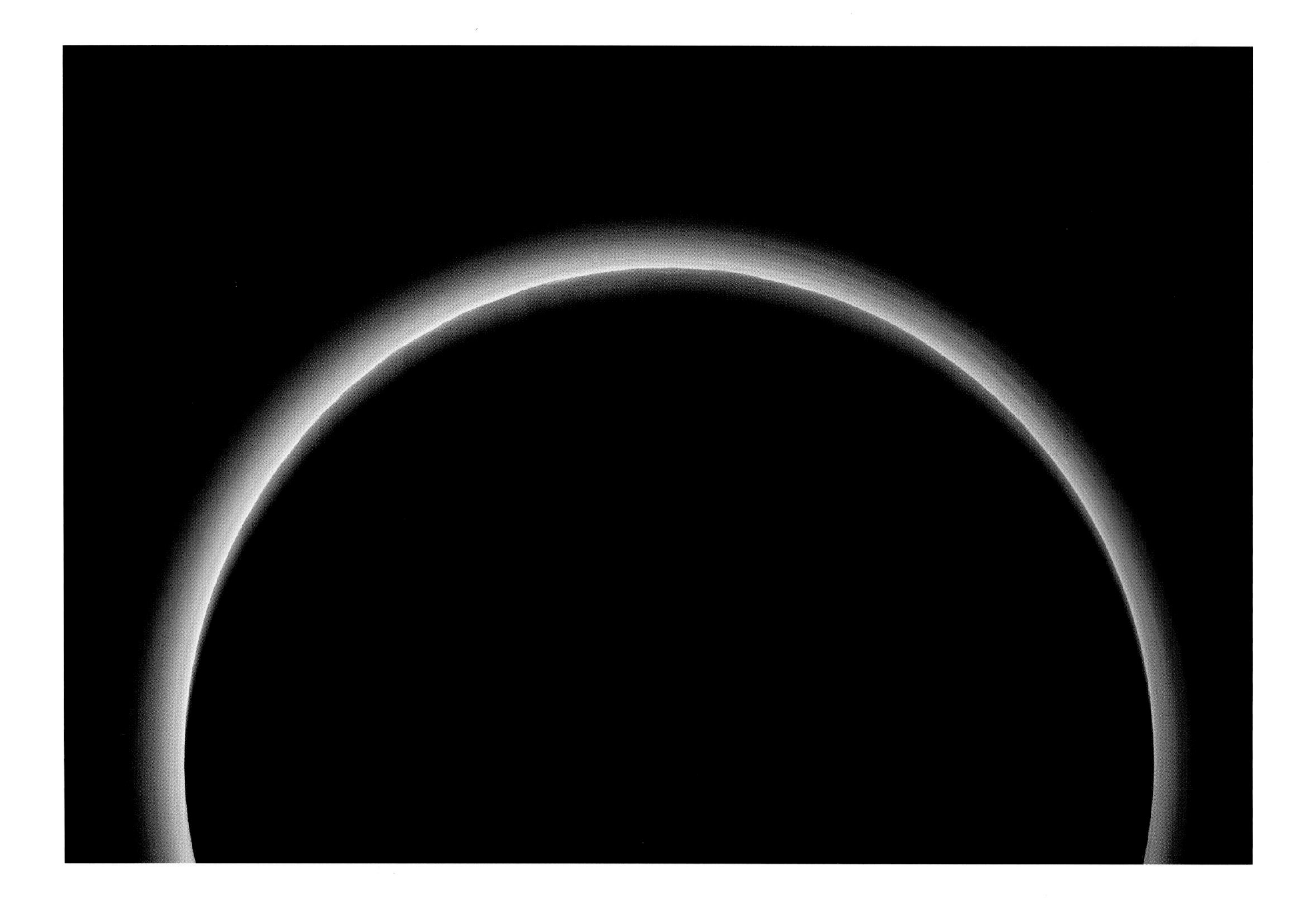

a large, planetlike object orbiting out beyond Pluto. Eventually named Eris for the Greek goddess of strife and discord, this object is actually more massive than Pluto, although it is smaller in volume. Since that time, more than a dozen other planetlike objects have been found in the Kuiper belt, and astronomers expect to find many more.

PLUTOIDS

All of which brings us to the August 2006 meeting of the International Astronomical Union in Prague. On the last day of that conference, with only a little more than 400 of the 2,400 participants voting, Pluto was designated a dwarf planet. Later, all of the newly discovered Kuiper belt planets were put in a class called plutoids. The idea behind this switch is that all of the difficulties that arise when we try to lump Pluto with the inner planets disappear when we realize that Pluto belongs in an entirely different category. It is, in fact, the first plutoid rather than the last planet, the beginning of the end rather than the end of the beginning.

This image of Pluto taken by NASA's New Horizons spacecraft shows the spectacular layers of blue haze in Pluto's atmosphere. Scientists believe the haze is a photochemical smog resulting from the action of sunlight on methane and other molecules in Pluto's atmosphere.

NEW HORIZONS

The New Horizons spacecraft was launched from Cape Canaveral in 2006 and, after a circuitous journey through the solar system, made humanity's first flyby of Pluto on July 14, 2015.

At closest approach, the craft was just 12,500 kilometers (7,800 mi) above Pluto's surface. The

PLUTO

An icy world on the edge of the Kuiper belt, dwarf planet Pluto is the first to be visited and mapped by human spacecraft.

KEY FEATURES

1. **TOMBAUGH REGIO:** Region named for Pluto's discoverer, Clyde Tombaugh
2. **BURNEY CRATER:** Named for the 11-year-old girl who suggested the discovery be named Pluto.
3. **VOYAGER TERRA:** Area named for the Voyager spacecraft that explored the outer planets

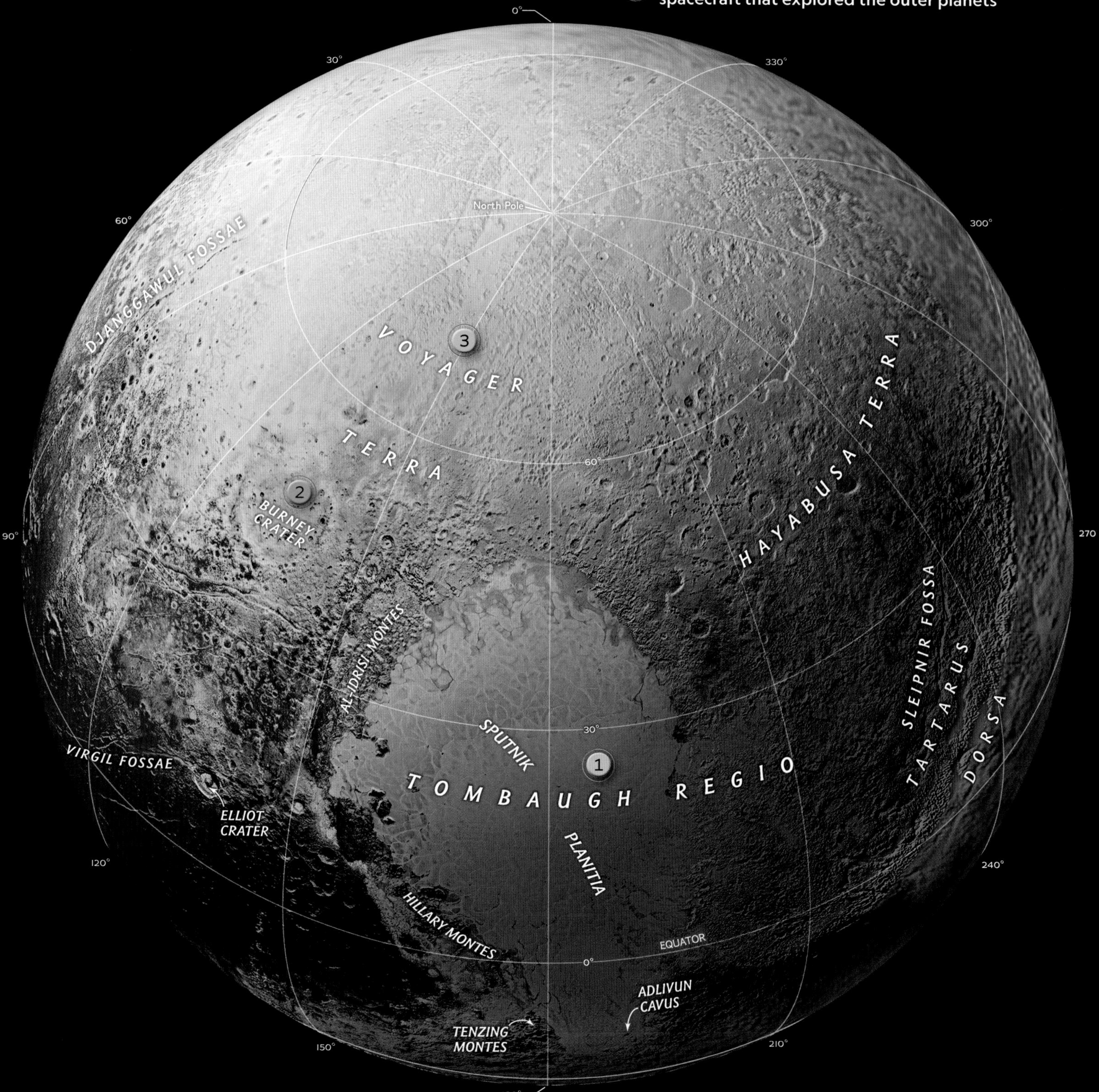

CARTOGRAPHER'S NOTE: The New Horizons probe provided all the data used to create this map of Pluto. Features are named after famous explorers, pioneering spacecraft, and mythological characters from the Underworld.

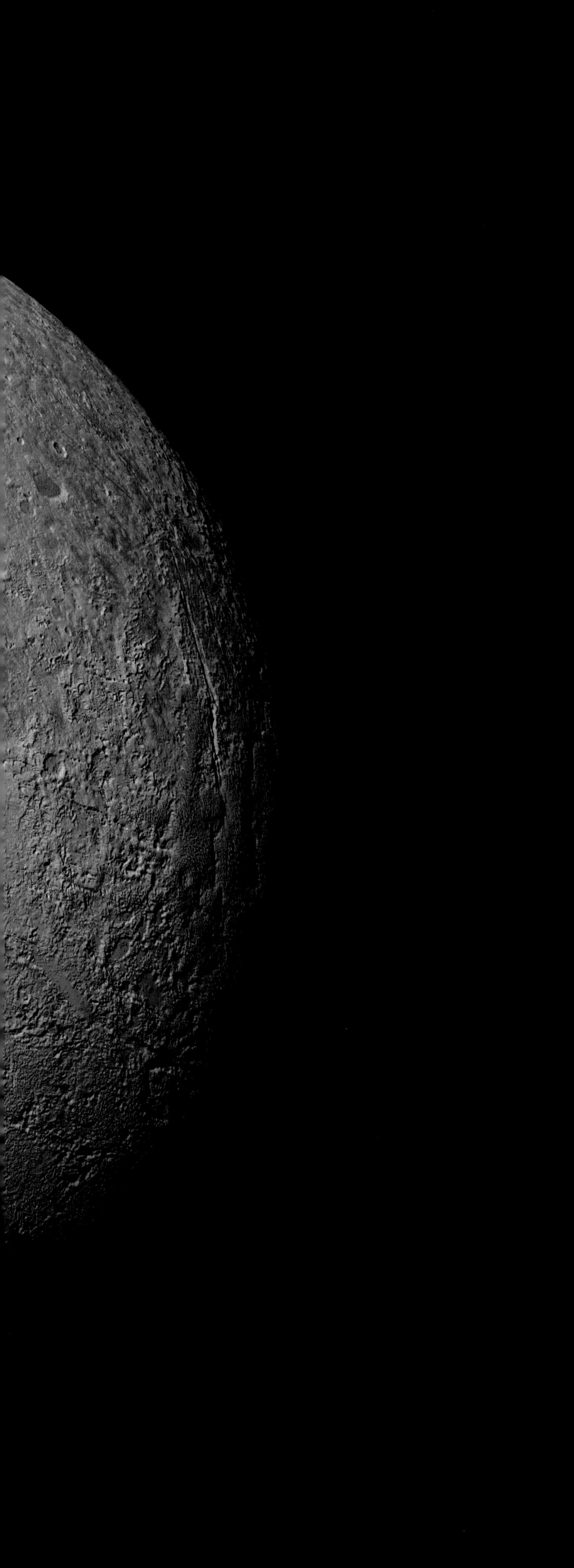

Pluto's surface as seen by the New Horizons spacecraft. The main feature shown is the large whitish heart-shaped surface, called Tombaugh Regio. It is believed to have been produced by a large meteorite impact in the past.

complexity of the dwarf planet was something of a surprise. Frozen nitrogen makes up most of the surface, but there are mountain chains made of water ice, hard as steel at this temperature. The main surface feature is a large heart-shaped structure tentatively named Tombaugh Regio, after Clyde Tombaugh. The western lobe of the "heart," tentatively named Sputnik Planitia, is made of nitrogen, carbon monoxide, and methane ices, and it is in this region that we find the water ice mountains.

Current hypotheses suggest that this structure was formed after a large impact created a deep crater on Pluto's surface—a crater that was partially filled in by liquid water welling up from the planetary interior. Current calculations suggest that, like the moons of Jupiter and Saturn, Pluto has a subsurface ocean of liquid water, perhaps 100 kilometers (67 mi) deep. How the water is kept from freezing is a question scientists are still asking.

Pluto also has a thin atmosphere, composed mainly of nitrogen with small admixtures of carbon monoxide and methane. The most interesting feature of this atmosphere is a layered haze, probably formed by interactions of its molecules with cosmic radiation. Along with its scientific cargo, New Horizons is carrying the usual sort of miscellaneous stuff—an American flag, a Florida state quarter showing a shuttle launch, and other mementos. In addition, as a result of what has to be one of the classiest decisions ever made by a government agency, the spaceship is carrying some of the ashes of Clyde Tombaugh, a Kansas farm boy far from home.

Comets have always been a problem. From ancient times, their appearance in the sky has been taken as an evil omen. There was, for example, a comet in the sky in 1066 when the Normans invaded England. (The comet may have portended disaster for the Saxons, but it certainly brought good luck to the Normans.) • Comets were also a problem in the scientifically minded 17th century, when Isaac Newton described a clockwork universe, with the orderly motion of the planets being the hands of a clock and the rational laws of nature the gears that drove them. There was no room in this universe for bodies that showed up at unpredictable times, stayed in the sky for a while, and then disappeared.

COMETS

VISITORS FROM THE OUTER LIMITS

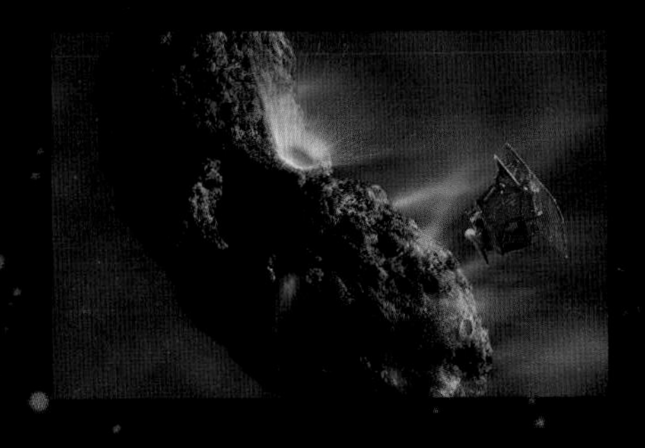

RECENT APPEARANCES
CAESAR'S COMET: **44 B.C.**
GREAT COMET OF 1577: **1577**
GREAT COMET OF 1744: **1744**
GREAT MARCH COMET: **1843**
GREAT SEPTEMBER COMET: **1882**
COMET WEST: **1976**
HALLEY'S COMET: **1986**
COMET HYAKUTAKE: **1996**
COMET HALE-BOPP: **1997**
COMET MCNAUGHT: **2007**

Comet Hale-Bopp. (Inset) Art of Deep Impact space probe at comet Tempel 1.

Our current understanding of comets began with a dinner between Newton and his friend Edmond Halley (1656–1742), who would eventually become Britain's Astronomer Royal. Halley apparently asked Newton what sorts of orbits comets would have if they were material objects subject to Newton's law of universal gravitation. Newton had actually worked the problem out, but he had not bothered to publish the result. He told Halley that comets would follow elliptical paths through the solar system. Armed with this knowledge, Halley examined the data on 26 comets. To his surprise, he found that three followed the same ellipse. The inference was obvious—these were not three different comets, but one comet coming back three different times. Halley established that some comets are in elongated orbits that bring them near Earth periodically. He predicted that this particular comet, which now bears his name, would return in 1758. When a German amateur astronomer sighted it on Christmas Eve of that year, it was a great triumph for Newton's clockwork universe.

Since then, historians have found sightings of Halley's comet in Chinese and Babylonian records going back to 240 B.C. The comet last visited Earth's neighborhood in 1986 and will come around again in 2061.

THE DIRTY SNOWBALL

The best picture we have now of comets was put forward by the American astronomer Fred Whipple (1906–2004) in the early 1950s. His ideas have acquired the evocative name dirty snowball, which turns out to be a pretty good description.

The central body of a comet, which can measure anywhere from a few hundred meters to some 10s of kilometers across, is called the nucleus. Most of the nuclei that have been analyzed contain dust and mineral grains embedded in water ice, along with a surprising array of trace constituents like methane, ammonia, and even, in some cases, complex molecules like amino acids.

When a comet is far away from the sun, the nucleus is basically frozen solid by the cold of space. As it approaches the sun, however, it

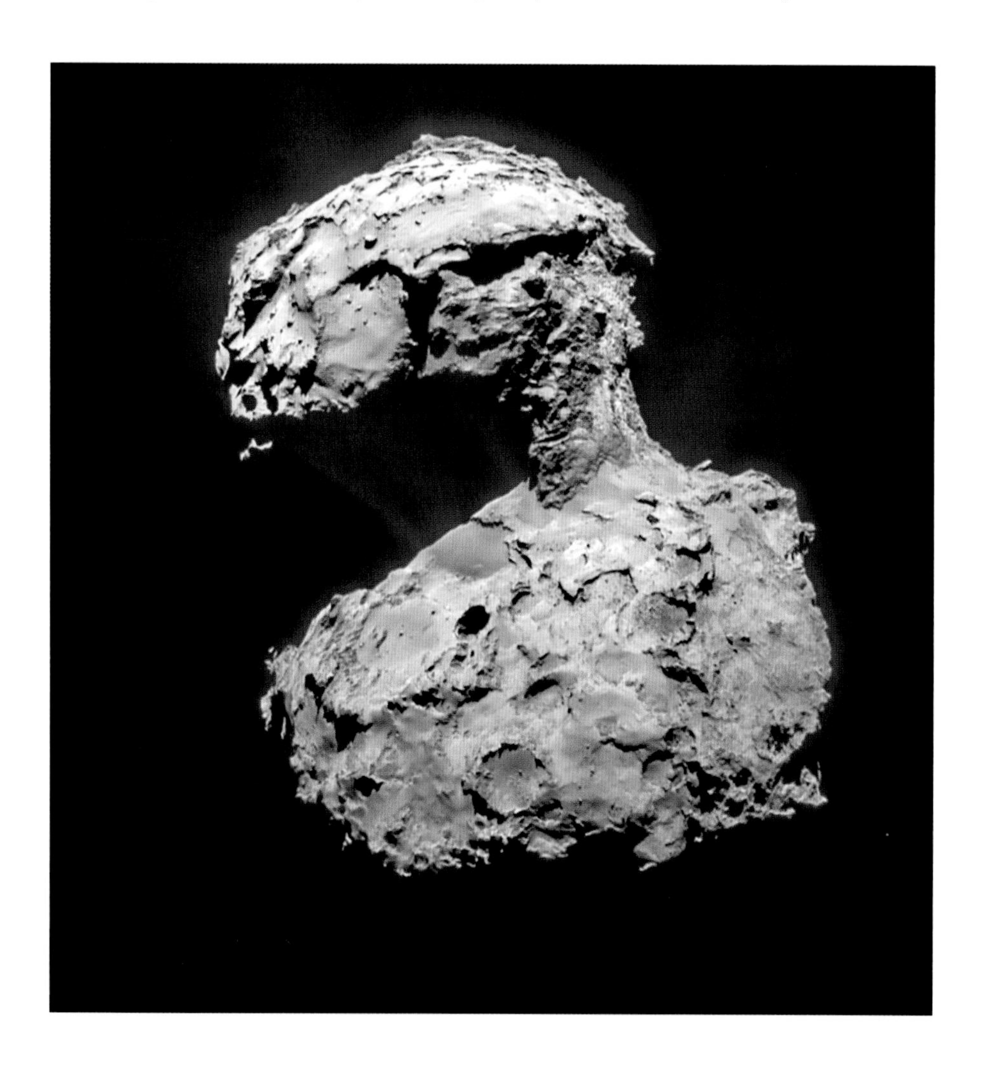

A closeup view of the comet 67P/Churymov-Gerasimenko taken by the European Space Agency's Rosetta spacecraft from a distance of only about 85 kilometers (57 mi) from the surface. The dumbbell shape is thought to be unusual.

heats up and the volatile materials in the nucleus begin to boil off. This gives rise to two structures—a thin atmosphere surrounding the nucleus (called a coma) and a tail—the thing we most associate with a comet. In fact, every comet has two tails. One is made up of the gases that boil off the nucleus; the solar wind—particles streaming out from the sun—exerts a pressure on these gases, blowing the tail so that it always points away from the sun. The other tail is a trail of dust that has drifted off the surface. This dust pretty much stays put along the track of the comet, like the trail a muddy dog leaves on a living room carpet.

Comets are usually divided into long-period and short-period bodies. Short-period comets complete their orbits in less than 200 years. They are thought to originate in the region outside the orbit of Neptune called the Kuiper belt (see pages 200–203). Long-period comets—and we know of some that take thousands of years to complete an orbit—are thought to originate still farther out, in an icy collection of distant bodies known as the Oort cloud (see pages 204–205).

Astronomers have cataloged thousands of comets, and roughly one a year passes close enough to Earth to be visible to the naked eye. Most of these, however, are faint and, frankly, pretty unremarkable. To be noticeable, a comet has to pass close to Earth after its tail has grown to its maximum brightness—not an easy set of requirements to satisfy.

SPACECRAFT AND COMETS

In the late 20th century spacecraft visited comets for the first time either flying by or, in one case, bringing material from the comet's tail back to Earth. There have been close to a dozen such encounters, but let me just describe two, the Deep Impact and Stardust missions.

In 2005 the Deep Impact spacecraft (later renamed EPOXI) released a probe that blasted a crater in the surface of comet Tempel 1. By monitoring the material that came off, scientists were able to establish that most of the water ice in a comet is beneath the dust layer at the surface.

The Stardust spacecraft was launched in 1999 and flew through the tail of comet Wild 2 (pronounced vilt) in 2004. Material from the tail was absorbed in special materials in the spacecraft and returned to Earth in a capsule in 2006. The results from this mission caused a stir in the astronomical community because they showed that the comet contained grains of materials that could have formed only at high temperatures, lending support to the idea that the formation of the solar system (see pages 55–7) must have been more turbulent than had been thought.

Some scientists think that comets, with their high water content and brew of organic molecules, may have played an important role in shaping the early Earth. Some have suggested that comets provided the water for Earth's oceans and perhaps even brought in the molecules from which the first life developed.

The most dramatic visit to a comet was made by the European Space Agency's Rosetta spacecraft. Named for the famous Rosetta Stone that allowed us to decipher Egyptian hieroglyphics, it was launched in 2004 and arrived at a comet named 67P Churyunev-Gerasimenko in 2014. After matching speeds with the comet as it moved toward the sun—a maneuver some commentators compared to matching speeds with a rifle bullet—the spacecraft dropped a lander named Philae onto the comet's surface. A malfunction caused the lander to bounce several times before coming to rest at the base of a cliff. With its solar panels shaded, the lander was able to return data for only a couple of days before its batteries were drained.

I f this book had been written 20 years ago, our discussion of the solar system would end at this point, with Pluto as the outermost planet. Today, however, we understand that the planets are just the beginning and that the actual solar system extends far out into space—farther than we ever imagined. • To understand what is meant by this, we're going to have to change our scale of perspective. Scientists usually use something called the astronomical unit, or AU, to talk about distances in the solar system. The AU is the distance between Earth and the sun—roughly 150 million kilometers (93 million mi), or about 8 light-minutes. In terms of this unit, Mars is about 1.5 AU from the sun, Jupiter about 5.2 AU, and Neptune about 30 AU.

KUIPER BELT & OORT CLOUD

THE ICY OUTSKIRTS

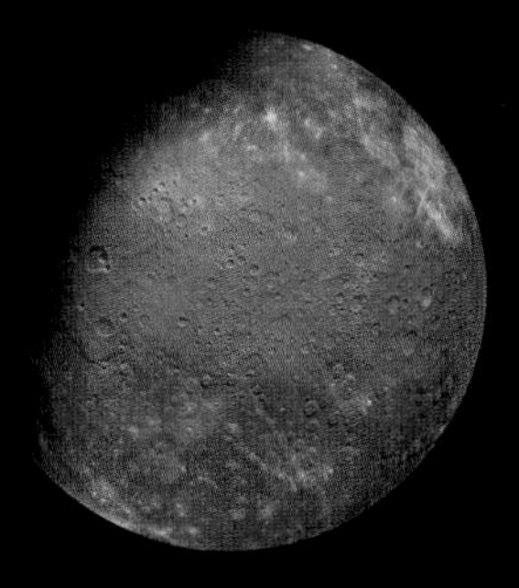

KUIPER BELT DISCOVERERS: **DAVID JEWITT AND JANE LUU**
CONFIRMED DISCOVERY DATE: **1992**
DISTANCE FROM SUN: **30–55 AU**
OORT CLOUD PROPONENTS: **JAN OORT AND ERNST ÖPIK**
DATE FIRST PROPOSED: **1932**
DISTANCE FROM SUN: **5,000–100,000 AU**

KNOWN PLUTOIDS AND AVERAGE DISTANCE FROM SUN:
PLUTO: **39 AU**
HAUMEA: **43 AU**
MAKEMAKE: **46 AU**
ERIS: **68 AU**

Artwork shows the disk and shell of the Oort cloud around the sun. (Inset) Artist's conception of the plutoid Eris.

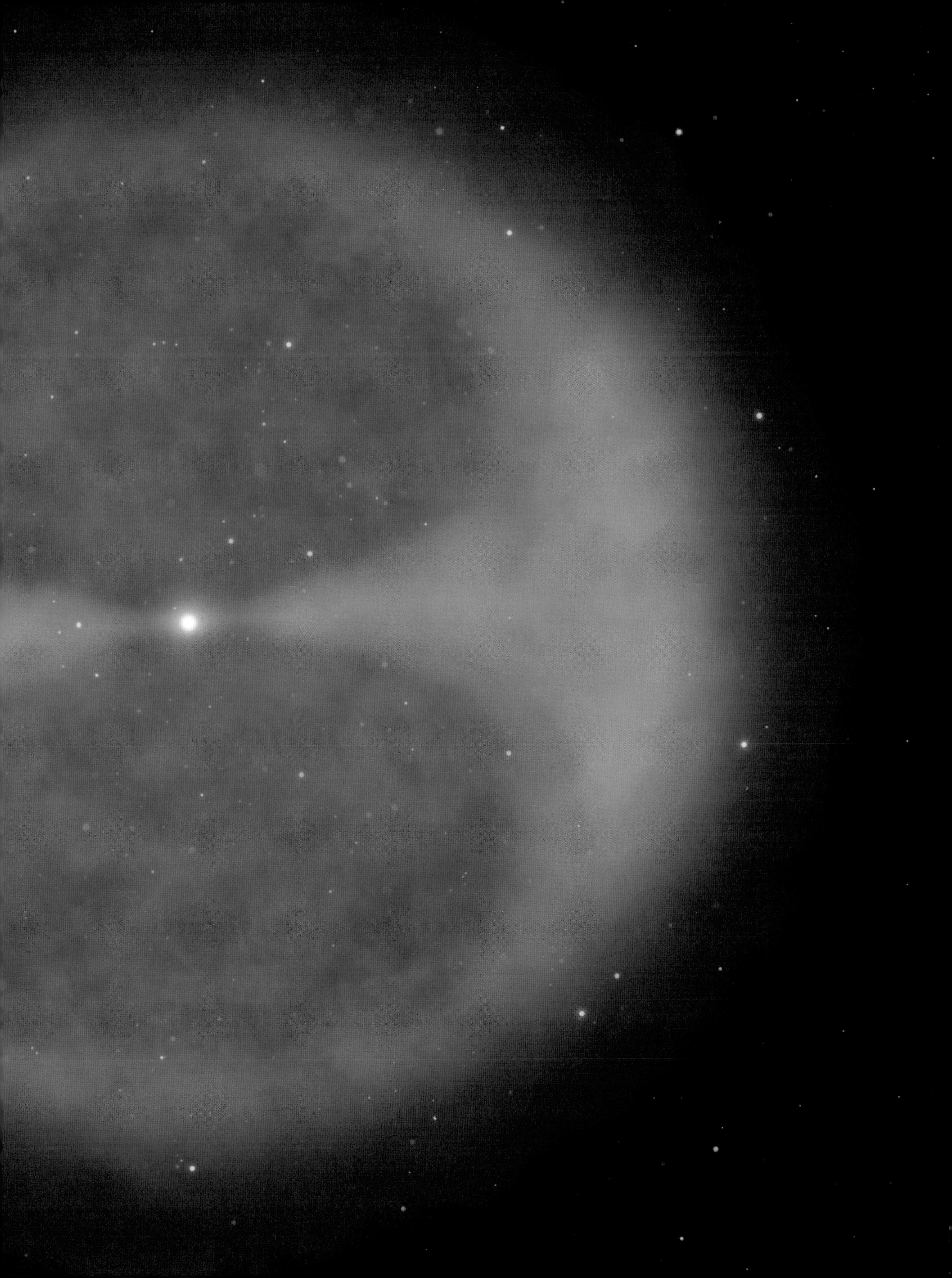

From Neptune's orbit to about 55 AU, the solar system extends outward in a giant doughnut-shaped structure known as the Kuiper belt. (Think of the planets as fitting into the hole in the doughnut.) The structure is named after the Dutch astronomer Gerard Kuiper (rhymes with "viper"), one of the scientists who did an early calculation of the belt's properties in 1951.

The Kuiper belt consists primarily of material that appears to be planetesimals left over from the formation of the solar system—indeed, one author referred to it as a "reservoir of rejects." Most likely the belt represents the remnants of the protoplanetary disk from which the solar system formed—remnants that survived the migration of Uranus and Neptune to their present orbits (see pages 55–7).

Today it is just a shadow of its former self, comprising no more, in aggregate, than 10 percent of the mass of Earth. Since 1992, telescopic exploration of the Kuiper belt has uncovered more than 1,000 Kuiper belt objects (KBOs), and astronomers expect to find many more. After leaving Pluto, the New Horizons spacecraft will perform a flyby of a smallish KBO named 2014MU69 with an anticipated arrival in early 2019. Pluto is now seen as the first KBO rather than as the last planet, and Neptune's moon Triton is considered a captured KBO.

Telescopic exploration continues with a survey called Pan-STARRS (Panoramic Survey Telescope and Rapid Response System). Started in 2008 with a telescope located on Mount Haleakala in Hawaii, its primary mission is to locate asteroids and comets that might impact Earth. As a side benefit, however, it is producing a detailed map of faint objects in the sky, including KBOs.

Outside the Kuiper belt "doughnut" is a sparsely populated region of objects in highly eccentric orbits, called the scattered disk, extending out to about 100 AU. Unlike objects in the Kuiper belt proper, which are in stable orbits, objects in the scattered disk have orbits that bring them in to 30 AU, where they can be affected by the gravitational pull of Neptune. It is believed that most short-period comets (see page 205) had their ultimate origin in the scattered disk.

DWARF PLANETS

We can't leave the Kuiper belt without mentioning a couple of objects found out beyond the orbit of Neptune. One is Eris, larger than Pluto and actually part of the scattered disk. Discovered in 2005 by astronomer Mike Brown and his team at Palomar Observatory, the dwarf planet was first nicknamed Xena after the TV character. Right

now Eris is about 97 AU from the sun, well outside the Kuiper belt proper, and is the farthest known member of our solar system. Almost a dozen planet-size objects have been found in the Kuiper belt, and astronomers expect to find more. In addition, some astronomers have proposed that there is an as yet undiscovered addition to the Kuiper belt, an object dubbed "Planet 9" that could be as much as 10 times as massive as the Earth.

Another strange discovery is the dwarf planet Sedna. Found in 2003 by Brown's team, it is smaller than Pluto but has a very unusual orbit. Circling 88 AU from the sun right now, Sedna's orbit never takes it closer than 76 AU. Thus, it orbits well outside both the planets and the Kuiper belt. Astronomers calculate that its farthest distance from the sun will be an astonishing 975 AU, far beyond anything we've discussed so far. There are no plans to send a spacecraft to Sedna, and astronomers estimate that it would take a spacecraft upward of 25 years to get there even at its closest approach to the sun. Sedna's remarkable distance from the sun has led Brown to suggest that Sedna may not be a scattered disk object at all, but the first member of the Oort cloud—which brings us to the last part of our story.

An artist's view of the types of objects we might find in the Kuiper belt. To get all the objects in a single picture, the artist has put the objects much closer together than they actually are—like the asteroid belt, the Kuiper belt is mostly empty space.

OORT CLOUD

In 1950 the Dutch astronomer Jan Oort (rhymes with "sort") suggested that somewhere in the outskirts of the solar system there had to be a reservoir of comets. His argument was simple: Comets can't last forever—they lose some of their mass every time they get near the sun and are subject to the gravitational effects of the planets. Since we still see comets today, Oort argued, there has to be a process by which new comets can be formed, some reservoir of comets out beyond Pluto.

Today this reservoir is thought to consist of a huge cloud at the very edge of the solar system. Named the Oort cloud, it extends from a few thousand AU out to at least 50,000 AU, and perhaps farther. It has two parts—an inner doughnut-shaped section that extends out to about 20,000 AU and a thinly populated outer sphere. The Oort cloud is thought to be the remains of the original protoplanetary disk. Objects in the disk probably formed closer to the sun but moved out to their present location

SEPARATING THE SCIENCE FROM SCIENCE FICTION

It used to be thought that the asteroid belt was the remains of a planet that had exploded – this hypothesis was behind the story of the fictional planet Krypton, home of Superman. In fact, the belt is just the opposite – it represents the remains of a planet that never formed. In the early years of our solar system, the planet Jupiter moved closer to the sun, and its gravity ejected most of the material in what is now the asteroid belt into space. What remains is only about 10 percent of the mass needed to form a planet.

We should also note that the picture of the asteroid belt shown in science fiction movies – the one in which a spacecraft has to thread its way through a densely packed field of boulders – is completely misleading. In fact, the asteroid belt is almost all empty space – as with the Kuiper belt. In fact we have sent many probes through the asteroid belt en route to the outer planets without encountering so much as a grain of sand.

following the great reshuffling of the solar system four billion years ago.

The Oort cloud may be the point of origin for what are known as long-period comets—comets that have a period of more than about 200 years. The comet Hale-Bopp that visited Earth in 1997 is a recent example of a long-period comet. Oddly enough, so is Halley's comet. Even though it now has a period of roughly 72 years, it is believed to have originated in the Oort cloud and been pulled into its present short orbit by the gravitational attraction of the planets. Most comets with a period of less than 200 years—the so-called short-period comets—are thought to originate in the scattered disk.

A rendering of the plutoid Sedna in the Kuiper belt shows the distant sun as a bright star. The reddish color of Sedna's surface is what we see in our telescopes. The moon the artist has depicted was thought to be present when Sedna was first discovered, but no evidence for its existence as been found since then.

As befits a region so remote and mysterious, there are many imaginative explanations for the origin of larger Oort cloud candidates, including the idea that they were captured into the Oort cloud from a passing star. Thus, our exploration of our own cosmic backyard—our first "universe"—brings us to the realm of the stars. Our next stop is our own galactic neighborhood, the Milky Way, a dynamic, churning spiral of activity.

THE MILKY WAY

Seen from Earth, the central plane of our galaxy, the Milky Way, forms a dense stream of stars across the sky.

Our planet, as magnificent as it is, is just one planet circling an ordinary star in a low-rent section of our galaxy. The Milky Way isn't just a passive collection of stars, however, but a roiling, dynamic place. Stars are born in the collapse of gigantic clouds of dust and debris. They keep themselves alive by consuming the primordial hydrogen of the universe. They use nuclear processes to create the heavier elements, including the carbon that makes up some of your body. Eventually stars run out of fuel and die, returning their heavy elements to the interstellar clouds from which new stars and planets are made.

THE G

When stars die they leave behind a strange menagerie of objects. Some leave white dwarfs, the dying embers of stars like the sun. Others become pulsars, incredibly dense objects 10s of kilometers across. A few collapse into black holes, representing the ultimate triumph of gravity.

In recent decades astronomers discovered that, just as the inner planets are actually a small part of the solar system, the mighty pinwheel of the Milky Way is just a small part of the galaxy. In fact, the spiral arms of the Milky Way are enclosed in a mysterious substance known as dark matter. Puzzling out the nature of dark matter remains a major research project today.

ALAXY

I THE MILKY WAY

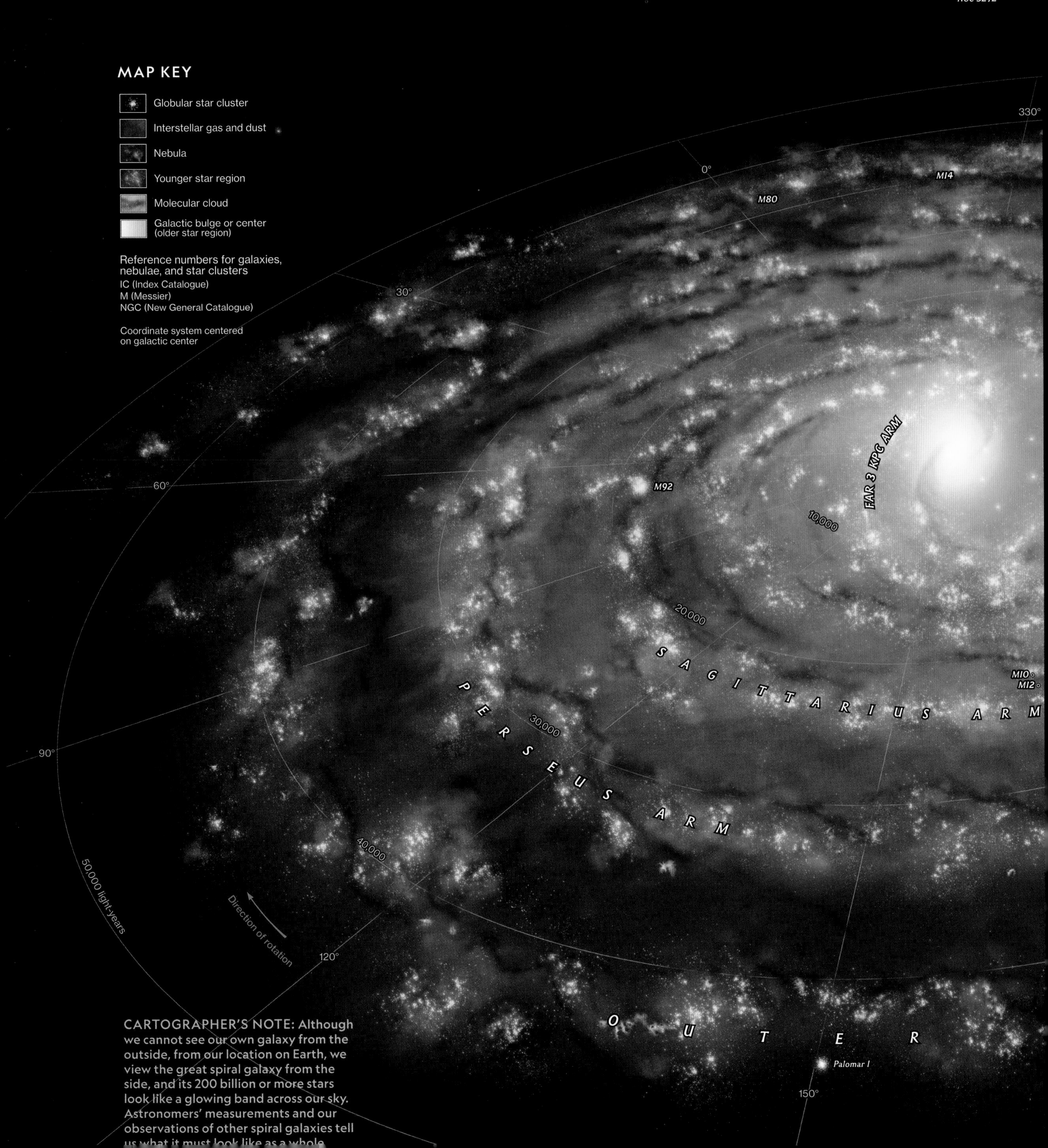

CARTOGRAPHER'S NOTE: Although we cannot see our own galaxy from the outside, from our location on Earth, we view the great spiral galaxy from the side, and its 200 billion or more stars look like a glowing band across our sky. Astronomers' measurements and our observations of other spiral galaxies tell us what it must look like as a whole.

An artist's conception of the Milky Way galaxy shows its dense central core and flat disk of spiral arms, whose brightest regions contain young stars. Our own solar system is located about 25,000 light-years out on one of these arms and rotates with the entire structure, making a complete circuit every 250 million years or so. The galaxy's crowded center contains a black hole millions of times more massive than the sun.

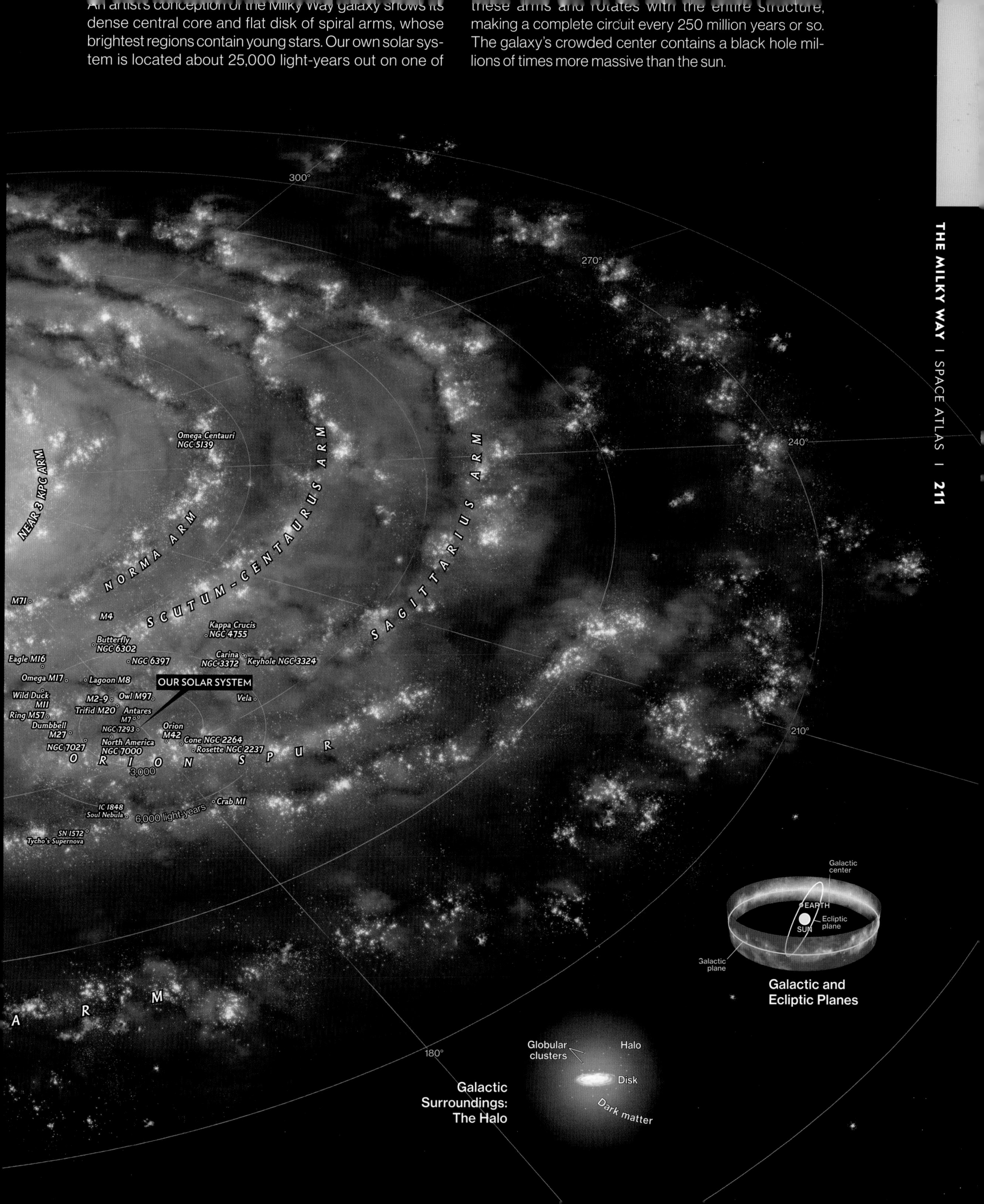

When we venture out of the confines of our own solar system to explore the Milky Way galaxy, the first thing we have to do is reset our thinking about distances. Consider this analogy: If the sun were a bowling ball in the center of a city on the U.S. East Coast—Washington or New York, for example—then all of the planets (including Pluto) would be found within a dozen or so city blocks and the very outermost reaches of the Oort cloud would be somewhere around Saint Louis. You would then have to travel to Hawaii to get to the nearest star, and the rest of this giant city of stars we call our home galaxy would take you right off the planet Earth.

SIZING UP THE MILKY WAY

HOW FAR ARE THE STARS?

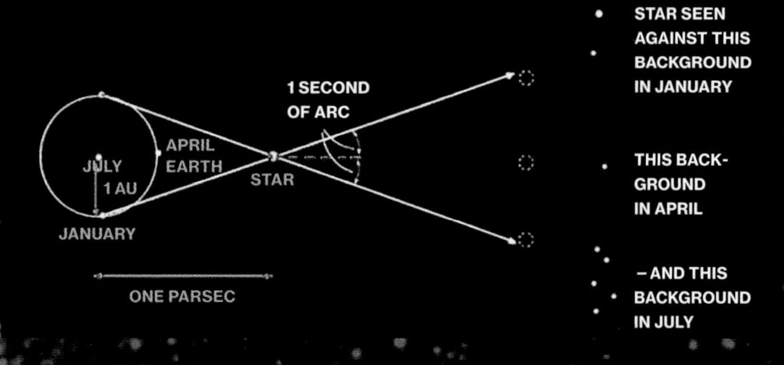

DISCOVERER: **UNKNOWN**
DISCOVERY DATE: **PREHISTORIC**
DISTANCE TO MILKY WAY CENTER: **28,000 LIGHT-YEARS**

DIAMETER: **100,000–120,000 LIGHT-YEARS**
THICKNESS: **1,000 LIGHT-YEARS**
ROTATION: **250 MILLION YEARS AT SUN'S DISTANCE**
MASS: **ABOUT 1 TRILLION (10^{12}) SOLAR MASSES**
NUMBER OF STARS: **300 BILLION ± 100 BILLION**
AGE: **13.2 BILLION YEARS**
GALAXY TYPE: **BARRED-SPIRAL**
MAIN SATELLITES: **LARGE AND SMALL MAGELLANIC CLOUDS**

Bright stars of the southern sky include Alpha Centauri and Beta Centauri (at left, left and right). (Inset) Diagram of measurement by parallax

W e introduced the astronomical unit (AU)—the distance from Earth to the sun—as a convenient unit for dealing with our solar system since it is comparable to the distance between planets. In the same way, we need a new distance measure in our new universe. After all, measuring the distances between stars in terms of the AU is a little like measuring the distance between cities in inches—you could do it, but it would be inconvenient at best. Consequently, astronomers use a new yardstick in this universe—the light-year. This measurement is defined to be the distance light travels in one year: 9.5×10^{12} kilometers, 6×10^{12} miles, or 63,000 AU. (For technical reasons, astronomers also use a unit called the parsec, which is 3.3 light-years.) Roughly speaking, stars in the Milky Way are separated by a few light-years—the nearest star to us, for example, is a little farther at 4 light-years away. The Milky Way itself is about 100,000 light-years across and about 10,000 light-years thick at the center. How do astronomers determine these distances? After all, when you look at the sky, what you see is a two-dimensional display, lights on an inverted bowl. The third dimension—the distance to each object—is not immediately apparent. A star might appear faint either because it really is faint and close or because it is bright but far away. Over the millennia, a lot of effort has been devoted to putting the third dimension into our picture of the heavens. Astronomers have developed a variety of techniques to measure distances to celestial objects, depending on how far away they are. In this section we look at the two most important of these: parallax (or triangulation) and standard candles.

PARALLAX

Hold out your finger and look at it with one eye closed, then the other. Notice how your finger appears to move against the background? This, basically, is what parallax is all about. Your finger appears to move because you

HENRIETTA LEAVITT

"A comparison [of the plates] led immediately to the discovery of an extraordinary number of new variable stars."

Educated at Oberlin College and what is now Radcliffe, Leavitt joined the staff of the Harvard College Observatory in 1893 as a "computer." (In those days, computers were human beings who carried out long, tedious calculations with pencil and paper.) Analyzing countless photographs of the night sky on glass plates, she noticed the connection between a variable star's brightness and its cycle time. In 1908 Leavitt published the results of her painstaking study of some 1,777 variable stars, establishing the basis for the Cepheid distance scale. Eventually she was appointed head of photometry at the observatory, a position she held until her death in 1921. Because her work proved to be the foundation of so many important later advances in astronomy, her name has been given to an asteroid and a crater on the moon.

are looking at it from two different positions, separated by the distance between your eyes.

Here's how you can use this effect to measure the distance to an object out of reach: Suppose you want to find out how far away a flagpole is, but you can't actually get to the flagpole itself. (Imagine it being on the other side of a river, for example.) You could find the distance by looking at the flagpole from two different spots on your side of the river and measuring the angles between your two lines of sight and the line connecting the two observation points. If you then measure the distance between the two observation points—what is called the baseline of the measurement—you have one side and two angles of a triangle, a triangle whose apex is the flagpole. Some simple geometry will then give you the distance you want.

This technique, also called triangulation, gives us a way of adding that third dimension to our picture of the heavens, provided that we can actually measure the difference between the two angles. And there's the rub, because the farther away the object is, the harder it is to use this method.

Imagine that our flagpole is being moved farther and farther away from us. Depending on the kind of instrument we are using to measure the angles, there will eventually come a point where the flagpole is so far away that we won't be able to tell the difference between the two angles we are measuring. As far as we are concerned, the two lines are parallel. At this point, the parallax measurement has run out of oomph, and we can no longer measure the distance we want. We now have two alternatives:

1. Increase the baseline, thereby increasing the difference between the two angles so that it becomes detectable with the instruments we have
2. Get better instruments, so that we can detect the difference in the angles while keeping the same baseline

VARIABLE STAR V1 (INSETS) IS FOUND IN THE ANDROMEDA GALAXY (BACKGROUND).

Being confined to the planet Earth, we are somewhat limited in our ability to increase our baseline. Nevertheless, the most obvious baseline is the diameter of Earth itself. We could take simultaneous measurements from opposite sides of Earth, or we could take a measurement from one place, wait 12 hours until the rotation of Earth has carried us around a half turn, then take the second measurement. In either case, the baseline of the measurement is limited to about 13,000 kilometers (8,000 mi), the diameter of our planet.

WITHIN THE MILKY WAY

After Eratosthenes estimated the circumference of Earth in about 240 B.C. (see sidebar page 90), celestial distance measurements ran out of steam for almost two millennia. The reason is simple: The next available longer baseline is the diameter of Earth's orbit, which you can use by taking angle measurements six months apart. The problem is that to use this baseline you need to know the distance from Earth to the sun, and you can't get that if you are restricted to using Earth's diameter as a baseline and don't have telescopes. It wasn't, in fact, until 1672 that French astronomers, using the best telescopes available, were able to get a reasonably accurate measurement of the distance between Earth and Mars and, from this, using some simple math, the distance between Earth and the sun. Even with this much larger baseline, it still took more than a century for telescopes to get good enough to measure the distance to a star.

This feat was accomplished in 1838 by the German scientist Frederick Bessel, who measured the way the star 61 Cygni appeared to move against the background of more distant stars and determined that its distance was 10.9 light-years. This discovery established once and for all that the universe is a much bigger place than had ever been imagined.

As telescopes improved, astronomers used the triangulation method to measure the distance to stars some 10s of light-years away. This process ran into a roadblock, however, because fluctuations in Earth's atmosphere set a limit on our ability to measure angles. Then, in 1989, the European Space Agency launched the satellite Hipparcos. From above the atmosphere, it accumulated a massive amount of data and pushed the limit on parallax measurements out beyond 130 light-years.

In this century, astronomers using the extreme accuracy available to radio telescopes have been able to determine the parallax of objects called pulsars (see pages 259–61) and have pushed this limit past 500 light-years. Since then, the Hubble Space Telescope has recorded triangulations of a few objects as much as 20,000 light-years away, and in 2013 the European Space Agency (ESA) launched the GAIA (Global Astrometric Interferometer for Astrophysics) satellite, which is in the process of creating the most complete catalog of stars and stellar distances ever produced. In 2016, ESA published a catalog of over a billion stars. These impressive achievements, however, are still not enough to get us out of the Milky Way galaxy. To go farther, we need a new measurement technique.

STANDARD CANDLES

A standard candle is an object for which we know the total energy output. A good example of a standard candle is a 100-watt lightbulb, since all we have to do to find out how much energy it is putting out is to read the label on the bulb. With this information, we can measure how much energy actually reaches us from the bulb (for example, by using the light meter on a camera). Then, knowing the way that distance dilutes energy, we can find out how far away the bulb is. For example, if we knew we were looking at a 100-watt

If we know how bright an object is intrinsically, then we can tell how far away it is by measuring how much light we receive. Such an object is called a standard candle and is used to measure the distance to the stars.

bulb and were getting 10 watts in our detector, we could use standard equations to find the distance to the bulb. If we had a way to do that with a distant star, then, we could find out how far away it is.

The trick, of course, is to "read the label" on a star—to find out how much energy it is actually sending into space. This was first done by a rather remarkable person: Henrietta Leavitt (1868–1921), America's first woman astronomer (see sidebar page 214). While working at the Harvard College Observatory, she noticed something interesting about a certain class of stars known as Cepheid variables, and she exploited this insight to produce what is now known as the Cepheid distance scale.

Most stars shine with a constant light—their brightness doesn't vary much except on astronomical timescales. Some stars, however, do not share this property. Watch them for weeks or months, and you will see them get brighter or dimmer, sometimes on a regular cycle. These are called, appropriately enough, variable stars.

The point about these stars is that their brightness varies regularly, first brightening, then dimming, then brightening again. We now understand that this behavior is due to processes in the outer atmospheres of certain kinds of stars as they reach the end of their life. What Leavitt found was that the time it took this type of star to go through its cycle depended on how much energy the star was pouring into space—the longer the cycle took, the more energy was being emitted. Watch a Cepheid variable go through its bright-dim-bright cycle, in other words, and you have, in effect, read the label on the lightbulb. It is then a simple matter to measure how much light you are actually receiving in your telescope and figure out how far away the star is. This means that as long as we can see the variable star, we can get its distance. As we shall see in chapter 3, it was Leavitt's work that allowed Edwin Hubble, several years later, to establish both the existence of other galaxies and the expansion of the universe.

Every day on Earth begins when one representative of the Milky Way galaxy, one very ordinary star, pokes its nose above the eastern horizon, and every day ends when that same star disappears in the west. The sun plays such a central role in life on Earth that it is easy for us to forget that it is, after all, just one more star in a galaxy full of them. The proximity of the sun does have an advantage, though—it allows us to study a star up close. In fact, everyday experience can lead you to one of the great scientific questions that occupied astronomers when they began thinking seriously about the sun. Stand outside on a summer day and you will feel warmth on your face. Energy—in the form of infrared radiation—is coming to you from the sun.

THE SUN

THE STAR NEXT DOOR

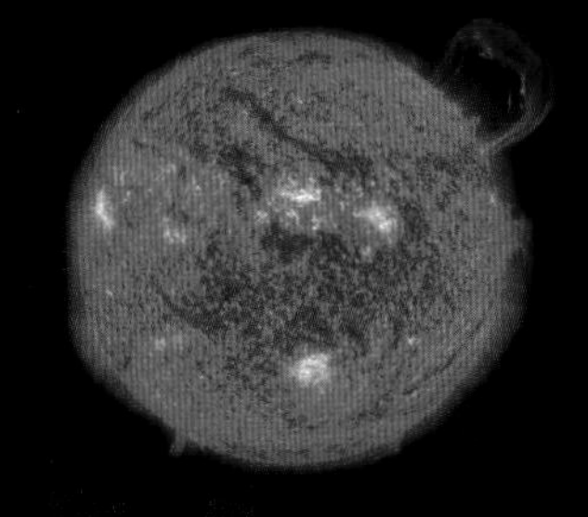

AGE: **4.567 BILLION YEARS**
BECOMES RED GIANT: **5.5 BILLION YEARS FROM NOW**
MASS (EARTH = 1): **333,000**

DIAMETER: **1,392,000 KILOMETERS (865,000 MI)**
ROTATION: **25.1 DAYS (EQUATOR), 34.4 DAYS (POLES)**
CORE TEMPERATURE: **16,000,000 KELVINS (16,000,000°C/29,000,000°F)**
SURFACE TEMPERATURE: **5,800 KELVINS (5500°C/10,400°F)**
COMPOSITION: **HYDROGEN 74.9 PERCENT, HELIUM 23.8 PERCENT**
ENERGY: **TURNS 400 MILLION TONS HYDROGEN INTO HELIUM PER SECOND**
SURFACE GRAVITY (EARTH = 1): **28**
VISIBLE DEPTH: **ABOUT 160 KILOMETERS (100 MI)**

Sunset in Indonesia. (Inset) The sun, with a large, handle-shaped prominence.

THE SUN

A cutaway view shows the dynamic structure of the sun, our nearest star.

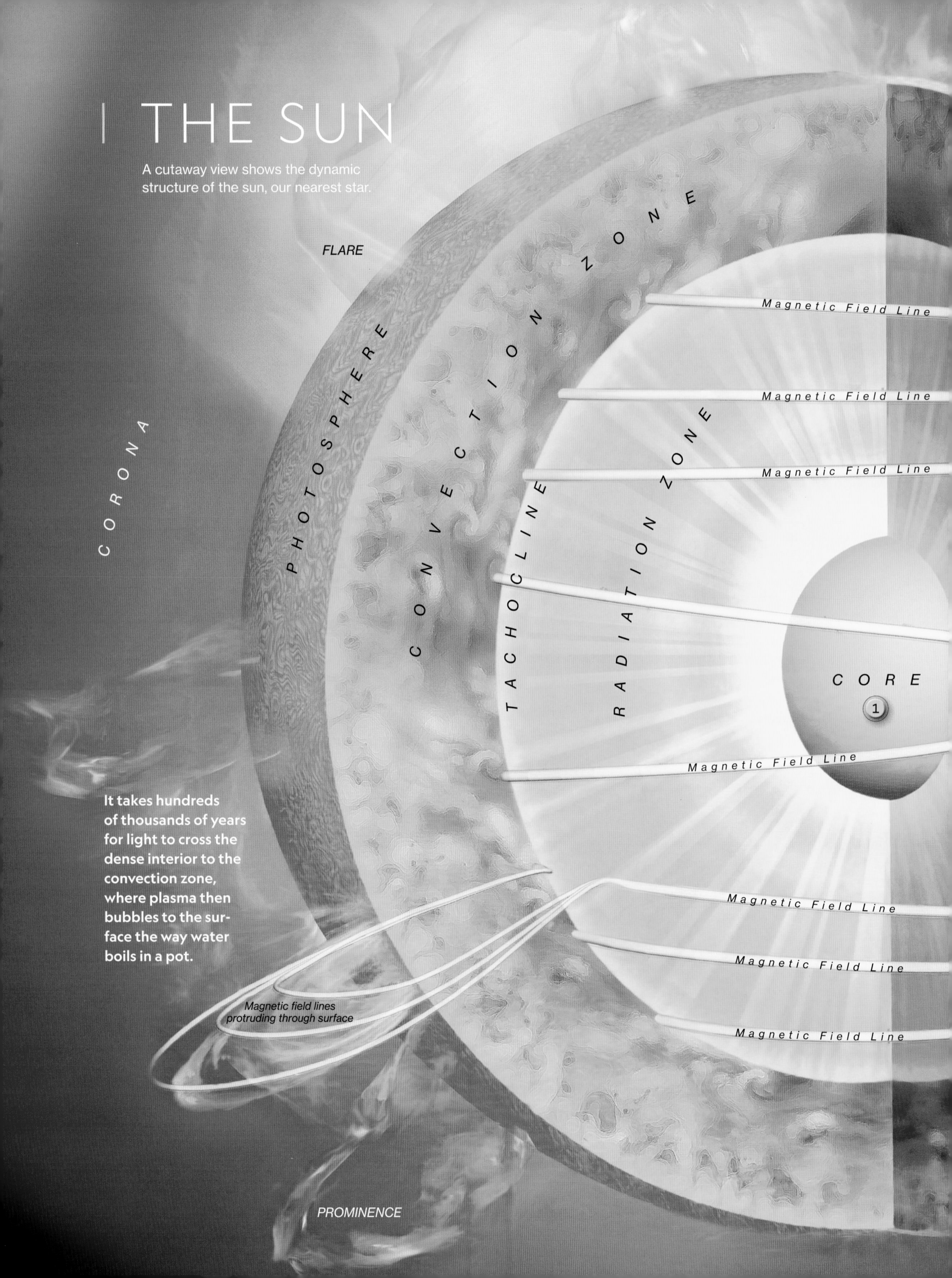

FLARE

CORONA

PHOTOSPHERE

CONVECTION ZONE

TACHOCLINE

RADIATION ZONE

CORE

1

Magnetic Field Line

Magnetic Field Line

Magnetic Field Line

Magnetic Field Line

Magnetic Field Line

Magnetic Field Line

Magnetic Field Line

It takes hundreds of thousands of years for light to cross the dense interior to the convection zone, where plasma then bubbles to the surface the way water boils in a pot.

Magnetic field lines protruding through surface

PROMINENCE

KEY FEATURES

1. **SOLAR CORE:** Center of fusion and the sun's energy
2. **SOLAR CORONA:** The sun's superheated outer atmosphere
3. **SUNSPOTS:** Cool spots on the sun's surface where magnetic lines break through

SUNSPOTS
3
TACHOCLINE
PHOTOSPHERE
CHROMOSPHERE
CORONA
2

STAR POWER

The sun's core is a thermonuclear reactor, fusing hydrogen into helium. Because of the intense heat, these gases exist in an electrified state of matter called plasma.

MASSIVE STORMS

A flare explodes when magnetic field lines become overloaded with electrical current, not unlike a fuse blowing. More menacing to Earth, a coronal mass ejection erupts when field lines snap in a way that lets billions of tons of plasma lift off. Moving as fast as five million miles per hour, the plasma cloud can expand to a width of tens of millions of miles.

MAGNETIC DYNAMO

Magnetism is key to solar behavior. A north-south magnetic field is generated in the tachocline, then is pulled into an east-west pattern as different layers of the star rotate at different speeds. The stretching adds energy to the lines, which break through the surface as sunspots or soar into the corona as loops and prominences.

SUN

Average surface temperature:	5,500°C
Average core temperature:	16,000,000°C
Rotation period:	24.6 days
Equatorial diameter:	1,392,000 km
Mass (Earth=1):	332,950
Density:	1.41 g/cm^3
Surface gravity (Earth=1):	28.0

Until the middle of the 19th century, the fact that the sun gives off energy would have been an unremarkable observation. It was about that time, however, that what we now call the law of conservation of energy was discovered. That law tells us that energy is neither created nor destroyed but just shifted from one form and place to another. The energy that warms your face, in other words, has to come from some source inside the sun, and the fact that you feel the warmth means that that energy has left the sun forever. The warmth of a summer day contains within it one of the most important messages about the Milky Way—every star must sooner or later run out of energy. Stars are not forever but are born and die like everything else.

Once this fact sank in, a whole constellation of theories as to the source of the sun's energy appeared. One late 19th-century astronomy textbook, for example, devoted several pages to calculating how long the sun could burn if it was made of the best fuel known at the time, anthracite coal. (The answer is about 10 million years.) In fact, it wasn't until the early 1930s that the young German-American physicist Hans Bethe showed that the energy source of stars was the process of nuclear fusion.

THE POWER AT THE CORE OF THE SUN

To understand the basis of Bethe's breakthrough, we have to go back to page 56, where we talked about how the sun and the solar system formed from the gravitational collapse of an interstellar dust cloud. In that discussion we concentrated on the small amount of material that went into the formation of the planets, but the fact of the matter is that more than 99 percent of that dust cloud was incorporated into the body we call the sun. Let's look at what happened as that part of the dust cloud contracted.

LITTLE NEUTRAL ONES

When we look at the sun, we can see down only 160 kilometers (100 mi) or so. No light reaches us directly from the interior because the solar material absorbs it. There is, however, an elusive particle called a neutrino that can travel to Earth from the sun's center without being absorbed. Produced in nuclear reactions in the sun's core, neutrinos give us a way of seeing into the heart of our star.

Neutrinos (the name means "little neutral one" in a combination of Latin and Italian) have no electric charge, almost no mass, and interact very weakly with matter. Billions of them have passed through your body since you started reading this sentence, for example, without disturbing a single atom. The only way to detect neutrinos is to put a large target in their way and measure the occasional interaction.

The first attempt to make such a measurement took place a mile underground in a South Dakota gold mine in the late 1960s (the overlying rock shielded the apparatus from cosmic rays). The detector was basically a tank full of carbon tetrachloride, or cleaning fluid. Once a day or so a neutrino from the sun would convert one of the chlorine atoms to a detectible atom of argon. At first scientists were mystified – too few neutrinos were being detected – but eventually they realized that during their transit from the sun, some of the neutrinos changed form and could no longer produce argon atoms in the tank. Today there are neutrino detectors in many places around the world, and they are producing results consistent with our notions of the power at the core of the sun.

One of the basic rules of nature is that as objects contract, they heat up. The rule applies to the cloud that became the sun, so as the great gas cloud contracted, it got warmer. The particles and atoms that made up the cloud moved faster and faster, and the collisions between them became more and more violent. Eventually they became so violent that electrons started to be torn loose from their atoms (this actually doesn't take all that much energy—it happens all the time in fluorescent lightbulbs, for example). The material in the sun became what physicists call a plasma—a collection of loose, negatively charged electrons and their positively charged nuclei moving around independently of each other.

When the temperature in the core of the sun reached the range of millions of degrees, particles in the center of the newly forming star began moving very fast indeed. Eventually, protons (the positively charged nuclei of hydrogen atoms) started moving so rapidly that they overcame their mutual electrical repulsion and got close enough to initiate a nuclear reaction. In a series of reactions, four hydrogen nuclei fused together to make the nucleus of a helium atom (two protons and two neutrons) and a miscellaneous spray of fast-moving particles. The mass of the final particles was less than the mass of the four initial particles, and the difference was converted, via Einstein's famous formula $E = mc^2$, to energy. This energy streamed outward, creating a pressure that balanced the inward force of gravity, and the sun stabilized.

THE SUN'S STRUCTURE

Since that time, about 4.5 billion years ago, the sun has been burning hydrogen at the rate of more than 400 million tons per second and will continue to do so for another 5.5 billion years. The core of the sun, where it's hot enough for nuclear reactions to occur, extends outward about a quarter of the way to the surface. Moving outward from

NEUTRINO DETECTOR IN HOMESTAKE GOLD MINE, SOUTH DAKOTA

SOLAR ACTIVITY PEAKS EVERY
11
YEARS

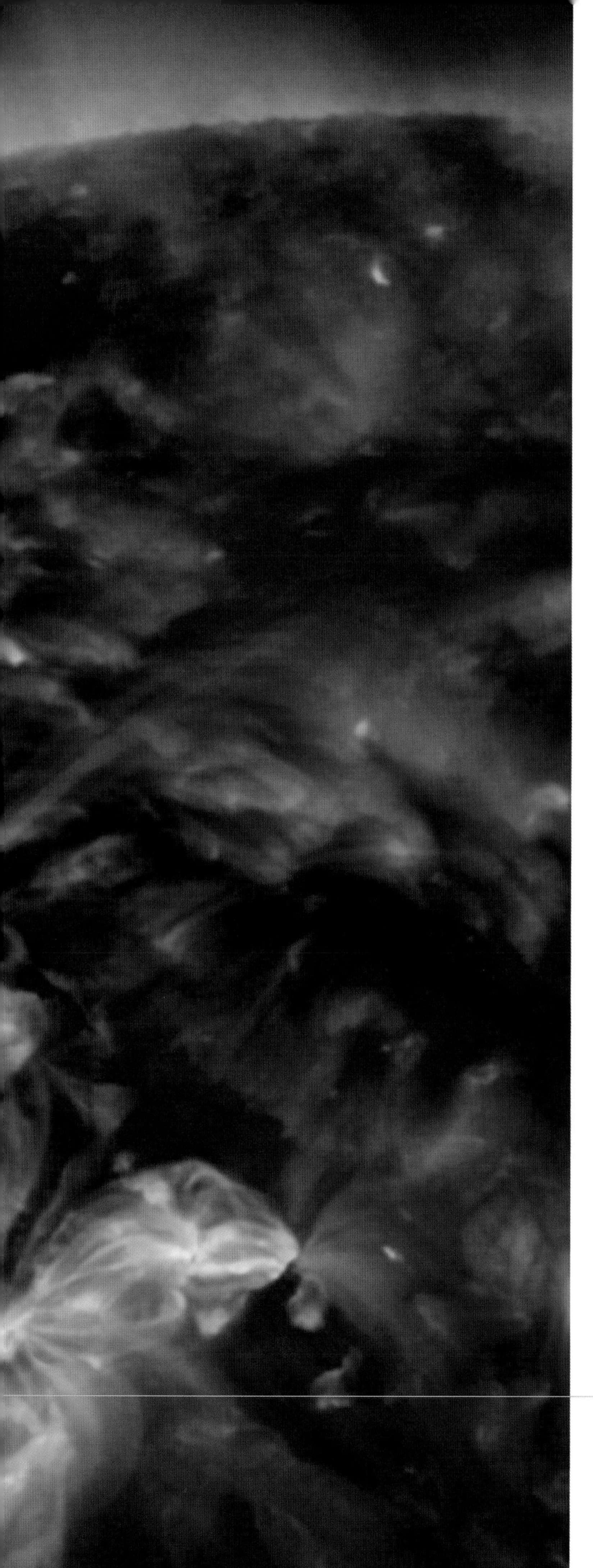

the core about 70 percent of the distance to the surface, we encounter what is called the radiative zone. In this region, where matter is still very dense, the fast-moving particles streaming out from the core suffer a series of collisions—think of the process as being like a giant pinball machine—as the energy streams through. Beyond that, the density of matter becomes too low for these kinds of collisions to impede the energy flow, and the sun actually boils, like water on the stove. This convective zone extends almost all the way to the surface.

Looking at the sun from the outside is a little like looking down into murky water. We can see only about 160 kilometers (100 mi) down into the sun, and this thin outer layer, called the photosphere, is what presents itself to us. Above the photosphere are tenuous layers of atmosphere such as the corona (visible during eclipses) and the heliosphere, which actually extends out past the orbit of Pluto.

Because the sun is not solid, different parts of it rotate at different speeds. The poles, for example, complete a revolution in about 35 days, while the equator takes about 25 days to do the same. This difference, together with the constantly churning convection beneath the photosphere, causes the magnetic field of the sun to be constantly twisted, distorted, and pulled around. This gives rise to phenomena like sunspots (dark spots that move across the face of the sun) and solar flares (in which huge numbers of particles are thrown into space). Sunspots go through an 11-year cycle, in which the number of observed spots increases and drops in a regular pattern. The solar cycle, and particularly solar flares, can have an effect on things like satellite operations and radio transmissions on Earth.

A false-color image shows the sun emitting a large solar flare (white spot at the center). The flare was associated with a coronal mass ejection, a fast-moving cloud of electrical particles.

One of the grand questions we ask about our galaxy is, Are there other life-forms out there—or are we alone in the universe? Science fiction is full of encounters with other intelligent beings—but there are lots of possible life-forms besides Klingons. For most of its history, for example, life on our planet consisted of little more than, essentially, green pond scum. Much of the motivation behind the search for the life-nurturing Goldilocks planet comes from the fact that we want to know if life developed on those planets as it did on ours. • The problem is this: Every living thing we know about works the same way, through a DNA-based chemical code. In effect, all life on Earth is the result of one experiment.

THE ORIGINS OF LIFE

ARE WE ALONE IN THE UNIVERSE?

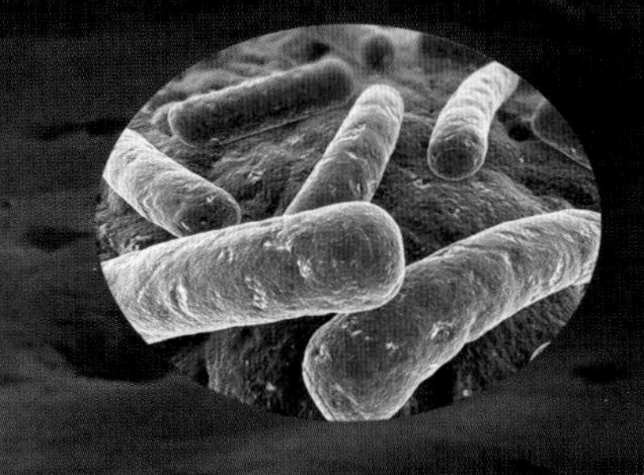

SOLAR SYSTEM'S AGE: **4.6 BILLION YEARS**
EARTH'S AGE: **4.56 BILLION YEARS**
OLDEST MINERAL (ZIRCON): **4.404 BILLION YEARS OLD**
FIRST OCEANS: **4.4 TO 4.2 BILLION YEARS AGO**
OLDEST DATED ROCKS: **4.031 BILLION YEARS OLD**
OLDEST GEOCHEMICAL SIGNS OF LIFE: **3.8 BILLION YEARS OLD**
OLDEST FOSSIL STROMATOLITES: **3.45 BILLION YEARS OLD**
ATMOSPHERE TURNS OXYGEN-RICH: **2.4 BILLION YEARS AGO**
"SNOWBALL EARTH" GLACIAL ERA: **2.3 BILLION YEARS AGO**
FIRST MULTICELLULAR FOSSILS: **2.1 BILLION YEARS AGO**
FIRST ANIMAL FOSSILS: **580 MILLION YEARS AGO**

Algae color a hot basin at Yellowstone National Park. (Inset) Bacteria are an ancient form of life.

With only one data point, there is no way we can tell whether we are the result of some wildly improbable cosmic coincidence (as some have argued) or the result of normal chemical processes that will occur many times in the galaxy. Obviously, one important step in solving this puzzle would be to understand how life developed on our own planet.

The watershed experiment in this field took place in the basement of the chemistry building at the University of Chicago in 1952. Harold Urey, a Nobel Prize–winning chemist, and his then student Stanley Miller decided to try a rather unusual experiment. They attempted to set up a miniature model of what they thought Earth was like when it was very young. In a closed system they had water (to simulate the ocean), heat (to simulate the action of the sun), an electric spark (to simulate lightning), and a collection of gases—hydrogen, ammonia, methane, carbon dioxide—that they thought were present on the early Earth. They sealed the system off, turned on the heat and the sparks, and watched to see what happened. After a few weeks, the water turned a brownish maroon color, and on analysis they found that it contained amino acids, one of the basic building blocks of living systems.

The philosophical impact of the Miller-Urey experiment was more important than the chemical details. What the two men showed was that it is possible to start with quite ordinary substances and, by normal chemical processes, produce the molecules that make up living systems. In effect, they moved origin-of-life studies from philosophy into serious science. The consensus today is that they had the wrong mixture of materials in their simulated atmosphere, but it really doesn't matter. Organic molecules of the type they produced have been found in meteorites, in comets, and even in interstellar dust clouds. It seems to be easy, in other words, to produce the molecules of life by normal chemical means. With this understanding, origin-of-life research moved from the question of how you make the basic building blocks of life to the question of how those blocks were assembled into a living cell.

PRIMORDIAL SOUP

The picture that developed after the Miller-Urey experiment was one in which the chemical process they discovered (or, alternatively, the impact of organic-rich meteorites or comets) turned the oceans into a thin broth of the molecules present in living systems, a broth that was given the wonderful name of primordial soup. Once this broth formed, the argument went, a chance collection of just the right molecules would eventually produce a primitive cell that could take in energy and reproduce. Locales for this chance assembly varied according to the theory, but tidal pools, oily drops in the ocean, and clays on the ocean floor have all been suggested.

Hypotheses like this, in which life arises through a random occurrence, go by the name of frozen accident theories, since they rely on the idea that a set of molecules, assembled first by random interactions, lock in the chemistry of all of their descendants.

Recently, an alternative explanation for the origin of life on Earth has been gaining favor in the scientific community. Because of tectonic activity (see pages 88–90), there are places on the Earth where hot material from the mantle is being brought to the surface, usually on an ocean floor. This produces what are called deep sea vents or, more technically, hydrothermal vents. In 1977, scientists using deep sea diving equipment made an astonishing discovery. Around these vents, feasting on the rich broth of energy-laden materials rising from the planet's interior, were complex ecosystems composed of many kinds of life-forms.

Once this fact was confirmed, scientists began to argue that the sea floor is actually a good place for life to develop. Not only is there an ample source of energy,

Photo of a hydrothermal vent where hot, mineral-rich material from Earth's interior is brought to the surface, supporting a complex ecosystem.

but the overlying water would shield the molecules necessary for life from harmful ultraviolet radiation from the sun. In these scenarios, life on Earth developed first at deep sea vents and only later migrated to the surface.

Once we realize that life need not begin on a planetary surface, the discovery of all those subsurface oceans in the outer solar system takes on a new significance. Places like Europa suddenly become prime locations for the possible development of life.

METABOLISM FIRST

There is, however, another way that life could have developed, and that is to imagine that there are normal chemical processes, which we are still in the process of discovering, that will drive certain kinds of reactions in the chemical stew of the early ocean—reactions that will lead directly to a primitive living system without the need of complex molecules. This is called the metabolism-first school of thought. According to the metabolism-first approach, life could have started with simple chemical reactions and evolved its present complexity over billions of years.

Whichever path life took on our planet, we know that it developed quickly, at least on a geological timescale. A half billion years after the end of the Late Heavy Bombardment (see page 59) made it possible for life to develop without interruption, we find fossils of a complex bacterial ecosystem—the green pond scum mentioned previously. The complexity of these organisms suggests that the earliest, simplest cells must have appeared fairly quickly after the end of the bombardment, and this, in turn, suggests that the chemical reactions that produced living systems on our planet could operate with equal speed elsewhere.

The idea that planetary systems circle other stars is an old one. Only in the last few decades, though, has the search for extrasolar planets, or exoplanets, become a major feature of galactic astronomy. The reason for the delay is simple: Planets shine by reflected light and are thus much dimmer than stars. Furthermore, they are located close to stars, so whatever light they send out is swamped by light from their parent star. One astronomer likened the problem of seeing an exoplanet directly to detecting a birthday candle next to a searchlight in Boston by using a telescope in Washington, D.C.! Consequently, the discovery of exoplanets had to wait for new kinds of detection techniques.

EXOPLANETS

| OTHER WORLDS, OTHER EARTHS |

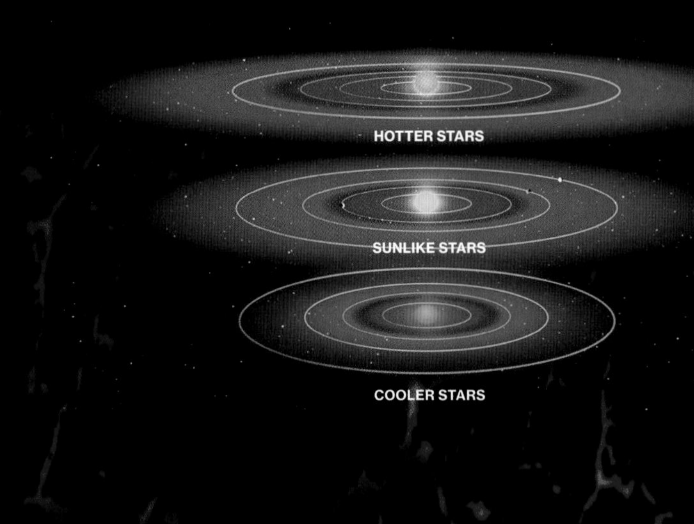

YEAR FIRST EXOPLANET DISCOVERED: **1992**
NAME OF FIRST DISCOVERED EXOPLANET: **PSR 1257+12B**
NUMBER OF EXOPLANETS KNOWN: **4,000 (AND COUNTING)**

Artist's conception of ringed exoplanet and its moon.
(Inset) Habitable zones (green) around different kinds of stars.

TWO PLANETARY SYSTEMS

The habitable zone of our solar system (bottom) compared with that of the newly discovered Kepler-22 system (top), which has a near-Earth-size planet, dubbed Kepler 22b. It is the first planet discovered that orbits within the habitable zone of a sunlike star.

KEY FEATURES

1. Earth's orbit in the sun's habitable zone
2. A planetary orbit in Kepler-22's habitable zone
3. Kepler-22's habitable zone

CARTOGRAPHER'S NOTE: Because the Kepler-22 star is smaller than the sun, its habitable zone is closer to the star. Although its planet Kepler-22b orbits within that zone, its orbit alone does not guarantee the presence of life, as can be seen with Mars in our own system.

The first successful detection technique is known in astronomical jargon as radial velocity measurement. To see how it works, imagine that you are an observer looking at our own solar system from a distance of many light-years. We are accustomed to thinking of the sun as being stationary while planets circle in their orbits, but in fact the sun moves around in response to the planets' gravitational pull. For example, if Jupiter lay between you and the sun when you were making your observation, the sun would be pulled slightly in your direction. On the other hand, if Jupiter were behind the sun, the sun would be pulled slightly away from you. Over a 10-year period, then, you would see the sun moving toward you for a while, then away from you. This motion can be detected by observing the Doppler shift in the light the sun emits—blue as it moves toward you, red as it moves away. So, although you couldn't see Jupiter directly, you would know that it's there because of its effect on the sun.

Oddly enough, the first detection of an exoplanet, in 1992, involved a rare case, a planet orbiting a pulsar (see pages 259–61). This planet must have

An illustration depicts the Kepler space observatory, launched in 2009, which has detected more than a thousand possible exoplanets. Kepler's discoveries are changing our view of how planetary systems can be organized.

formed after its parent star became a supernova, and the discovery was certainly unexpected. It was followed quickly in 1995, however, by the more conventional discovery of a planet circling a star (in this case in the constellation Pegasus), and it was this that ushered in the modern era of exoplanet detection. At first, the discoveries came in slowly, a few a year, but as techniques improved, the pace picked up. We now know of more than four thousand possible planetary systems around other stars, and some astronomers predict that the number will climb into the 10s of thousands once data from the Kepler probe, launched by NASA in 2009, are analyzed.

KEPLER MISSION

Kepler is a spacecraft weighing in at a little more than a ton. It is equipped to provide continuous monitoring of the brightness of more than 150,000 stars in our immediate galactic neighborhood. Because a satellite in low earth orbit can have up to half the sky blocked by Earth's disk, and because Kepler needs to watch the sky continuously, the spacecraft is actually maintained in an orbit around the sun, not the Earth. You can think of it as trailing along behind Earth like a miniature planet.

The basic planet-hunting technique used by Kepler is simple to describe, but it requires sophisticated equipment to make it work. The central point is that if a planet passes in front of a star, the brightness of that star will

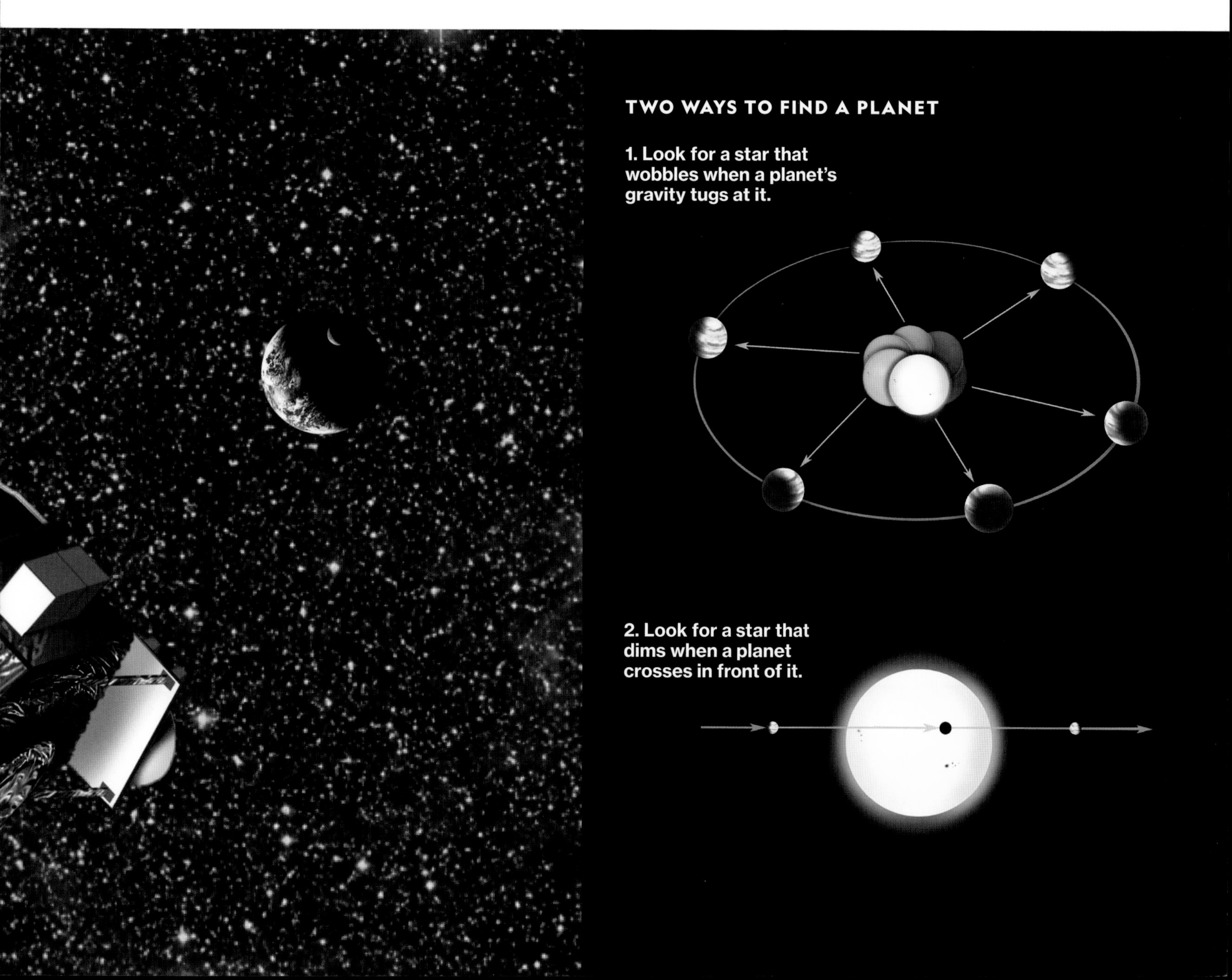

Artist's conception of sunrise on an Earth-type exoplanet. This planet circles two stars instead of one—the second star is the white dot to the right of the star that is rising.

drop while the planet is making its transit, then pick up again as the planet moves around its orbit. Of course, this so-called transit method of detection works only if the orbit of the planet lies in the line of sight to Earth. (For example, if someone were observing our solar system, Jupiter could be detected by the transit method if the observer were located on the same plane as the planetary orbits, but not if he or she was above or below it.) This means that only a fraction of planetary systems can be detected by transit techniques.

On the other hand, Kepler scientists are quick to point out that the transit method has one important advantage over the older radial velocity technique—it will detect every planet, no matter its size or distance from its star. For a planet to jerk a star around enough to produce a detectable Doppler shift, the planet has to exert a large gravitational attraction on the star. This means that the radial velocity method is most likely to detect large planets orbiting close to their stars—the so-called hot Jupiters. Indeed, most of the planets detected before the Kepler launch were of this type. The transit method, on the other hand, detects any planet that affects its star's brightness, regardless of its size or distance from the star.

THE KEPLER LEGACY

Two astonishing facts have emerged from exoplanet searches, particularly from the Kepler satellite:

- There are more planets than stars in our galaxy, and
- There are more planets not attached to stars than there are planets in stellar orbits.

Let's look at these facts one by one. Twenty years ago, astronomers could have serious debates about whether other planetary systems existed at all. Now we know that they are everywhere. We also know that planetary systems come in a wide variety of configurations and that very few of them resemble our own solar system.

One example illustrates this fact. Before the launch of the Kepler satellite, the only way to find exoplanets was to use the radial velocity method described above. If you think about the way this system works, you will realize that it is most likely to detect large planets close to their stars, simply because such planets will exert the greatest gravitational force. And indeed, that is what astronomers found. Initial searches seemed to indicate that we lived in a galaxy full of what came to be called hot Jupiters—huge planets closer to their stars than Mercury is to ours.

If you think back to our best theories about the formation of our solar system (see pages 56–9), you will see why the existence of hot Jupiters was puzzling. Planets close to their stars are supposed to be small and rocky with gas giants like Jupiter forming farther out. Were our theories really that wrong?

As it turned out, we needn't have worried. The Kepler satellite quickly established the fact that hot Jupiters constituted a small fraction of the planets out there and probably represented systems where giant planets formed far out and later moved closer to their stars. In fact, the proper lesson to draw from the hot Jupiter episode is that we are likely to find all sorts of planetary systems out there, and most of those systems will be quite different from ours.

The star TRAPPIST-1, about 40 light-years from Earth, has no fewer than seven Earth-size planets in orbit, including three in the continuously habitable zone (CHZ). The system is shown here in comparison to our own.

LIFE AT THE EXTREMES

No discussion of exoplanets or life elsewhere in the universe would be complete without a discussion of strange life right here on planet Earth. Over the past 50 years, scientists have discovered life in unexpected places. These newly discovered forms of life are called extremophiles, from the Latin extremus and the Greek philia, and the word can be translated roughly as "loving extremes."

The first extremophiles were discovered in the 1960s in the hot springs of Yellowstone National Park. The temperature of the water – at or above the boiling point – would have killed ordinary bacteria, but the extremophiles thrived. In fact, they are responsible for some of the spectacular coloring of those pools. Since that time, life has been found in more than a dozen unlikely environments – in highly acidic and highly salty environs and at the unimaginable pressure and temperatures of deep-sea vents, to name a few. Experimenters in Japan even found microbes that could thrive in centrifuges producing gravity 400 times that of Earth!

These discoveries have had a profound effect on science. Some biologists have proposed, for example, that life on Earth began with extremophiles at deep-sea trenches and only later migrated to the surface. Astrobiologists who think about life on other planets have been cautioned not to be too restrictive in their definitions of where life might develop. And, of course, there is always the possibility that surprises might be waiting for us here, on our home planet. In the words of physicist Paul Davies, "Life might be under our noses ... or in our noses."

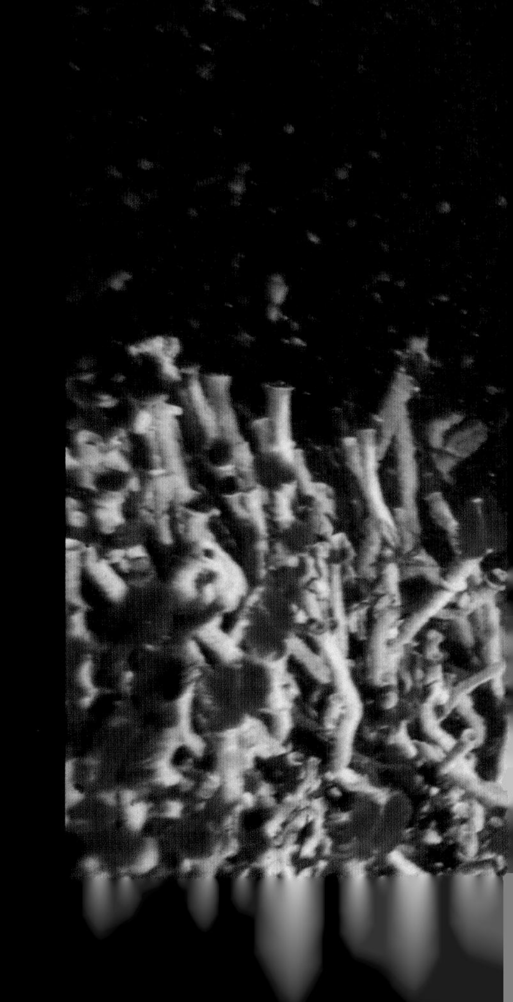

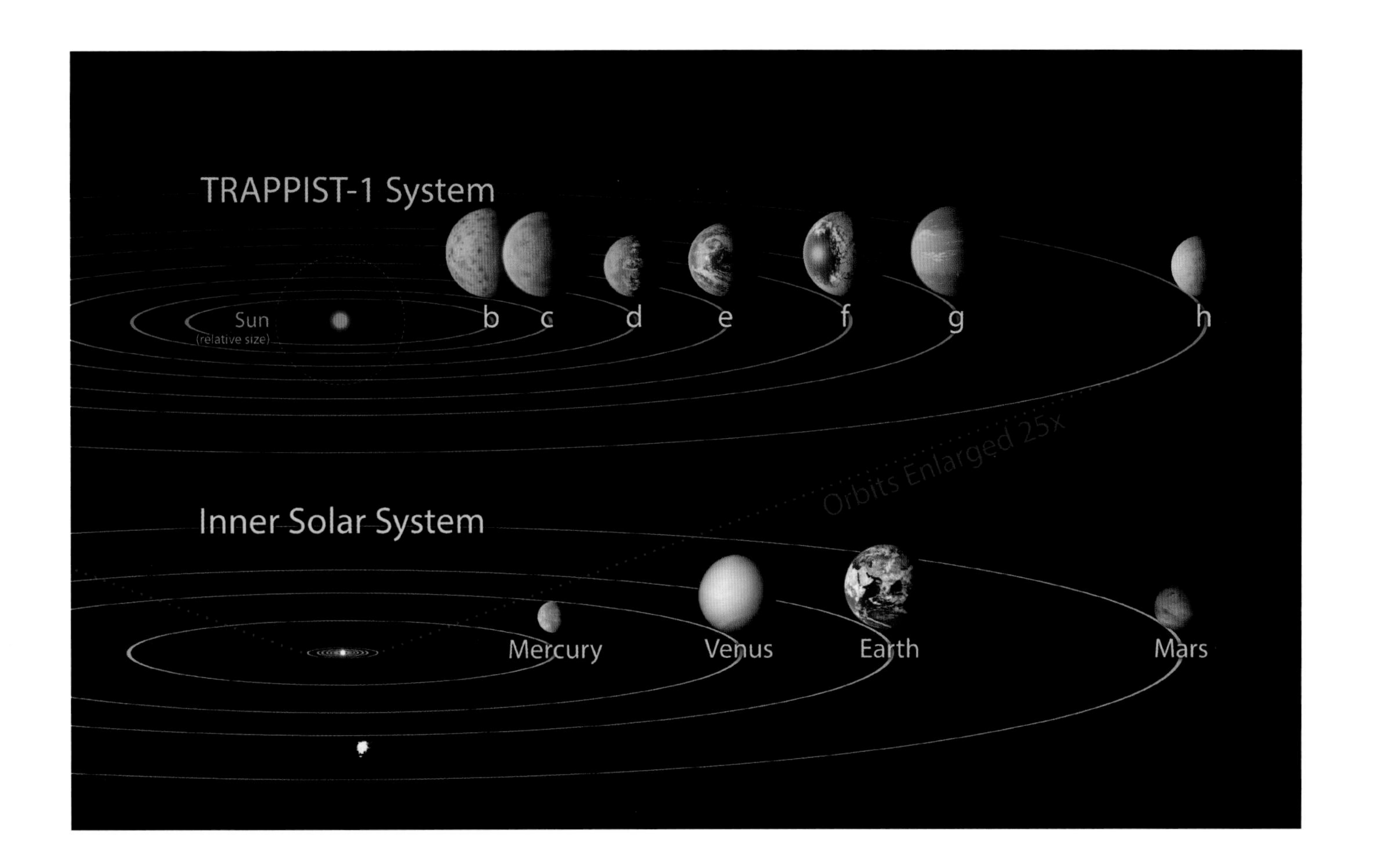

BLACK SMOKER NEAR HYDROTHERMAL VENTS

GOLDILOCKS PLANETS AND THE CHZ

Nowhere does this point have more relevance than when we consider the search for extraterrestrial life. Since life began in surface oceans on Earth, the assumption was that we were most likely to find life on planets that have had surface oceans for long periods. This idea led scientists to define something called the continuously habitable zone (CHZ), which is the region around a star in which planetary temperatures can remain between the freezing and boiling points of water for billions of years. In our system, for example, only the Earth is in the CHZ of the sun. Earth-size planets located in the CHZ of their star are often referred to as "Goldilocks planets" because, like the porridge in the well-known children's tale, they are not too hot and not too cold, but juuust right. Many such planets have been found; the first was a planet known as Kepler 186f (the name indicates that it is the fifth planet found around the

186th star in the catalog of exoplanets discovered by the Kepler satellite). These planets will most likely be the first that will be examined by the advanced detection systems described below.

AN IMPRESSIVE ARRAY OF EXOPLANETS

To get some sense of the variety of exoplanets out there, consider a couple of examples:

- A planet whose orbit is so close in that it can be considered to be actually inside its star
- A planet so hot that rocks on the starward side are vaporized. When the planet rotates, these materials solidify and "snowflakes" made of solid rock fall from the sky
- Planets covered with oceans a hundred miles deep

But the most amazing discovery, alluded to above, has been the detection of what are called "rogue planets"—planets wandering in the dark of space far from any star. If you go back to our discussion of the formation of our own solar system (see pages 56–9), you will recall that in its early evolution many planet-sized objects were ejected into space. These bodies didn't disappear but created a large body of rogue planets. It has even been argued that since such planets would

At left is an artist's interpretation of a rogue planet, wandering through space without a star. Above, a humorous poster advertises an exoplanet adventure in the form of a trip to the TRAPPIST-1 system.

retain their internal heat sources, they might even support life. Some authors compare them to a house whose lights have been turned off but whose furnace is still running.

New and exciting discoveries about exoplanets are being made rapidly. Two such discoveries have caught the attention of the public in recent years:

- The discovery of an Earth-sized planet in the CHZ of Proxima Centauri B, the closest star to Earth. The star is only about four light-years away, which means that it is likely to be the first exoplanet visited by human spacecraft. But even if we could build a spaceship capable of traveling at 10 percent of the speed of light, a round trip to this planet would take 80 years
- In 2017 scientists at NASA and the ESA announced the discovery of seven Earth-size planets in orbit around a small star called TRAPPIST-1, about 40 light-years away. Three of those planets are in the star's CHZ, and the others are close enough to have liquid water on their surfaces, given the right atmospheric conditions. Thus, the TRAPPIST-1 system will be a major focus of research in the search for signs of life on an exoplanet

It's not often that the birth of a scientific field can be dated precisely, but SETI—the Search for Extraterrestrial Intelligence—is an exception. It began with a paper by physicists Giuseppe Cocconi and Philip Morrison in 1959 and came to full fruition at a conference in the mountains of West Virginia in 1961. In their paper, the physicists pointed out that with the new availability of radio telescopes, it had become possible to scan the radio bands and see if anyone out there was trying to contact us. "The chance of success if we try is small," they argued, "but the chance of success if we don't try is zero."

SETI

ARE WE ALONE?

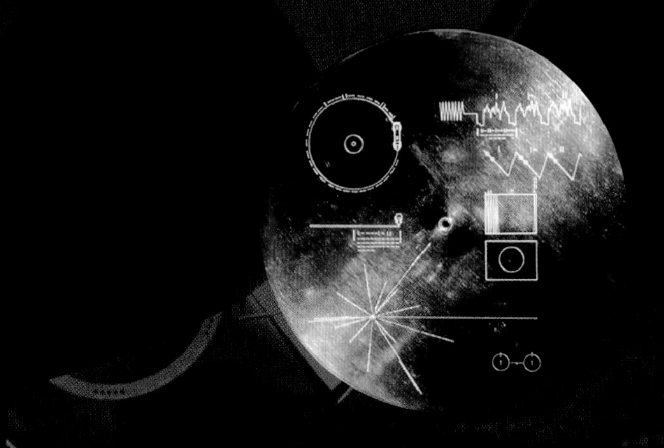

CLAIM THAT MARTIANS DUG CANALS: **PERCIVAL LOWELL, 1895**

RADIO PROPOSED TO FIND EXTRATERRESTRIALS: **NIKOLA TESLA, 1896**

FIRST LISTENING FACILITY FOR MARS RADIO SIGNALS: **U.S. NAVAL OBSERVATORY, 1924**

MONITORING WITH RADIO MICROWAVES: **PHILIP MORRISON AND GIUSEPPE COCCONI, 1959**

FIRST RADIO SEARCH: **PROJECT OZMA, HEADED BY FRANK DRAKE, 1960**

SEARCHING WITH ANTENNA ARRAYS: **PROJECT CYCLOPS, 1971**

MULTICHANNEL SPECTRUM ANALYZER: **PAUL HOROWITZ, 1981**

NASA DROPS SETI FROM BUDGET: **1981**

SETI@HOME BEGINS: **1999**

ALLEN TELESCOPE ARRAY COMES ON LINE: **2007**

Allen Telescope Array, California. (Inset) Diagrams on Voyager golden record cover.

The conference at the National Radio Astronomy Observatory in Green Bank, West Virginia, gathered 11 scientists to talk about this new possibility. The conferees eventually summarized their estimates of the number of extraterrestrial civilizations in a compact notation that has come to be called the Drake equation, after Cornell astronomer Frank Drake, one of the organizers of the conference. The equation estimates the value of N, the number of extraterrestrial (E.T.) civilizations trying to communicate with us right now, as

$$N = R f_p n_e f_l f_i f_c L$$

where the symbols represent, from left to right, the rate at which new stars are forming in the galaxy, the probability that the new star will have planets, the number of those planets capable of supporting life, the probability that those planets will actually develop life, the probability that life will develop intelligence, the probability that intelligent life will develop a technology capable of interstellar communication, and the length of time that that communication, once started, will continue. Obviously, as we move from left to right through these terms, we go from fairly well known astronomy to sheer guesswork. Nevertheless, the Drake equation is a useful way to organize our knowledge (and our ignorance) on the subject of extraterrestrial intelligence.

In 1961 the Green Bank attendees argued that *N* could be as high as 200 million or as low as 4 but came up with a most probable estimate of around a million. This was a time, remember, when scientists still harbored hopes of being able to find life on Mars and other places in our solar system. The idea of a vast, intercommunicating Galactic Club made up of thousands of intelligent species entered the public consciousness, fueling countless science-fiction scenarios.

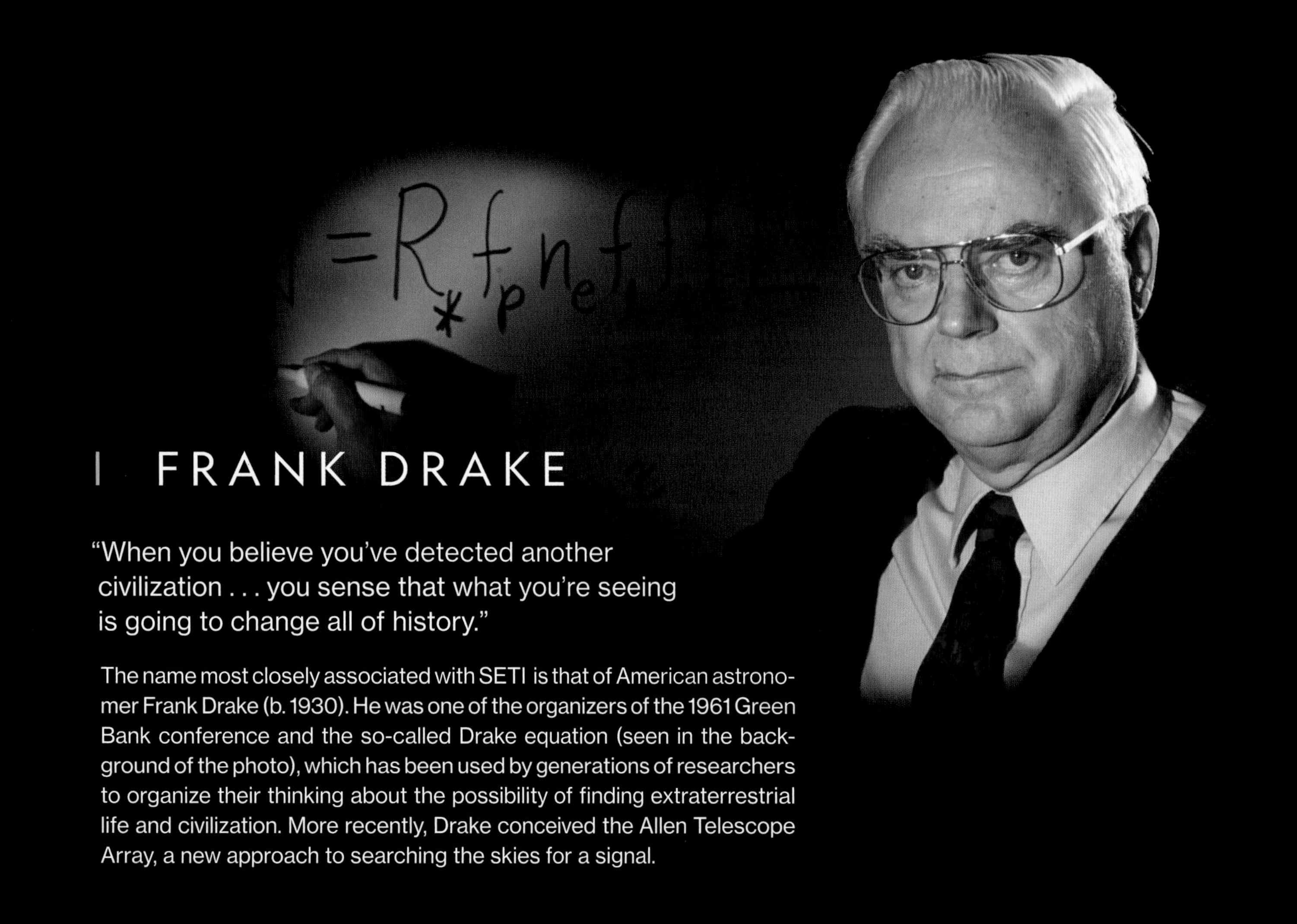

FRANK DRAKE

"When you believe you've detected another civilization . . . you sense that what you're seeing is going to change all of history."

The name most closely associated with SETI is that of American astronomer Frank Drake (b. 1930). He was one of the organizers of the 1961 Green Bank conference and the so-called Drake equation (seen in the background of the photo), which has been used by generations of researchers to organize their thinking about the possibility of finding extraterrestrial life and civilization. More recently, Drake conceived the Allen Telescope Array, a new approach to searching the skies for a signal.

Unfortunately, this is one case where science didn't keep up with science fiction. In the beginning, there was modest federal support for SETI programs, but when the searches failed to turn up anything interesting, that source of funding waned. Today, the SETI program is being carried forward using private funding. For example, in 2015, Russian investor Yuri Milner committed $100 million over a 10-year period to support SETI.

When you think about the problems involved in the search, you can see why it is taking so long. There are billions of stars out there, and, as we saw on page 235, a good proportion of them have planetary systems. With each star, you have to follow a slow process: You don't know what frequency the aliens might be using for their broadcasts, so, like someone listening to radio programs in a strange city, you slowly move through the dial, pausing long enough at each position to see if something interesting is coming through. Today SETI searches make use of modern fast electronics to monitor thousands of stars and thousands of frequencies at once as they sort through a mountain of data.

WHERE IS EVERYBODY?

But despite these efforts and widespread public interest, the unfortunate fact is that no contact has yet been made with extraterrestrials. One of the engaging things about SETI is that it is tailor-made for "armchair science"—amiable speculation that doesn't require rigorous evidence to support it. Here, for example, are a few of the explanations that have been given for the lack of a SETI signal:

The zoo hypothesis—our solar system has been declared a kind of galactic wilderness area, off-limits to ETs.

The gloom and doom hypothesis—any species aggressive enough to win the evolutionary battle and develop technology will wipe itself out with nuclear weapons, thereby producing a low value of L in the Drake equation.

The magic frequency hypothesis—we're looking at the wrong frequencies in our searches, although there is some other magic frequency, recently discovered by the proponents of this hypothesis, that we should be monitoring instead.

And so on. You can undoubtedly add to this list yourself.

The most interesting SETI argument, though, was made early on by the Italian American physicist Enrico Fermi. When presented with the kind of arguments that eventually led to the idea of the Galactic Club, he is supposed to have thought for a while, then asked, "So where is everybody?" Fermi, a genius at seeing through to the heart of complex problems, argued as follows: Modern science is only a few hundred years old—less than the blink of an eye on astronomical timescales. It's pretty likely that in another few hundred years—another eyeblink—we will have solved the problem of interstellar travel and be colonizing the stars ourselves. If there are really millions of other civilizations out there, some must have passed this stage already, so they should be here already. So, he asked, "Where is everybody?"

The point of the argument, of course, is that we shouldn't be looking for ET life out there, we should be looking right here. Many authors (myself included) have used the Fermi argument to propose that N is quite small, possibly even 1, and that humans might well be alone in the universe.

The point about SETI, though, is that no matter how the search turns out, it's worth doing. Are there ETs out there? Fantastic! Are we alone in the galaxy? Even more fantastic! There aren't many other kinds of scientific activities with this kind of payoff.

All stars begin life the way our sun did, as a condensing cloud of interstellar dust. From then on, the star is essentially devising strategies to ward off the eternal inward pull of gravity. We discussed the first of these strategies in the birth of our own sun—the initiation of fusion reactions in the core, which sets up an outward pressure to stop the cloud from collapsing. Almost every star you see is in this hydrogen-burning phase—astronomers call them main-sequence stars. Our sun has been in this phase for roughly 4.5 billion years.

STARS IN OLD AGE

WHAT HAPPENS WHEN THE FUEL RUNS LOW?

LIFETIMES OF STARS OF DIFFERENT MASSES

0.1 SOLAR MASS: **6 TO 12 TRILLION YEARS**
1 SOLAR MASS: **10 BILLION YEARS**
10 SOLAR MASSES: **32 MILLION YEARS**
100 SOLAR MASSES: **100,000 YEARS**

FATES OF STARS OF DIFFERENT MASSES

0.1 SOLAR MASS: **RED DWARF**
1 SOLAR MASS: **RED GIANT, THEN WHITE DWARF**
10 SOLAR MASSES: **SUPERNOVA, THEN BLACK HOLE**
100 SOLAR MASSES: **SUPERNOVA, THEN BLACK HOLE**

Stars in the globular cluster Omega Centauri. (Inset) Typical stars Sirius A and the sun, hot blue star Rigel, and red supergiant Antares (in the background).

Obviously, hydrogen burning can't go on forever. Sooner or later the hydrogen fuel in the star's core will run out and the star will have to develop a new way of countering gravity. How long that takes depends on how big the star is. There are two competing effects here. On one hand, bigger stars have more fuel to start with. On the other hand, bigger stars also generate a stronger gravitational force and have to burn fuel faster to overcome it. It turns out that the second effect wins, and that the bigger a star is, the shorter the time it can spend burning hydrogen. If you imagine the Milky Way galaxy being a year old, for example, a star like the sun might have fuel enough for 10 months or so, whereas very large stars may last as little as half an hour.

RED GIANTS

So what will happen to a typical star, like our sun, when the hydrogen in its core runs out? Obviously, the nuclear fires will start to fade and the pressure that has held off the forces of gravity for billions of years will grow weaker. Gravity will take over again, the star will start to contract, and once again the contraction will cause the star's interior to heat up. This will have two effects: First, the temperature of the region just outside the core, which still has a lot of unburned hydrogen in it, will increase to the point where that hydrogen can start fusing into helium. Second, the temperature of the core itself goes up until the helium nuclei that were the end product of the original fusion reactions are moving fast enough to start a new cycle of fusion. In the end, three helium nuclei, each with two protons and two neutrons, come together in the core to produce a carbon nucleus (six protons and six neutrons) along with some extra energy. This sort of sequence, in which the ashes of one nuclear fire become the fuel for the next, is one of the major features of energy generation in aging stars.

In the end, for stars up to about six times as massive as the sun, the renewed fusion reactions will increase the star's energy output as well as cause its outer atmosphere to expand. The edge of our sun, for example, will eventually extend outside the current orbit of Earth. Because the star's energy is being sent out through a much larger surface, the color of that surface changes from white-hot (as in the sun today) to a cooler red. A star like this is called a red giant.

What will happen to Earth when the sun turns into a red giant 5.5 billion years from now? Clearly Mercury and probably Venus will be swallowed up. During the run-up to the red giant phase, the sun will throw an appreciable fraction of its mass into space, however, weakening its gravitational hold on the planets. Consequently, Earth's orbit will move outward. If this were the only thing happening, Earth would narrowly escape being engulfed, but recent calculations indicate that tidal effects may pull the orbit in enough to ensure destruction. Even if our planet is not engulfed, however, its oceans would evaporate and surface rocks would melt, ending any life that has survived to that time.

WHITE DWARFS

What's next? The pressures that stars like the sun can generate are not high enough to initiate any new nuclear fusion reactions with the carbon created earlier in the star's core, so that avenue of escape from gravity is closed off. Gravity takes over again and the collapse resumes.

We know that early in the collapse of the interstellar gas cloud that led to the sun, collisions between atoms became so violent that electrons were torn from their nuclei. Throughout the whole stellar life cycle, from main sequence to red giant, these electrons were basically spectators while nuclear events held center stage. Now it becomes their turn to shine.

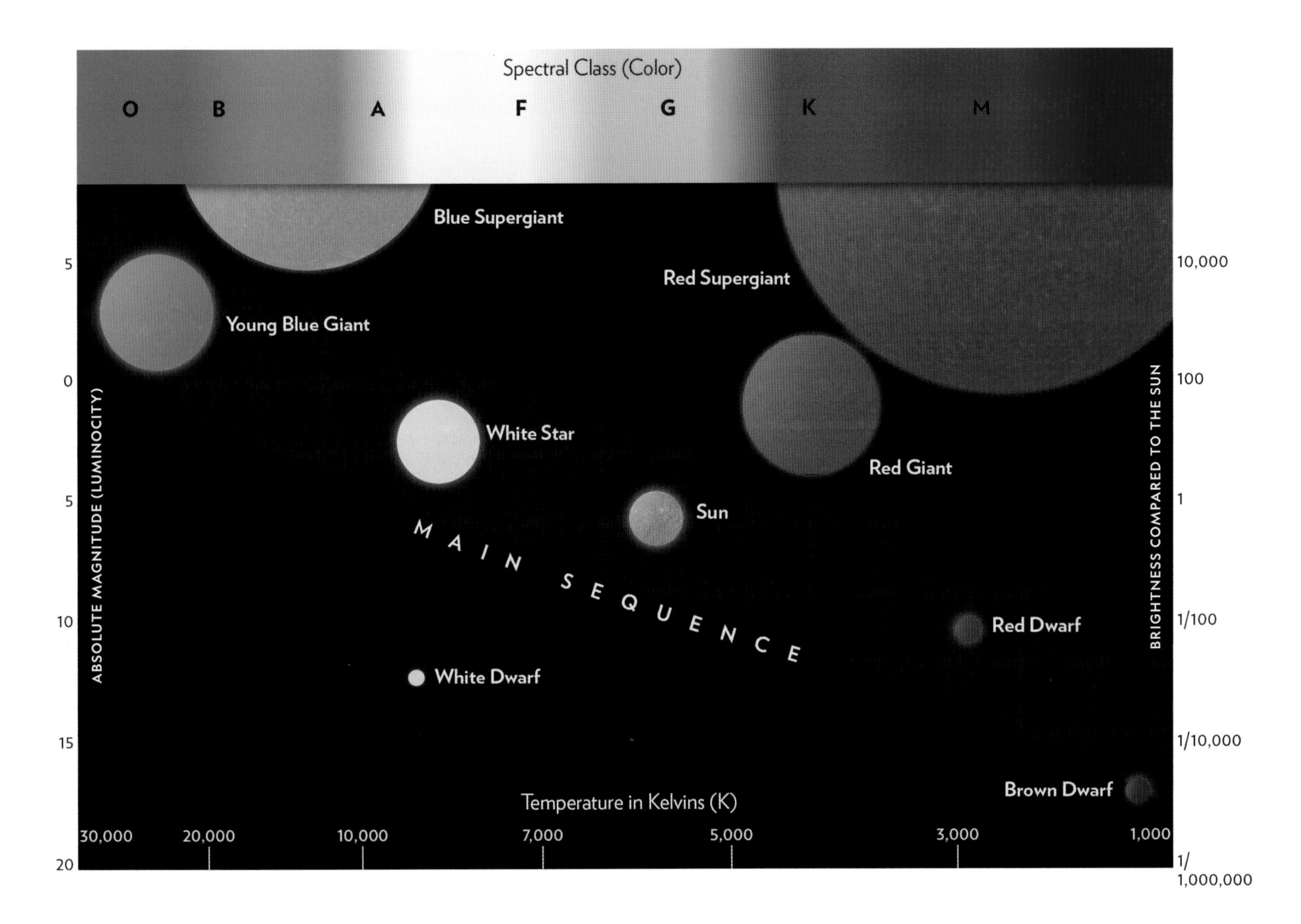

The connection between a star's temperature and its luminosity can be seen on a Hertzsprung-Russell diagram (above). From left to right, stars run from hottest to coolest; from bottom to top, dimmer to brighter. Our sun is a typical star on the central main sequence.

One fact about electrons becomes very important at this stage of the stellar life cycle. It's called the exclusion principle, and it says that no two electrons can be in the same state. Think of a crowd of people: You can push them close together, but eventually you reach a point where each person requires some minimum amount of elbow room, at which point you can't make the crowd any smaller. In the same way, the loose electrons in the sun are pulled in as the final collapse starts, but they eventually reach a point where they just can't be pushed together anymore. At this point gravity is pushing in while the electrons push out, and the star reaches a final point of stability. It will last forever.

For the sun, this new equilibrium will be reached when the star has shrunk down to about the size of Earth. The star is white-hot, like the ember of a recent fire. Astronomers call this sort of object a white dwarf. It is still radiating leftover energy produced during its long life cycle, but like a dying ember it will keep cooling and dimming. This is how our own star will end its life, along with many of the stars we now see in the sky.

This is not the only way for a star to die, however. Some life stories have a much more spectacular endings, as we shall see in the next section.

When you look out into the Milky Way, you see all kinds and sizes of stars. If the sun were the size of a bowling ball, for example, the galaxy would be littered with bowling ball-size stars, interspersed with every other conceivable size ball up to giant beach balls. As you might expect from a collection with this much diversity, not all stars have the same life story, although all of them, like the sun, start by burning hydrogen. Stars like the sun (roughly from golf ball to basketball size) go through the red giant–white dwarf sequence described in the preceding chapter. Larger stars have a much more spectacular endgame and play a much more important part in the history of life on Earth.

SUPERNOVAE

GOING OUT WITH A BANG

MOST FAMOUS SUPERNOVA: **1054 (CRAB NEBULA)**
DISTANCE TO 1054 SUPERNOVA: **6,500 LIGHT-YEARS**
LAST SUPERNOVA SEEN IN MILKY WAY: **1604**
DISTANCE TO 1604 SUPERNOVA: **14,000 LIGHT-YEARS**
SUPERNOVA CONCEPT BORN: **1931, BY WALTER BAADE AND FRITZ ZWICKY**
MOST RECENT BRIGHT SUPERNOVA: **1987 IN LARGE MAGELLANIC CLOUD**
DISTANCE TO 1987 SUPERNOVA: **160,000 LIGHT-YEARS**
SUPERNOVA BLAST-WAVE SPEED: **30,000 KM/SEC (19,000 MI/SEC)**
NUMBER OF SUPERNOVAE DISCOVERED EACH YEAR: **HUNDREDS**
MILKY WAY SUPERNOVA FREQUENCY: **ABOUT EVERY 50 YEARS**
SUPERNOVA TYPES: **IA, IB, IC, IIP, IIL**

False-color image of supernova remnant Cassiopeia A. (Inset) Shock waves heat material around Supernova 1987A.

A quick review: Every star begins as a collapsing cloud of interstellar dust. It reaches a first point of stabilization when the temperature in its core gets high enough to begin nuclear fusion, turning hydrogen into helium. When the hydrogen in the core runs out, the collapse continues, the interior heats up, and the helium ash is burned along with some previously unused hydrogen. For stars up to six times the size of the sun, the nuclear story ends here—there just isn't enough mass to push the temperature high enough to ignite any new nuclear fires. The star becomes a white dwarf.

For stars nine or ten times more massive than the sun, though, the story is different. These stars do have enough mass to push both the compression and the temperature high enough to keep the nuclear fires burning. As in the sun, the first nuclear step for these stars once the hydrogen in the core runs out is to burn the resulting helium, ultimately combining three helium nuclei to create a carbon nucleus. This process is accompanied by the burning of previously unburned hydrogen into helium in a shell surrounding the core. The star now has a carbon core surrounded by a helium shell that, in turn, is surrounded by unburned hydrogen. When the collapse starts again, the temperature in all these regions rises. In the innermost core, the carbon (six protons, six neutrons) combines with other nuclei to produce oxygen (eight protons, eight neutrons) and other, heavier products. The helium in the first shell burns to carbon, and the hydrogen in the next shell burns to helium.

This process, with the ashes of one fire becoming the fuel for the next, continues through the periodic table of the elements, with the star developing an onion-like shell structure as successively heavier elements are created by

NOVAE AND SUPERNOVAE

The word *nova* means "new" in Latin, and the term has long been applied to a specific type of astronomical event: A star suddenly appears in the sky where no star had been before. Today we realize that there are many different processes that can produce new stars, and we give them different names if they have different causes.

A nova occurs in double-star systems when one of the stars has gone through its from its partner. When this hydrogen accumulates to a depth of several feet on the dwarf's surface, nuclear reactions ignite. In essence, the layer goes off like a huge hydrogen bomb, temporarily brightening the sky in a nova. This process can be repeated as the layer builds up again.

Although people often confuse novae and supernovae, only those stars that undergo this temporary surface brightening are prop-

each collapse. Each new reaction adds another layer to the onion. As this process goes on, however, the star gets less bang for the buck with each new cycle. Indeed, some calculations suggest that the last stage of the process, which creates nuclei of iron in the core, buys the star only a few days' respite from the relentless pull of gravity.

CRITICAL MASS

Iron is the ultimate nuclear ash. You can't get energy from iron by splitting it, and you can't get energy from iron by adding to it. It just builds up in the core of the star like ashes in the grate of a woodstove, setting the stage for one of the most spectacular events in the universe. As the iron accumulates in the heart of the star, nuclear forces cannot keep it from collapsing under the influence of its own gravity. Electrons in the iron core for a time provide the pressure to counteract gravity. In effect, the core of the massive star becomes a kind of white dwarf, except that it is made of iron.

As more and more iron "ash" falls into the star's center, the mass of the core approaches a critical value. When the mass gets to be about 40 percent more than that of the sun, the electrons begin to combine with the protons in the iron nuclei to produce neutrons. As each electron disappears, the ability of the remaining electrons to counteract gravity decreases; in a very short time the core turns into a mass of neutrons that collapses catastrophically.

Depending on the mass of the star, the collapse will continue until the neutrons can't be pushed together any further, or it will go on to form a black hole. We'll talk about both of these possibilities in the next sections. For the moment, though, let's think about the rest of the star—that huge, onion-like envelope of heavy elements that the nuclear reactions have built.

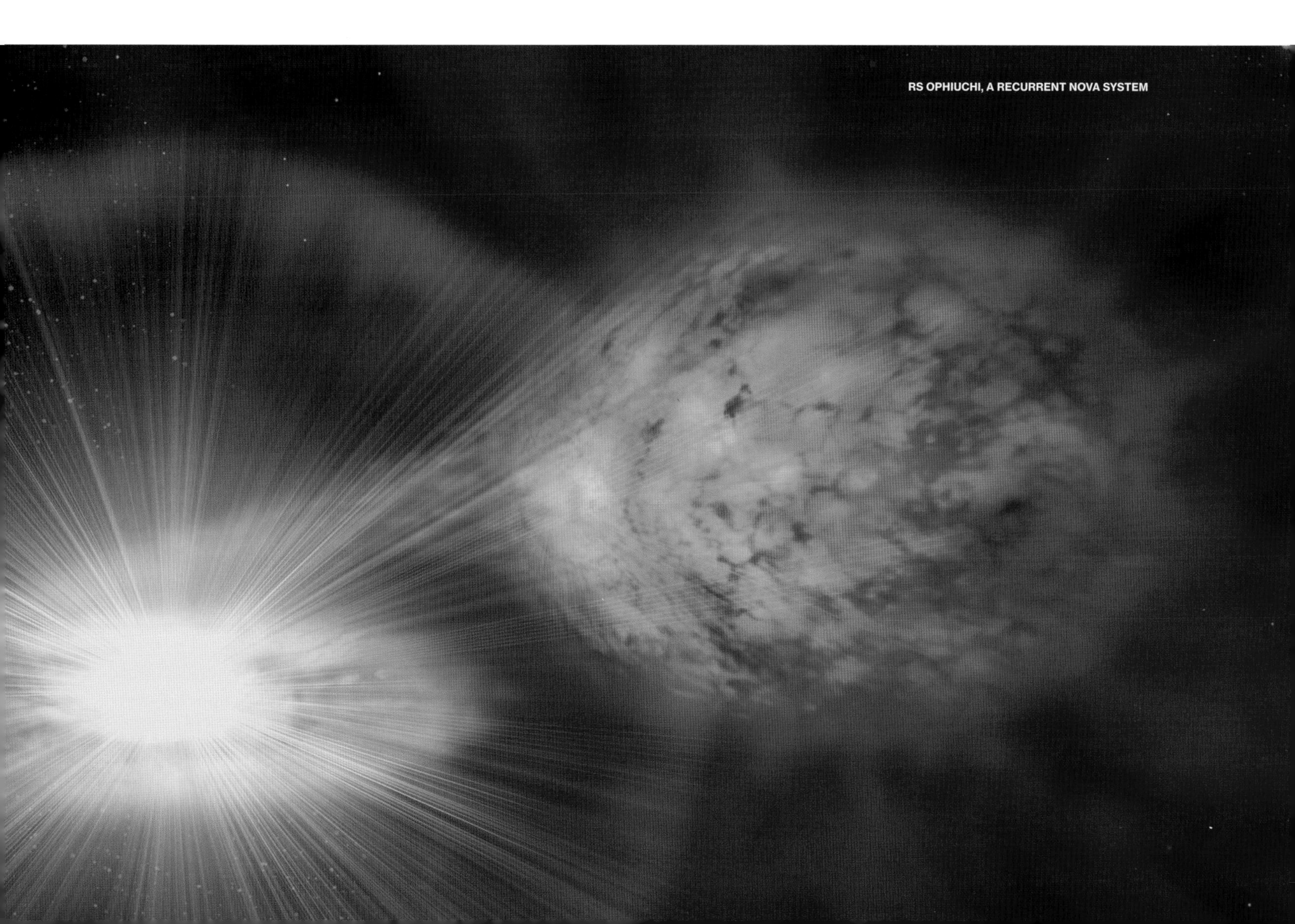

RS OPHIUCHI, A RECURRENT NOVA SYSTEM

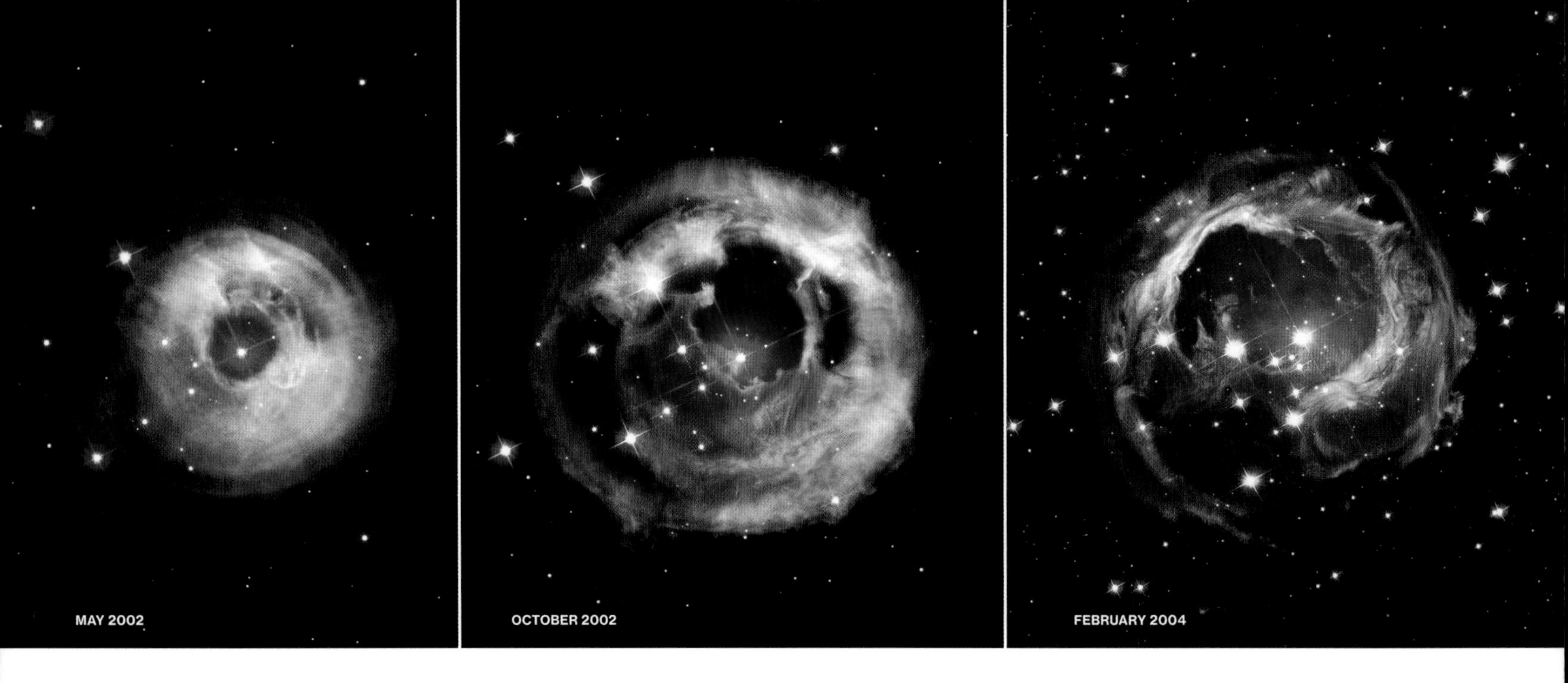

EXPLOSION

From the point of view of that envelope, the rug has just been pulled out from under its feet. The iron core that had supported the weight of the rest of the star has suddenly disappeared. In response, the envelope starts to collapse, falling inward until it hits the new neutron core, at which point it rebounds, producing massive shock waves—and the entire star explodes. In the resulting maelstrom, nuclear reactions produce all of the elements up to uranium. It is this titanic event that gives rise to a new star in the sky: a supernova.

The remnants of the envelope speed outward, carrying with them all of the star's heavy elements. Over the next few thousand years, these clouds of materials will cool and mix with the interstellar medium, becoming part of the dust clouds from which new stars and new planetary systems form. Our own solar system formed late in the history of the Milky Way, incorporating into its planets and even its life-forms the heavy elements made by long-dead stars.

TYPE I SUPERNOVAE

For historical reasons, the kind of event we've just described is called a type II supernova. A type I supernova operates differently, although it produces the same kind of massive explosion. Type I supernovae occur in double-star systems in which one star has gone through its life cycle and is a white dwarf. It can happen that this dwarf starts to pull hydrogen off of its partner, increasing its mass until it reaches that critical point, 40 percent higher than the mass of the sun. At this point the entire star explodes in a massive nuclear inferno.

Let's think about the dynamic picture the supernova story gives us of our home galaxy. It started off, billions of years ago, as a cloud made up primarily of hydrogen and helium left over from the big bang. As large, short-lived stars were born and became supernovae, heavier elements began to appear—a few at first, then more and more as time went on. One way to think about the Milky Way, then, is to picture it as a giant machine constantly taking in primordial hydrogen and churning out the rest of the chemical elements.

Oh, and in case you're interested, astronomers expect the star 1K Pegasi (in the constellation Pegasus), 150 light-years from Earth, to become a type I supernova sometime in the next few million years or so.

Beginning in 2002, the Hubble Space Telescope captured a remarkable sequence of images as the star V838 Monocerotis suddenly swelled, heating and illuminating a shell of surrounding dust clouds.

SEPTEMBER 2006

When we described supernovae in the preceding section, we concentrated on events in the exploding stellar envelope. Now it's time to go back and see what happened to the collapsing core. You may recall that the envelope suddenly collapsed because electrons were being forced to combine with the protons in the iron nuclei of the core, turning most of the core into a mass of neutrons. Because neutrons have no electrical charge and therefore do not repel each other, there was nothing to counter the force of gravity in the heart of the star, so the material in the core went into free fall. This fall continued until some force could be found to overcome the inward pull.

NEUTRON STARS & PULSARS

MAGNETIC SPINNERS

FIRST PULSAR DISCOVERED: **NOVEMBER 28, 1967**
DISCOVERERS: **JOCELYN BELL BURNELL AND ANTONY HEWISH**
IDENTIFIED AS NEUTRON STARS BY: **THOMAS GOLD AND FRANCO PACINI**

PERIOD OF FIRST PULSAR: **1.33 SECONDS**
LONGEST PERIOD PULSAR: **8.51 SECONDS (PSR J2144-3933)**
FIRST BINARY PULSAR: **PSR 1913+16**
FIRST DOUBLE-PULSAR BINARY SYSTEM: **PSR J0737–3039**
FIRST X-RAY PULSAR: **CENTAURUS X-3**
FIRST PULSAR WITH PLANETS: **PSR B1257+12**
FASTEST ROTATION: **716 TIMES A SECOND (PSR J1748-2446AD)**
CLOSEST TO EARTH: **510 LIGHT-YEARS (PSR J0437-4715)**

Illustration of pulsar in supernova remnant.
(Inset) A pulsar in the Crab Nebula.

Neutrons, like electrons, can't be crowded too closely together, so if the star isn't too big, eventually the pressure of the crowded neutrons will balance gravity. The laws of quantum mechanics tell us, however, that the heavier a particle is, the less elbow room it needs and the more closely it can be packed. Since neutrons are almost 2,000 times heavier than electrons, this means that the object formed from the collapsing core will be much smaller than the white dwarfs discussed on pages 248–9. In fact, most neutron stars are thought to be less than about 16 kilometers (10 mi) across—small enough to fit inside the city limits of many urban areas. When I want to make this point to my students in the suburbs of Washington, D.C., I point out that a neutron star would fit comfortably inside the Capital Beltway.

PROPERTIES OF A NEUTRON STAR

An object like this has several amazing properties. In the first place, a supernova's iron core doesn't start to collapse until it is significantly more massive than the sun (typically, between 40 and 200 percent more massive). If you cram that much material into something the size of a small city, the material is going to be incredibly dense. In fact, a tiny drop of neutron star stuff would easily outweigh the Great Pyramid at Giza. Because the mass is so concentrated, the force of gravity at the surface is huge—perhaps a hundred billion times stronger than on the surface of Earth.

The second amazing feature has to do with rotation. All stars rotate—we saw, for example, that the sun spins on its axis about once a month. The iron core of the supernova would have this sort of rotation rate as well. But just as an ice-skater's rate of spin increases when she pulls in her arms, the rate of rotation of the core will increase during the collapse. Sometimes, the spin can get very fast indeed. Some neutron stars rotate at nearly a thousand times a second!

Finally, the collapse will produce an extremely strong magnetic field in the neutron star. Stars normally have moderate magnetic fields—the field of the sun, for example, is about half that of Earth. This field is locked in to the matter of the star, however, so it is concentrated by the collapse of the core. The magnetic fields in some neutron stars may be a million billion times greater than that on Earth.

We can't study neutron stars close up, of course, but there are fairly solid theoretical models of what they must be like. A neutron star 16 kilometers (10 mi) in diameter is thought to have a solid crust—nuclei crushed into a kind of lattice—about 1.5 kilometers thick. Because of the intense gravity, the atmosphere (composed of atomic nuclei and electrons) is less than a meter high, and the surface is extremely smooth, with a maximum irregularity less than the thickness of a dime. In this model, the interior of the star would be a kind of nuclear liquid made up mostly of neutrons.

Putting all of this together, we get a compelling picture of the star. It is a compact, rapidly spinning object with a very strong magnetic field. In general, we would expect the star's north magnetic pole to be different from the geographic North Pole, just as it is on Earth. (Remember that the north magnetic pole of our planet is in Canada, not at the geographic North Pole.) This means that as the star rotates, the magnetic field is swept around in a circle. Because of the field's intensity, the neutron star emits a beam of radio waves that travels outward along the direction of the star's magnetic axis. Like a lighthouse beam sweeping around in a circle, then, the rotating neutron star sends out a radio beam that sweeps through space. It's what happens if Earth happens to be in the path of that beam that is interesting.

THE STELLAR LIGHTHOUSE

Imagine standing on a shore near a lighthouse. You will see a flash of light when the beam is in your direction, followed by a period of darkness as the beam travels around, and then another flash. In the same way, if you have a radio receiver in the path of the beam from a rotating neutron star, you will see a pulse of radio waves when the magnetic axis points in our direction, a period of no signal, then another pulse, and so on.

In 1967 scientists at a radio telescope in England first observed these sorts of regular pulses. No one had anticipated anything like this, and at first the scientists referred to them, jokingly, as the LGM signals—LGM standing for "little green men." Once the signal was understood to be from a rotating neutron star, the word "pulsar" was coined for these objects.

Since that first discovery, a couple of thousand pulsars have been discovered in the Milky Way. Their periods of rotation vary from somewhat less than 10 seconds down to a few milliseconds. Pulsars have also been discovered that emit pulses in the x-ray and gamma-ray regions of the spectrum; a few even have planets. The closest pulsar to Earth is 280 light-years away in the constellation Cetus.

PULSAR SCIENCE

Pulsars come in many varieties and have figured prominently in a number of different investigations. We will talk about only a few of them here.

JOCELYN BELL BURNELL

"Finding the first [pulsar] was disturbing – scary – because we weren't sure what it was."

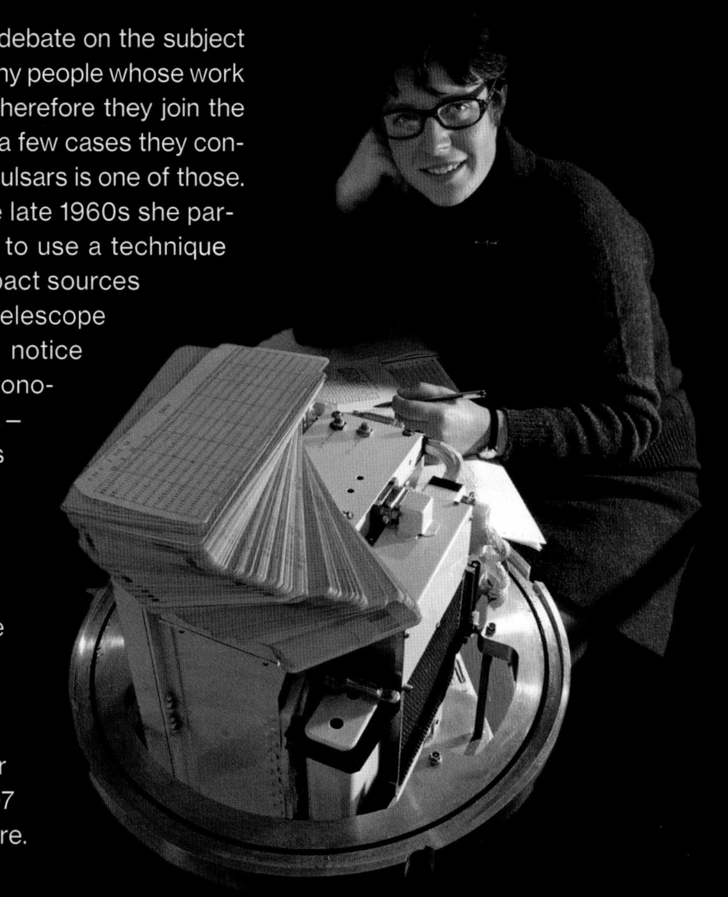

Every year when the Nobel Prizes are announced, there is a mini-debate on the subject of who was left out. In any scientific discovery, there are always many people whose work was important but who do not get the ultimate recognition, and therefore they join the ranks of the "also-rans." These debates are usually low-key, but in a few cases they continue for decades. The story of Jocelyn Bell and the discovery of pulsars is one of those.

Bell was a graduate student at Cambridge University. In the late 1960s she participated in building one of the first radio telescopes designed to use a technique suggested by her thesis adviser, Antony Hewish, to detect compact sources of radio waves in the sky. Bell was in charge of operating the telescope and doing the first analysis of the data. In 1967 she began to notice signals she called "scruff"; in the face of hostility from senior astronomers, she showed that these signals – regularly timed radio pulses – were real, and not caused by man-made interference. The signals were originally dubbed LGM, for "little green men," because of the possibility that they came from extraterrestrials. It was quickly realized, however, that they were what we now call pulsars (see above).

For this discovery, Hewish and radio astronomer Martin Ryle shared the Nobel Prize in physics in 1974, the first physics prize awarded in astronomy. The exclusion of Bell triggered protests from many prominent astronomers – though not from Bell (now Jocelyn Bell Burnell) herself, who is still active as an astronomer and has won many other awards in her career, including a 2007 appointment as Dame Commander of the order of the British Empire.

THE CRAB NEBULA SUPERNOVA
EXPLODED ALMOST
1,000
YEARS AGO

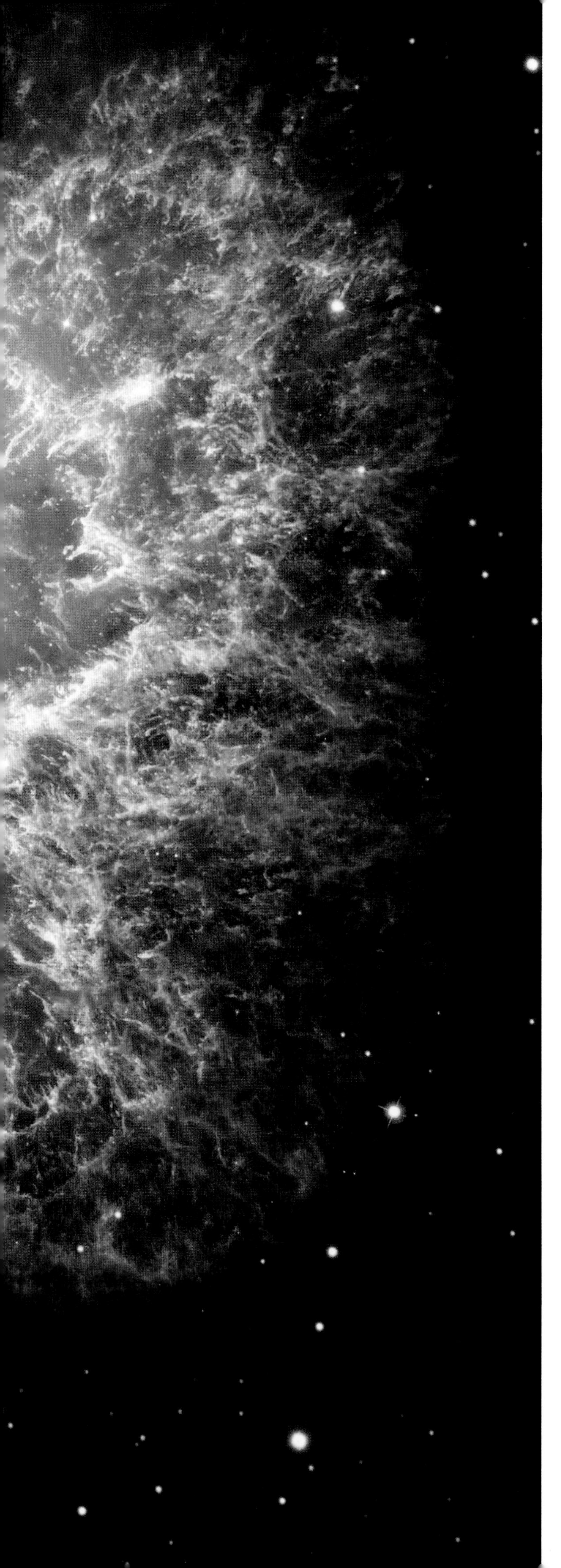

The Crab Nebula is the remains of a supernova whose light was first recorded by Japanese and Chinese astronomers in 1054. The remnants now form a gaseous cloud six light-years across. Its bluish glow is powered by the magnetic field of a rapidly spinning neutron star in the center, the collapsed core of the exploded star.

Sometimes the pulsar's rotation will speed up. Remembering the analogy of the ice-skater, we can conclude that the size of the neutron star must have decreased slightly. This is interpreted as a "starquake" — a rupturing and crunching of the star's crust.

In 1974 Russell Hulse and Joseph Taylor at Princeton University discovered a pulsar in orbit around another star. Using the precise timing of the pulses, they were able to document a slow decay in the pulsar's orbit. The energy loss turned out to be precisely that predicted by general relativity—basically, the system is giving off a type of radiation known as gravitational waves. (The direct discovery of these waves was made in 2016 and is discussed on pages 268–71. For this discovery, Hulse and Taylor were awarded the Nobel Prize in physics in 1993.)

When neutron stars are part of binary systems, in fact, they provide scientists with a fascinating natural laboratory. For instance, while studying a neutron star in orbit around a larger companion in the constellation Volans, astronomers were able to observe how light from the companion star was bent and redshifted through the extreme gravity of the neutron star's centimeter-high atmosphere.

Some astronomers have suggested using the precise rotation rate of pulsars to define a new time standard to improve on the current atomic clocks. Atomic clocks are accurate to "only" 13 decimal places, and pulsar timing is one system being proposed to push that limit to 15 decimal places. Given that time standards began with observations of heavenly bodies, there would be a certain philosophical rightness to returning those standards to the sky.

There is probably no object in the sky that has been taken up so wholeheartedly in fiction and common usage as the black hole. It is certainly the most exotic object we know. • The short definition of a black hole is that it is an object so massive, so compact, that nothing can escape the gravitational pull at its surface. Light that falls into it never comes out. We expect that stars that begin life with 30 times the mass of the sun and go through the supernova process will wind up as black holes. The gravitational force exerted by the core of such a star is so huge that it will overcome the attempts of the neutrons to oppose it.

BLACK HOLES

THE TIMELESS END OF THE BIGGEST STARS

FIRST PREDICTED: **1783, BY JOHN MICHELL**
FIRST MODERN THEORY: **1916, KARL SCHWARZCHILD**

MICRO BLACK HOLE SIZE: **UP TO ABOUT 0.1 MILLIMETER**
MICRO BLACK HOLE MASS: **UP TO ABOUT THE MASS OF THE MOON**
STELLAR BLACK HOLE SIZE: **ABOUT 30 KM (19 MI)**
STELLAR BLACK HOLE MASS: **ABOUT 10 SUNS**
INTERMEDIATE BLACK HOLE SIZE: **ABOUT 1,000 KM (600 MI)**
INTERMEDIATE BLACK HOLE MASS: **ABOUT 1,000 SUNS**
SUPERMASSIVE BLACK HOLE SIZE: **150,000–1.5 BILLION KM (93,200–932 MILLION MI)**
SUPERMASSIVE BLACK HOLE MASS: **100,000–A BILLION SUNS**

Gas jets out from the edges of a black hole in galaxy Centaurus A. (Inset) Illustration of matter sucked into a black hole.

Once these supermassive stars begin to collapse, the process takes a remarkable turn. Instead of forming a neutron star, the core of the dying star will continue its collapse until a stellar black hole is formed. There are other kinds of black holes, as we shall see, but before we get into that subject, let's consider what a black hole is in terms of Einstein's theory of relativity.

The best way to visualize the force of gravity as Einstein saw it is to picture a flexible plastic sheet marked off with a coordinate grid and stretched tightly over a frame. Roll a lightweight marble across the sheet and it will travel in a straight line. Now imagine putting a heavy object, like a bowling ball, on the sheet. The ball will drag down and distort the sheet; if you roll the marble again, its path will be warped around the slope created by the bowling ball. In Einstein's language, the mass of the bowling ball distorts the grid (he would say "space-time grid"), and what we interpret as the force of gravity is actually the result of this distortion.

Now imagine making the bowling ball heavier and heavier, pressing the distortion deeper and deeper into the sheet. Eventually, you might get to the point where the plastic just wraps itself around the ball and snaps off, isolating itself from the grid. In essence, you now have a black hole—a region of space that has cut itself off from the rest of the universe.

ACROSS THE EVENT HORIZON

German physicist Karl Schwarzchild predicted the existence of black holes in 1916, shortly after Einstein published his theory of relativity. For a long time, Schwarzchild's solution of the Einstein equations was regarded as an oddity—the duck-billed platypus of the astrophysical world. In fact, I can remember that when it was discussed in my general relativity class at Stanford in the 1960s, we were told that although black holes were possible in theory, they could never form in the real world. This was the prevailing orthodoxy for much of the 20th century.

What makes the Schwarzchild solution so strange is the existence of what is called the event horizon, or Schwarzchild radius. This marks a point of no return, a boundary in space that separates the interior of the black hole from the rest of the universe. Event horizons surround an incredibly small volume—you would have to pack all of the mass of an object like the sun into a sphere a little over a mile across to create one, for example. At the event horizon of a black hole, though, really strange things happen.

Here's an analogy that may help you understand the event horizon. Suppose you and a friend get into a couple canoes and start down a river. You communicate with people at your starting point by shouting regularly (for example, by sending sounds waves every minute by your clock) so that they can keep track of where you are. Suppose further that there is a waterfall downstream and that the speed of the water increases as you approach it. Finally, suppose that at some point the speed of the water exceeds the speed of sound. This will be a kind of event horizon. How will your trip downstream look to you and to your friends on the bank?

To them, your shouts will get farther and farther apart as you speed up. To be technical, they would see your clock (as measured by the arrival of your shouts) slowing down until, when you pass the event horizon, the shouts (and hence your clock) stop. You and your partner in the neighboring canoe, however, don't notice anything strange and can go on communicating normally. As far as you are concerned, nothing particular happens as you cross the event horizon.

In the same way, a distant observer watching an object falling into a black hole will see time on the object slow down and stop at the event horizon. An observer on the infalling object, however, sees no change in the clock that is traveling with her. Such is the nature of the event horizon.

FINDING BLACK HOLES

From our description of black holes, it's obvious that they cannot be detected in the normal way, by bouncing light off them, since any light sent into a black hole cannot, by definition, return. Black holes, however, do exert a gravitational force depending on their masses, and that can be used to find them.

In the 1980s scientists, observing the motion of objects in the gravitational field of an unseen body, found the first evidence for black holes in our universe. Researchers monitoring bodies near the center of our galaxy (the center being in the constellation Sagittarius) realized that objects there were in orbit around something extremely massive—what we call today a galactic black hole. The black hole at the center of the Milky Way has a mass several million times that of the sun.

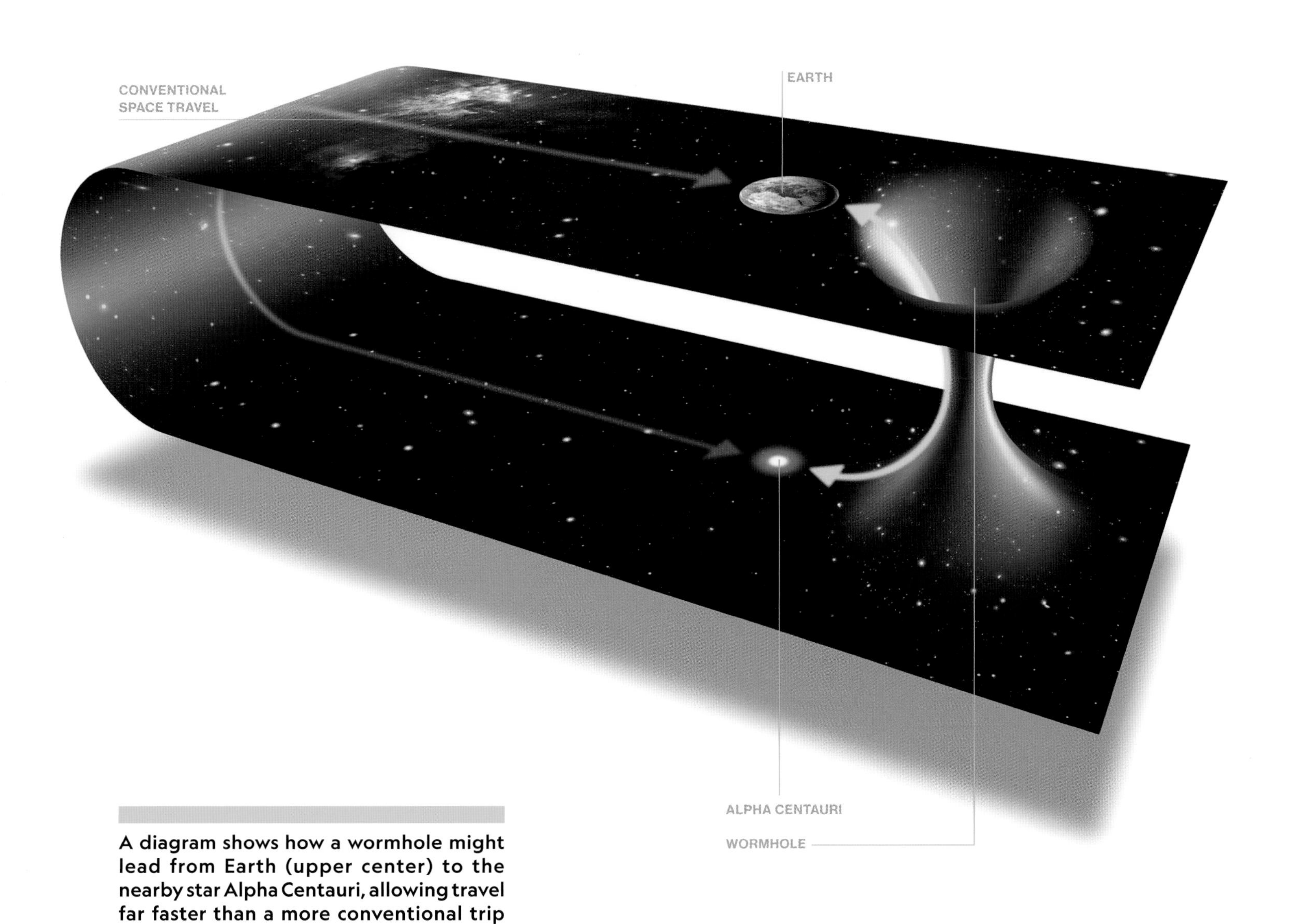

A diagram shows how a wormhole might lead from Earth (upper center) to the nearby star Alpha Centauri, allowing travel far faster than a more conventional trip (red arrow) at or below the speed of light. Wormholes, a theoretical possibility, are a favorite device in science fiction.

As impressive as this is, you have to remember that the galaxy contains many hundreds of billions of stars, so the black hole at the center represents a tiny fraction of the galaxy's mass. Astronomers think it is likely that almost all (and perhaps all) galaxies have black holes at their centers.

Another way to spot black holes is through radiation from their outskirts. Matter falling into a black hole tends to bunch up, forming what is called an accretion disk. Collisions in the disk heat it up, so that it emits energetic radiation. It is this effect that allows us to see the smaller black holes that result from the collapse of stars—the so-called stellar black holes. The best candidates we have for stellar black holes are double-star systems in which one of the partners has gone through its life cycle and evolved into a black hole. In this case, the black hole can pull material away from the other star and form an accretion disk, which would then heat up and emit radiation such as x-rays. The star Cygnus X-1, a strong x-ray source, is one of our best candidates for a stellar black hole.

The Cygnus X-1 system (below left) is found near star-forming regions of the Milky Way. It is thought that this double star contains a black hole of about 15 solar masses in orbit around a blue giant star (illustration, below right). Scientists believe the black hole pulls gas from its companion and ejects some of it in high-speed jets.

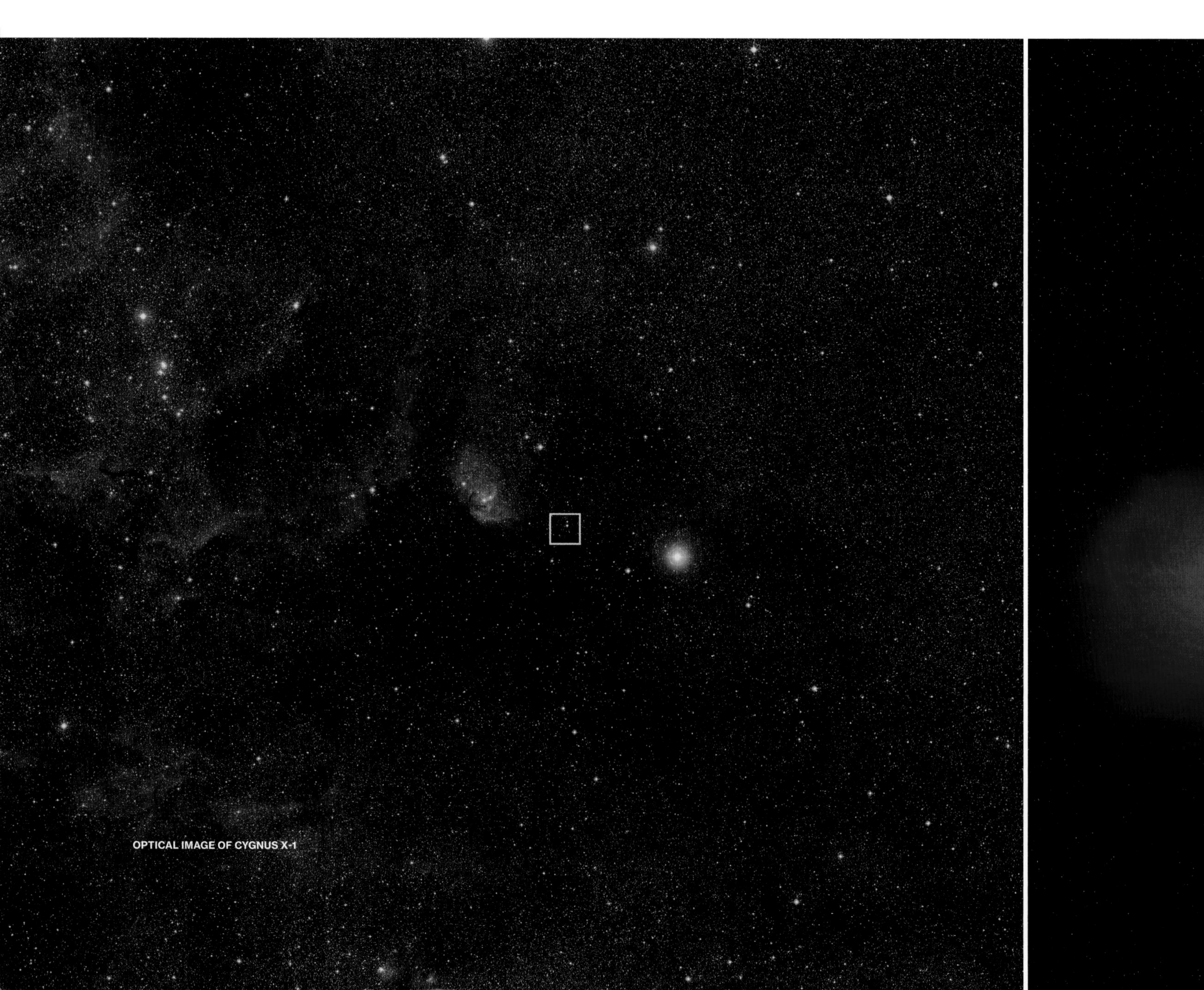
OPTICAL IMAGE OF CYGNUS X-1

INSIDE A BLACK HOLE

Up to this point we have talked only about observing black holes from the outside. We can have no direct knowledge of the interior of a black hole, for the simple reason that there is no way for information to get out. Our mathematical models, however, suggest that the interior may be very strange. The Schwarzchild solution, for example, predicts that at the center of the black hole may be a singularity—a place where the curvature of spacetime becomes infinite. The known laws of physics would break down at a singularity. In the case of electrically charged or rotating black holes, it is theoretically possible that someone entering a black hole at one point could exit in a different spacetime continuum. Such a path, beloved of science-fiction writers, is called a wormhole. The same models suggest that moving through a black hole could produce time travel.

Before you get too intrigued by these exotic results, though, you have to realize that the gravitational fields near a black hole are so strong that the difference in the force between your head and feet would be enough to stretch you out and tear you apart—a process astrophysicists call "spaghettification." Nothing, not even the sturdiest starship imaginable, would survive a trip through a singularity. Best to avoid black holes if you can!

ILLUSTRATION OF CYGNUS X-1 BLACK HOLE, CLOSE UP

On September 14, 2016, astronomers opened a new window on the universe. An enormous machine called LIGO (Laser Interferometer Gravitational Observatory) registered a gravitational wave passing by the Earth. These waves, which you can think of as ripples in the spacetime continuum, were predicted by Albert Einstein in 1916. He predicted that whenever immense objects moved, these waves would be created. The first wave detected resulted from the collision between two massive black holes in a galaxy billions of light-years away.

GRAVITATIONAL WAVES

A NEW WINDOW ON THE UNIVERSE

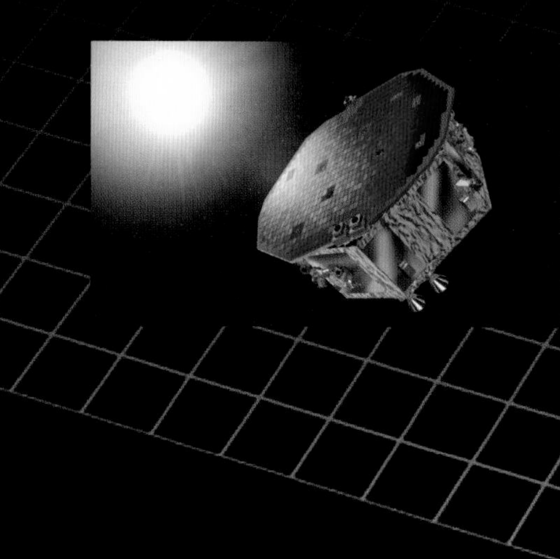

FIRST PREDICTED BY: **ALBERT EINSTEIN, 1916**
FIRST DETECTED BY: **LIGO, SEPTEMBER 14, 2016**

FIRST DETECTION EVENT: **COLLISION OF TWO MASSIVE BLACK HOLES**
ADDITIONAL DETECTIONS: **AT LEAST TWO MORE**
FIRST PHASE OF SPACE-BASED DETECTION: **ELISA PATHFINDER, 2016**

A drawing illustrating general relativity. A mass (the yellow ball) distorts the fabric of spacetime (the grid), deflecting the path of another object (the red ball). (Inset) Artist's depiction of an eLisa satellite, designed for space-based detection.

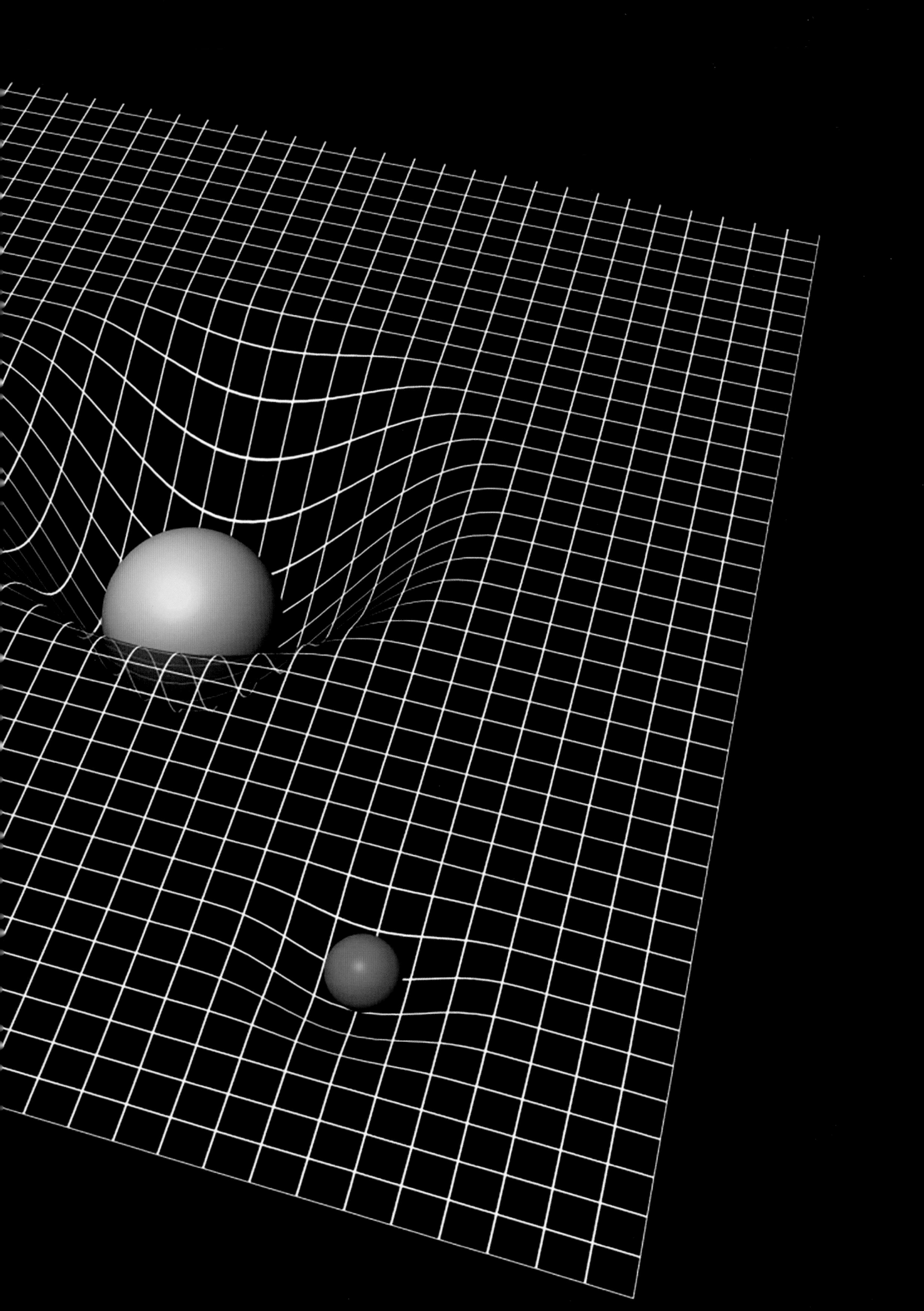

In 1916 a rising young theoretical physicist named Albert Einstein presented a paper to a group of scientists in Berlin. The work, which outlined the general theory of relativity, was formally published in 1917, and the theory quickly became—and remains—our best theory of gravity.

You can get a sense of how the theory works by thinking about its two central precepts:

- The presence of matter warps the fabric of space and time, and
- Objects travel on the shortest possible paths in the warped space.

An easy way to visualize the first precept is to imagine that you have a stretched-out sheet of rubber marked in a standard rectangular grid. Now imagine dropping a heavy object like a bowling ball on the sheet. The sheet will deform ("warp") because of the presence of the massive bowling ball (the first precept, and objects traveling on the sheet (imagine rolling a marble from one side to the other) will be deflected if they travel near the ball (the second precept).

The actual working out of general relativity involves complex mathematics, of course, but, as any scientific theory should, it makes predictions that can be checked against experiment and observation. For example, it predicts that light rays passing near the sun will be deflected by the warping discussed herein. It was the confirmation of this prediction by British astronomer Arthur Eddington during a solar eclipse in 1919 that catapulted Einstein to international fame. (The *New York Times* referred to him as "the suddenly famous Dr. Einstein"). Over the course of the 20th century, one prediction of the theory after another was confirmed by a series of high-precision experiments. One of those predictions, however, remained unverified, and that was the prediction involving a phenomenon known as "gravitational waves."

The easiest way to visualize a gravitational wave is to go back to our example of the bowling ball on a rubber sheet and imagine grabbing the bowling ball and pulling it up and down. If we do this, then it's not hard to see that ripples will travel outward on the sheet. These are gravitational waves, rippling distortions in the fabric of space and time.

There are two problems we have to deal with if we want to detect these waves. One of them is that the predicted distortions are incredibly small. The second problem, related to the first, is that gravitational wave detectors have to be so sensitive that they will pick up all manner of extraneous signals. A car going by on a nearby street or even wind blowing against a building could easily generate vibrations that might mask the readings generated by a real gravitational wave.

ENTER LIGO

LIGO stands for "Laser Interferometer Gravitational Wave Observatory." This incredibly complex apparatus is one of the most ambitious science projects ever built. For the reasons outlined above, there are actually two detectors, one in Washington state and the other in Louisiana, the idea being that only events registered by both detectors will be accepted. Each apparatus consists of a pair of tubes at right angles to each other and 4 kilometers (~3 mi) long—think of them as forming a large L. The tubes are kept at a high vacuum, and at the end of each is a mirror. A laser beam is brought into the place where the tubes meet and split into two beams, each of which travels down one arm of the L. At the end of the arm, each beam is reflected back by the mirror, and the returning waves are recombined at the central point. In the jargon of physics, the two laser beams "interfere" and produce a very sensitive method of detecting small changes in the positions of the mirrors.

The LIGO apparatus in Livingston, Louisiana. After the first detection of a gravitational wave in 2016, many more waves have been detected. The first detection was recognized by a Nobel prize in 2017.

LIGO first went into operation in 2002, and between then and 2010, extensive engineering and design work were carried out. The apparatus was then shut down for a major upgrade and brought back on line in 2015 (it was then called "Advanced LIGO"). On September 14, 2016, the first gravitational wave was detected, followed by another detection on December 26, 2016. Both of these waves resulted from the coalescing of two black holes. The actual motion of the mirrors during these events amounted to a distance about one-thousandth of the size of a proton.

We can understand the connection between LIGO and black holes by noting that general relativity predicts that the biggest (and most easily detected) gravitational waves will be generated by the motion of the biggest masses. In a double-star system in which both members have evolved into black holes, the two black holes can spiral in toward each other, finally uniting as a cataclysmic explosion that creates a single massive black hole.

There are two important consequences of the success of LIGO. First, it is a major confirmation of general relativity. Second, it opens a new window on the universe, giving us access to some of the most violent events imaginable. We can only guess about the sorts of things we will learn from this newfound ability.

Although several countries have built and operated LIGO-like detectors, the real future of gravitational wave detection lies with a project on the drawing boards at the European Space Agency. Called eLISA (evolved Laser Interferometer Space Antenna), it would involve three satellites orbiting the sun and occupying the vertices of a triangle whose sides are millions of miles long. Scientists believe that eLISA would be capable of detecting the waves generated by the big bang itself.

One of the most startling discoveries of the late 20th century was that ordinary matter, the stuff we're made of, is just a small part of our universe. We'll come back to this subject when we discuss the expansion of the universe, but here we will talk about the discovery of dark matter, another insight into our own insignificance. • To do this, we'll have to think about how galaxies rotate. Our galaxy, for instance, rotates like a giant pinwheel over periods of hundreds of millions of years, with our sun making a grand circuit every 220 million years or so.

DARK

One way that astronomers study the structure of galaxies is to look at the details of galactic rotation. The main tool in this study is something called a rotation curve, in which astronomers plot how fast a star is moving as a function of how far away from the galactic center it lies.

To understand a rotation curve, we consider a simple example of a rotating object—a merry-go-round. If you are standing near the inside of the platform, you will be moving fairly slowly, but as you move outward to the periphery, you will be whirling faster and faster. The rotation curve for this situation shows the speed increasing steadily as you get farther from the center. In the jargon of astronomers, this kind of rotation is called wheel flow, since it is characteristic of any solid rotating object.

In the central regions of a galaxy like the Milky Way, you would find something similar. Stars in the densely populated galactic center, locked together by gravity, exhibit wheel flow. But as you move outward, at some point wheel flow stops. Beyond this point, we find that all the stars are moving at the same speed, regardless of how far they are from the center. They are like runners who have to stay in their lanes on a curving track. Those on the outside, because they have farther to go and yet must move at about the same speed as those on the inside, will start to fall behind. This is why the spiral arms of our galaxy bend as they do.

What happens to the galactic rotation curve when we move still farther out? Here's a little thought experiment: Imagine getting so far away from the galaxy that the whole pinwheel shrinks down to a single faint point of light in the distance. The galaxy will still exert a gravitational force on you, so you will still be orbiting that distant point. Your situation, though, will be analogous to the planets circling the sun in our solar system. Just as Jupiter moves more slowly in its orbit than Mars, once you are this far away from a galaxy, you expect the speed of the stars and dust clouds out there to get slower the farther away they are. This is called Kepler rotation, after Johannes Kepler (1571–1630), who discovered the law for planets in the solar system.

A GRAVITATIONAL MYSTERY

As astronomers traced the rotation curve into the far outskirts of other galaxies, they fully expected to see that downturn. The problem was that they didn't. In the early 1970s, Vera Rubin, then a young astronomer at the Carnegie Institution of Washington, began using advanced imaging instruments to measure the rotation curves of galaxies. Starting with the nearby Andromeda galaxy, she was amazed to observe that the curve stayed flat out to the limits of what she could measure—the stars kept moving at the same speed no matter how far they were from the galactic center. As she extended her search outward, galaxy after galaxy gave the same result, and by 1978 astronomers realized

that their expectations about the rotation of galaxies were simply wrong.

In fact, scientists quickly realized that the only way to explain the observed rotation of the galaxies was to say that the visible part of a galaxy—the stars and dust clouds we've been exploring—was enclosed in a giant sphere of matter that we cannot see, but whose effects we can observe. The term "dark matter" was quickly applied to this new material. Whatever it is, it does not emit or absorb light or other electromagnetic radiation, nor does it have any other kind of interaction with ordinary matter except for gravity. This means that although we can detect it by observing its gravitational effect on luminous matter (for example, the effect it has on the rotation curve), we cannot see it directly. Furthermore, calculations showed that for a galaxy like the Milky Way,

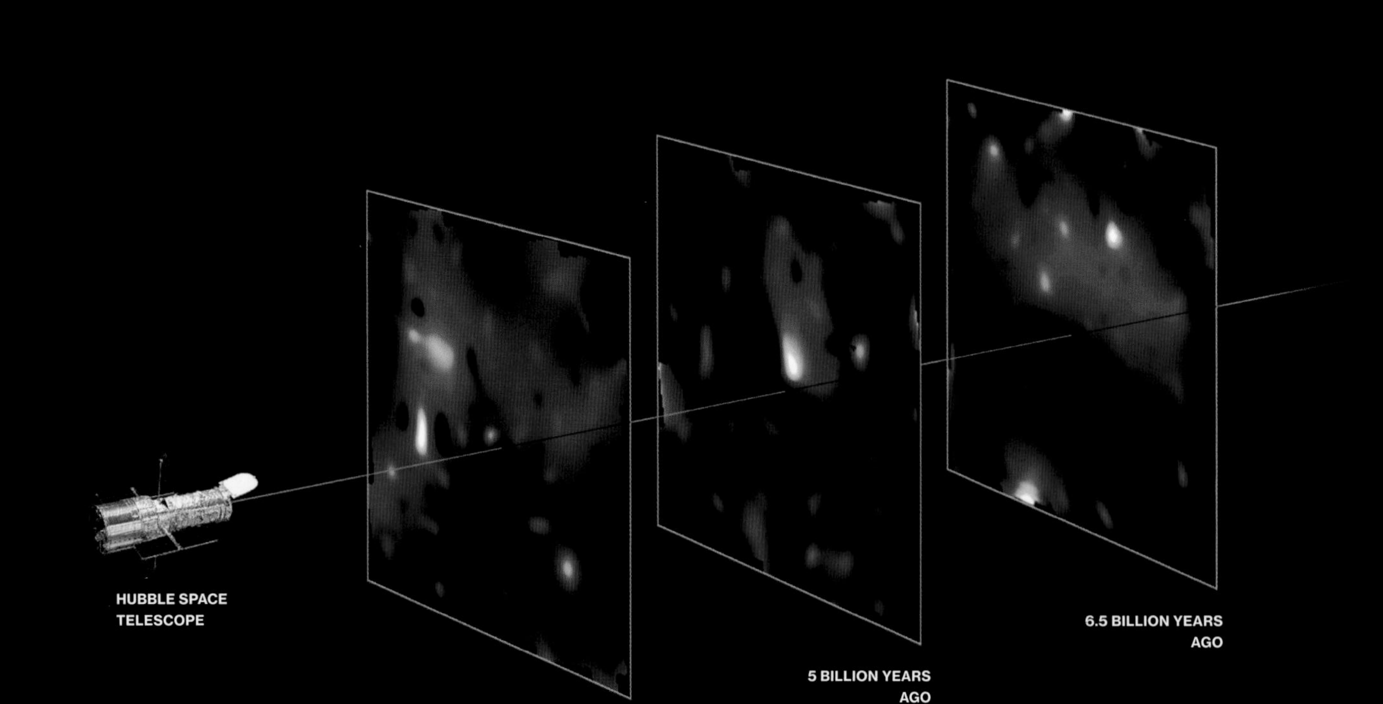

more than 90 percent of the galaxy's total mass had to be in this same new (and unexpected) form.

Once the idea of dark matter surfaced, evidence for its existence in other venues showed up quickly. For example, there are stellar clusters in which individual stars are moving too fast to be held in by the gravitational attraction of the other stars. In these cases, some extra gravitational oomph is needed, and that, of course, is what dark matter supplies. As we shall see on page 321 today astronomers believe that dark matter makes up 23 percent of the mass of the universe. (For reference, luminous matter like stars makes up a bit less than 5 percent.)

So if there's so much of this stuff around, what is it? Theorists have not been slow in producing hypothetical answers to this question. The most popular candidates for the missing matter are as-yet-undetected objects known as weakly interacting massive particles, or WIMPs. The hypotheses are interesting, but of course the only way we'll know for certain what dark matter is made of is to have someone identify the stuff in a laboratory.

SEARCHING FOR DARK MATTER

A number of dark matter searches are going on around the world right now. The fact that this new form of matter permeates the galaxy means that the particles of dark matter, whatever they are, have to be passing through us all the time, leaving no record of their passage (remember, they don't interact much with ordinary atoms). In fact, since Earth is constantly moving around the sun, there must be a "dark matter wind" blowing past us constantly, much as one can feel an apparent wind when a car drives through still air.

A three-dimensional map of dark matter in the universe. Since looking at objects a great distance away is equivalent to looking back in time, we can trace out the evolution of this distribution over billions of years.

Attempts to detect dark matter involve trying to see the rare events in which the dark matter wind jostles an atom or two in a detector. Many processes, particularly collisions with cosmic rays, can jostle atoms and mask the very faint dark matter signal. For this reason, dark matter searches tend to be located deep underground in mines or tunnels where the overlying rock provides a shield against this sort of interference. The most sensitive dark matter search is being carried out in an old gold mine in the Black Hills of South Dakota, in a chamber previously used for the detection of solar neutrinos (see page 222). The chamber is in what used to be the Homestake Gold Mine but is now known as the Sanford Underground Research Facility (after banker T. Denny Sanford, who contributed a sizeable sum to create the laboratory). The core of the experiment is a phone booth–sized container filled with liquid xenon. This so-called LUX detector (for Large Underground xenon) is the place where particles in the dark matter wind are supposed to interact occasionally with the nuclei of xenon atoms. When this happens, the xenon will produce a burst of light and an electron. Through a series of complex interactions, the electron will eventually produce a second burst of light. The signal for a dark matter event, then, will be those two bursts in a specific sequence.

The original search, which was run with 118 kilograms (260 lb) of xenon, ended in 2013 with no evidence for any dark matter interactions. An extended run with 368 kilograms (1,781 lb) ended in 2016 with the same result. These results call into question, but do not eliminate, the possibility that dark matter is composed of WIMPs. At the moment, theorists are busy proposing new and exotic candidates, but the best we can say right now is that although we know that dark matter exists, we do not yet know what it is.

HELIX NEBULA

An infrared image of the Helix Nebula, taken by the Spitzer Space Telescope. Located some 700 light-years away, the planetary nebula is the gaseous remains of a dying star, seen as a bright white dwarf in the center of the image.

Our current picture of the universe was born outside of Los Angeles at a newly built telescope atop Mount Wilson. It was the brainchild of one man: Edwin Hubble. Working at the new instrument in the late 1920s, Hubble established that matter in the universe was organized into galaxies like the Milky Way: other "island universes" outside our own galaxy. More important, he showed that those other galaxies were moving away from us—that the whole universe was expanding.

Given this fact, it's not hard to imagine running the film backward to a time when the entire universe was compacted into a single, unbelievably hot, dense point. This scenario, in which the universe began at a specific time in the past and has been expanding and cooling ever since, is known as the

THE UN

big bang. Confronted with this idea, three questions come to mind: Is this theory correct? How did the big bang begin? How will the universe end?

In this section we look at these questions, starting with a phenomenon known as the cosmic microwave background, which provides arguably the best evidence for the big bang. The second question will get us into one of the most exciting realizations in science—the idea that to study the largest thing we know about, the universe, we have to study the smallest things we know about, the elementary particles that constitute all of matter.

In recent years, we have come to realize that most of the mass of the universe is in the form of a mysterious substance known as dark energy, and the fate of the universe depends on what that dark energy is. At the moment, we have no idea.

IVERSE

I THE UNIVERSE

CARTOGRAPHER'S NOTE: From bottom right, our solar system is part of a stellar neighborhood, itself part of the Milky Way galaxy (bottom left), which belongs to the Local Group of galaxies (upper left). These, in turn, belong to a galactic supercluster (upper right). The largest structures known, the filaments and walls of superclusters are believed to vein the entire universe.

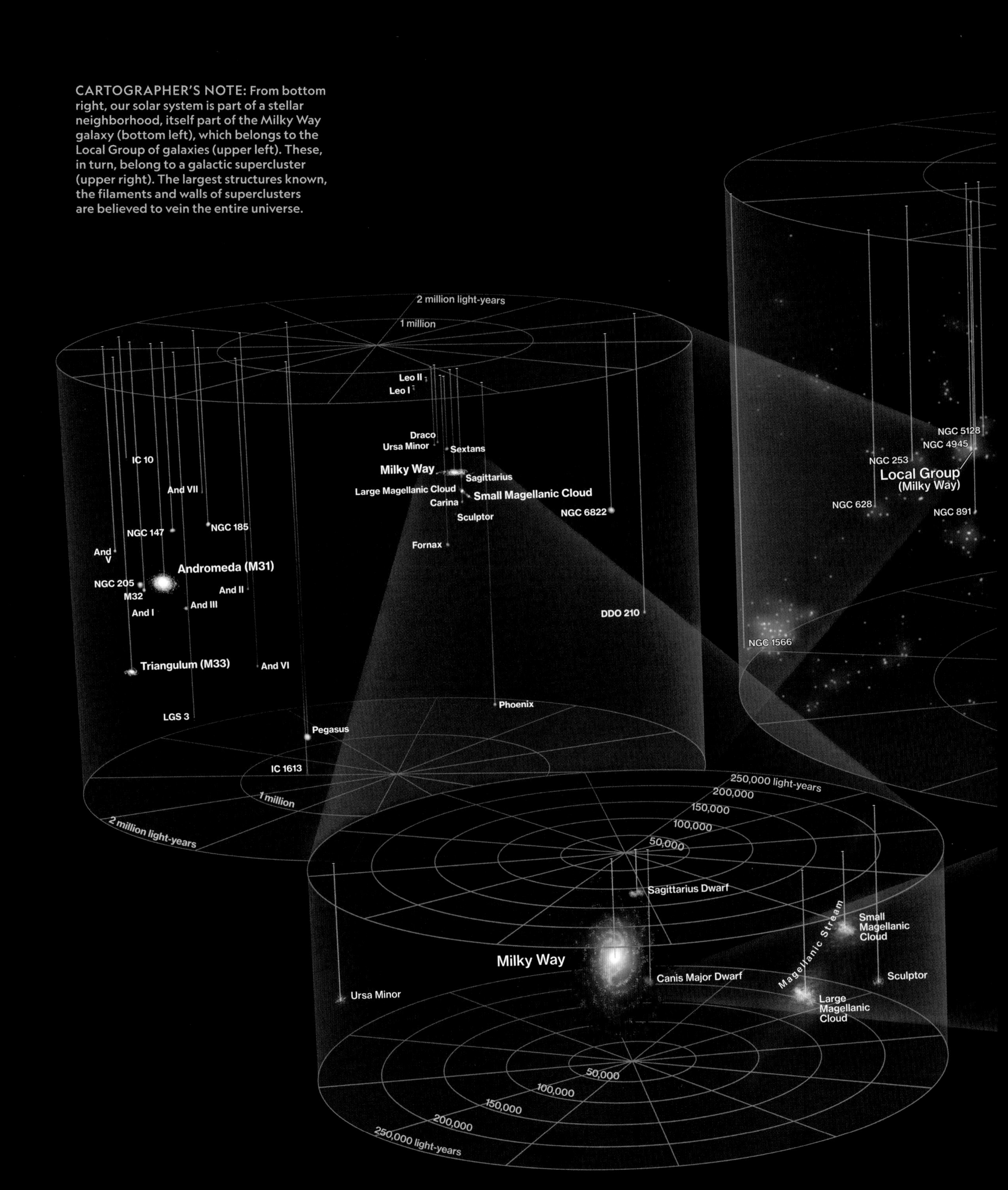

Our own solar system and its galaxy are just a small piece in the hierarchical structure of the universe. Gravity binds together galaxies as well as massive galaxy clusters, which can contain thousands of component galaxies. More than 100 billion galaxies altogether are gathered in these clusters throughout the universe—all of them flying apart as the universe is expanding.

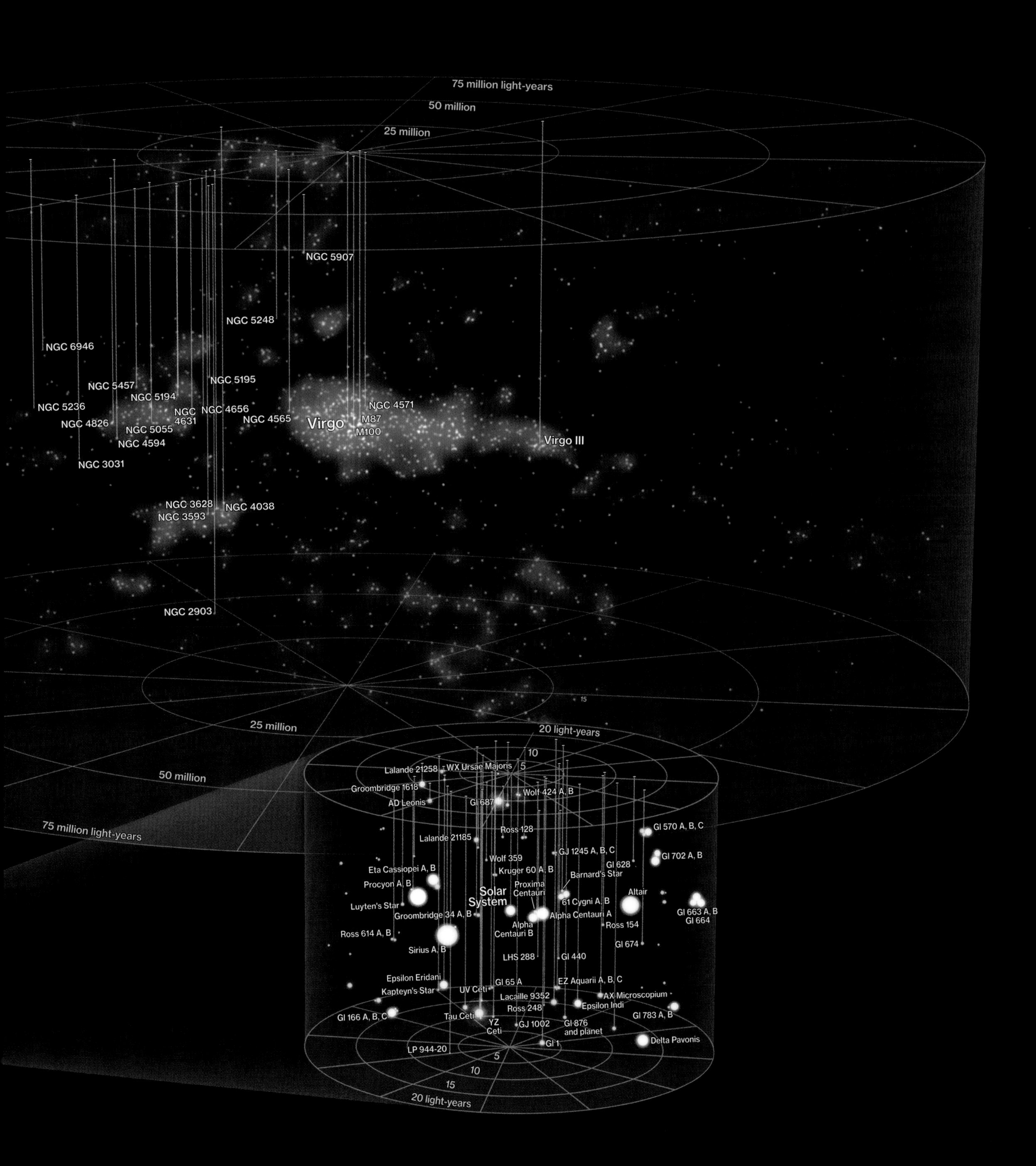

We are so used to thinking of our sun as one star among billions in the Milky Way galaxy, and of the Milky Way as one among billions of galaxies in the universe, that it is likely that few readers even noticed when these notions were introduced without explanation in the last section. Yet there was a time when the notion that matter in the universe was bunched up into what we call galaxies was hotly debated. • There are many possible ways a universe might be organized. Matter could be scattered randomly throughout space; one central clump of matter might be surrounded by emptiness; or matter could clump into galaxies, with the galaxies themselves scattered about randomly.

THE GREAT GALAXY DEBATE

EDWIN HUBBLE & ISLAND UNIVERSES

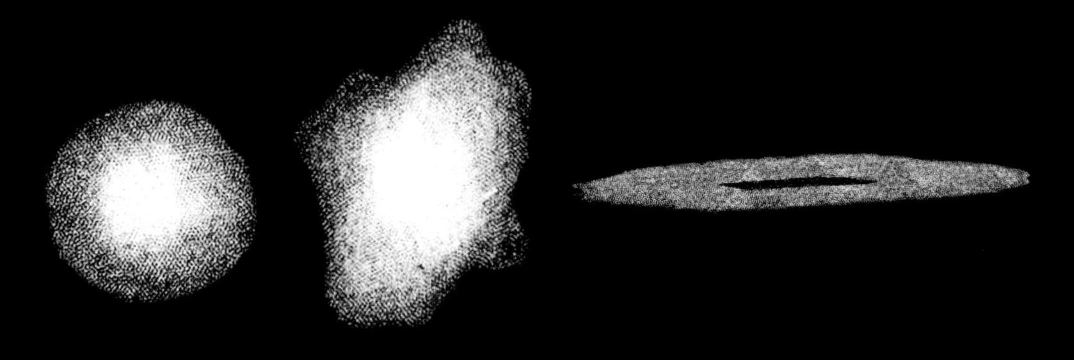

MILKY WAY IS FIERY VAPOR: **GREEK PHILOSOPHERS, 400S B.C.**
MILKY WAY IS MADE OF STARS: **GALILEO GALILEI, 1610**
SUN IS NOT CENTER OF MILKY WAY: **HARLOW SHAPLEY, 1918**
GALAXIES ARE MADE OF STARS: **EDWIN HUBBLE, 1920S**
GALAXIES ARE RECEDING: **EDWIN HUBBLE, 1929**
RADIO NOISE DETECTED FROM CENTER OF MILKY WAY: **KARL JANSKY, 1932**
"TUNING FORK" GALAXY CLASSIFICATION: **EDWIN HUBBLE, 1936**
DISCOVERY OF RADIO SOURCE CYGNUS A: **GROTE REBER, 1939**
MILKY WAY'S SPIRAL ARMS DETECTED IN RADIO: **1952**
DARK MATTER IN GALAXIES: **VERA RUBIN, 1970S**
HUBBLE SPACE TELESCOPE SURVEYS DEEP FIELD: **1990S**

Three galaxies known as ARP 274. (Inset) Three nebulae drawn by astronomer William Herschel.

Understanding what different imagined universes might be like is the job of theoretical astrophysicists. On the other hand, we can live in only one of those possible universes, and finding out which one we actually inhabit is the job of observational astronomers.

What I am calling the Great Galaxy Debate was triggered by the presence in the sky of objects called nebulae. Nebula means "cloud" in Latin, and the name comes from the fact that when nebulae were first observed by early astronomers, they looked like smeared patches of light—luminous clouds. The question was simple: Were nebulae just clouds of luminous stuff inside the Milky Way, or were they other "island universes," far beyond our own galaxy? To answer this question, astronomers needed two things: They needed a telescope with a resolution that would allow them to see individual stars within the nebulae, and they needed a way of measuring the distance to those stars.

By the early 20th century, the distance measurement problem had been solved by Henrietta Leavitt (see pages 216–17). Also, although it was far less obvious, the telescope problem was on its way to being solved as well.

A 100-INCH TELESCOPE

To understand the events that led to our current picture of the universe, we have to go back to the middle of the 19th century and meet one of the most remarkable men ever to appear on the American stage—Andrew Carnegie. Coming to America as a 10-year-old, he began his career delivering telegrams in Pittsburgh and wound up being one of the wealthiest men in America. Among other things, for example, he founded the company that eventually became United States Steel. Then, in what was an astonishing turn for a 19th-century robber baron, he wrote an essay known as "The Gospel of Wealth" in which he put forward the idea that once a man had acquired wealth, it was his duty to see that it was spent on solving important social problems. "The man who dies rich," he said, "dies disgraced." You have only to look at charitable trusts like the Bill and Melinda Gates Foundation to see the Carnegie legacy in operation today.

Andrew Carnegie founded the Carnegie Institution of Washington, which is dedicated to scientific research. One project in which he took a great personal interest was the construction of a major astronomical observatory on Mount Wilson, near Los Angeles. In the early years of the 20th century, this observatory housed the largest telescopes in the world. In 1919 Edwin Hubble (see sidebar opposite) joined the staff at Mount Wilson and proceeded to revolutionize our picture of the universe.

Using the observatory's magnificent new instrument, whose 100-inch mirror captured an unprecedented amount of light, Hubble began a systematic study of nebulae. The first thing he did was set up a classification scheme based on their appearance—a scheme still used today. More important, though, the new telescope could pick out individual Cepheid variable stars in nearby nebulae, which meant that Hubble could use Leavitt's standard candle method to find out how far away those nebulae were. The distances he measured turned out to be millions of light-years—much too great to have the nebulae included as parts of the Milky Way. By 1925 Hubble had established that we live in a universe in which matter is organized into galaxies.

Had this been the extent of Hubble's contribution, it would have ensured that his name would be in all the history books. There was another feature of his work, however, that led to the big bang picture of the universe—the picture that continues to dominate modern cosmology. To understand how he came to his theory, we have to make a slight diversion to discuss a phenomenon known as the Doppler effect.

REDSHIFT

You've had firsthand experience of the Doppler effect if you've ever listened to a car horn as the car drove past you on the highway. You probably noticed that the pitch of the horn dropped as the car went by. When the car is standing still, the crests of the sound wave move out uniformly in all directions, and everyone hears the same pitch. If the car is moving, however, the center for each outgoing wave is at the spot where the car was when that particular crest was emitted. This means that someone standing in front of the car will see the crests bunched up (that is, hear a higher pitch), while someone standing in back will see the crests stretched out (that is, hear a lower pitch). This shift in pitch is the Doppler effect, and it will be seen for any type of wave emitted by a moving source.

Astronomers before Hubble had noted that the light from the stars in nebulae was shifted to the longer wavelength (red) end of the spectrum, indicating that the crests were being stretched out. This means that

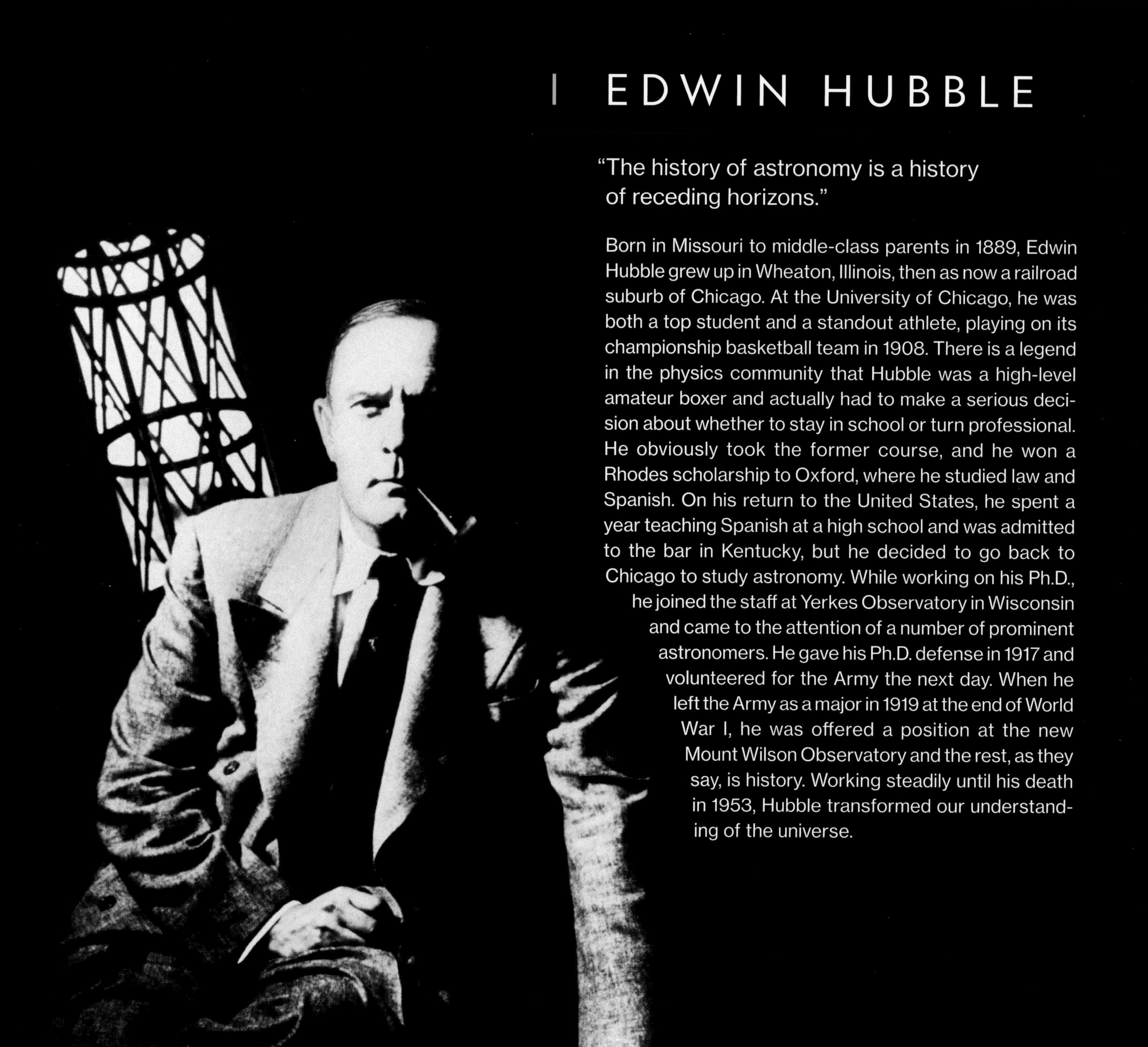

EDWIN HUBBLE

"The history of astronomy is a history of receding horizons."

Born in Missouri to middle-class parents in 1889, Edwin Hubble grew up in Wheaton, Illinois, then as now a railroad suburb of Chicago. At the University of Chicago, he was both a top student and a standout athlete, playing on its championship basketball team in 1908. There is a legend in the physics community that Hubble was a high-level amateur boxer and actually had to make a serious decision about whether to stay in school or turn professional. He obviously took the former course, and he won a Rhodes scholarship to Oxford, where he studied law and Spanish. On his return to the United States, he spent a year teaching Spanish at a high school and was admitted to the bar in Kentucky, but he decided to go back to Chicago to study astronomy. While working on his Ph.D., he joined the staff at Yerkes Observatory in Wisconsin and came to the attention of a number of prominent astronomers. He gave his Ph.D. defense in 1917 and volunteered for the Army the next day. When he left the Army as a major in 1919 at the end of World War I, he was offered a position at the new Mount Wilson Observatory and the rest, as they say, is history. Working steadily until his death in 1953, Hubble transformed our understanding of the universe.

DISTANT GALAXIES LOOK

REDDER

AS THEY RECEDE FROM US

the nebulae in question are moving away from us. Those astronomers didn't know how far away the various nebulae were, however, so they couldn't see any systematic relationships among the observed redshifts. Once Hubble figured out the distances, he was able to note that the larger the redshift (that is, the faster the galaxy was moving away from us), the farther away it was. This is usually expressed in an equation known as Hubble's law:

$$V = H \times D,$$

where V is the speed of the receding galaxy, D the distance to it, and H a number known as Hubble's constant. This equation tells us that if we look at two galaxies and one is twice as far away from us as the other, then the farther galaxy will be moving away from us twice as fast as the closer galaxy.

We'll discuss some of the astonishing consequences of Hubble's discovery in the next section.

The Doppler effect causes light from objects that are receding from us, such as galaxies, to be shifted to longer wavelengths (redshift, middle illustration). Light from approaching objects is shifted to shorter wavelengths (blueshift, bottom illustration).

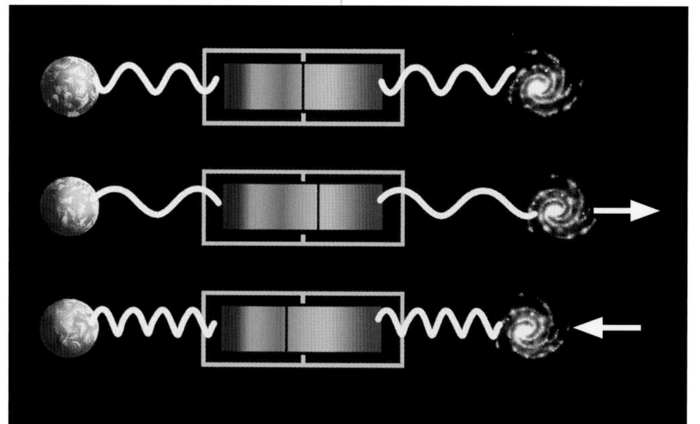

An artist's conception of the cosmic redshift depicts nearby galaxies in white light, but more distant galaxies in progressively redder light as they get farther away (and move more rapidly away from us).

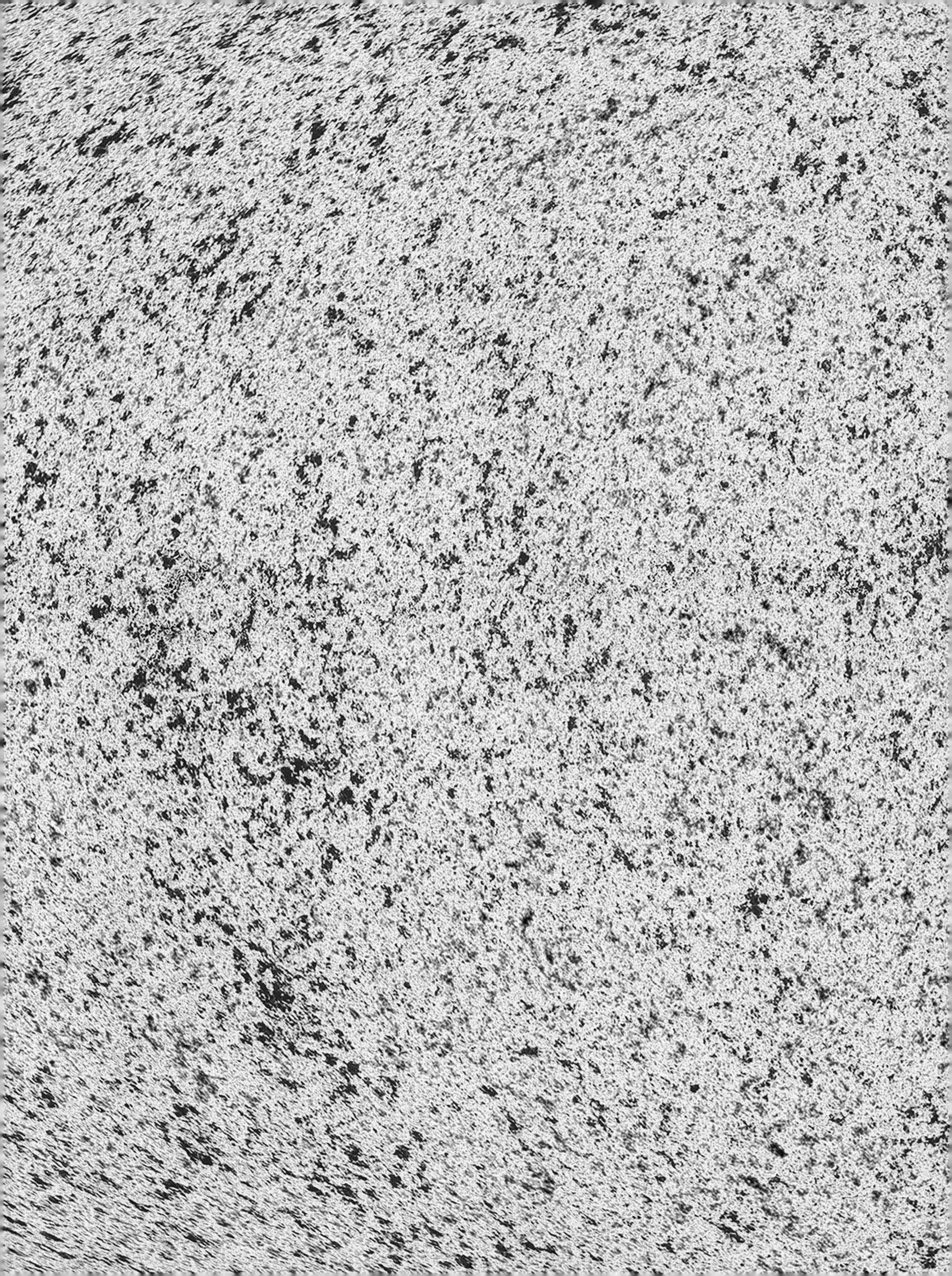

Sometimes the most important discoveries in science happen by accident. The discovery of the cosmic microwave background falls into this category. In the early 1960s transcontinental television transmissions were just becoming possible. The technologies were primitive by today's standards—someone who wanted to acquire a TV signal had to point a microwave receiver at the sky. And this raised the question of interference: Was there anything else out there that could also be sending microwaves into the receiver? • In 1964 two scientists at Bell Labs in New Jersey, Arno Penzias and Robert Wilson, began a survey of the microwave sky to resolve this issue. Using an old receiver to scan the sky, they systematically recorded the background microwave radiation that might interfere with TV reception.

COSMIC MICROWAVE BACKGROUND

MESSAGE FROM THE DAWN OF TIME

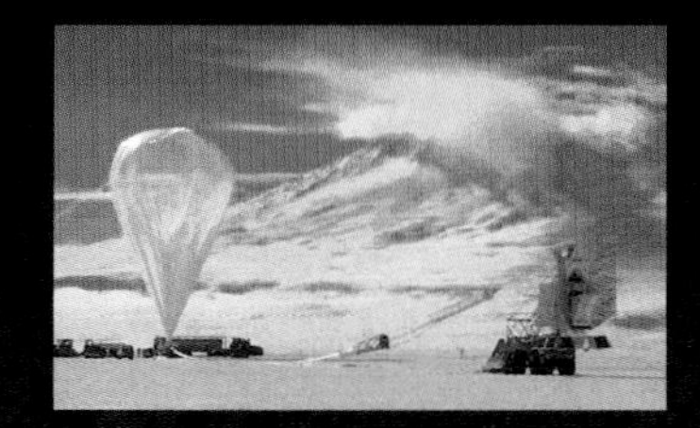

FIRST PREDICTED: **1948, BY RALPH ALPHER AND ROBERT HERMAN**
DISCOVERED BY: **ARNO PENZIAS AND ROBERT WILSON, 1964**
1989: **COSMIC BACKGROUND EXPLORER (COBE) PROBE LAUNCHED**
2001: **WILKINSON MICROWAVE ANISOTROPY PROBE (WMAP) LAUNCHED**
2009: **PLANCK SATELLITE LAUNCHED**
HOW DISCOVERED: **MICROWAVE RADIO TELESCOPE**

COSMIC BACKGROUND TEMPERATURE: **2.725 KELVINS (−270.425°C/−454.765°F)**
VARIATIONS IN COSMIC BACKGROUND: **1 PART IN 14,000**
BACKGROUND SHOWS UNIVERSE AT AGE: **380,000 YEARS**
CURRENT AGE OF UNIVERSE: **13.7 BILLION YEARS**
MILKY WAY'S VELOCITY RELATIVE TO BACKGROUND: **627 KM/SEC (390 MI/SEC)**

Detail of microwave map of early universe. (Inset) Launch of the BOOMERANG telescope, which measured the microwave sky.

I COSMIC MICROWAVE BACKGROUND

A map of background radiation across the sky reveals tiny differences in the temperatures and density of the early universe.

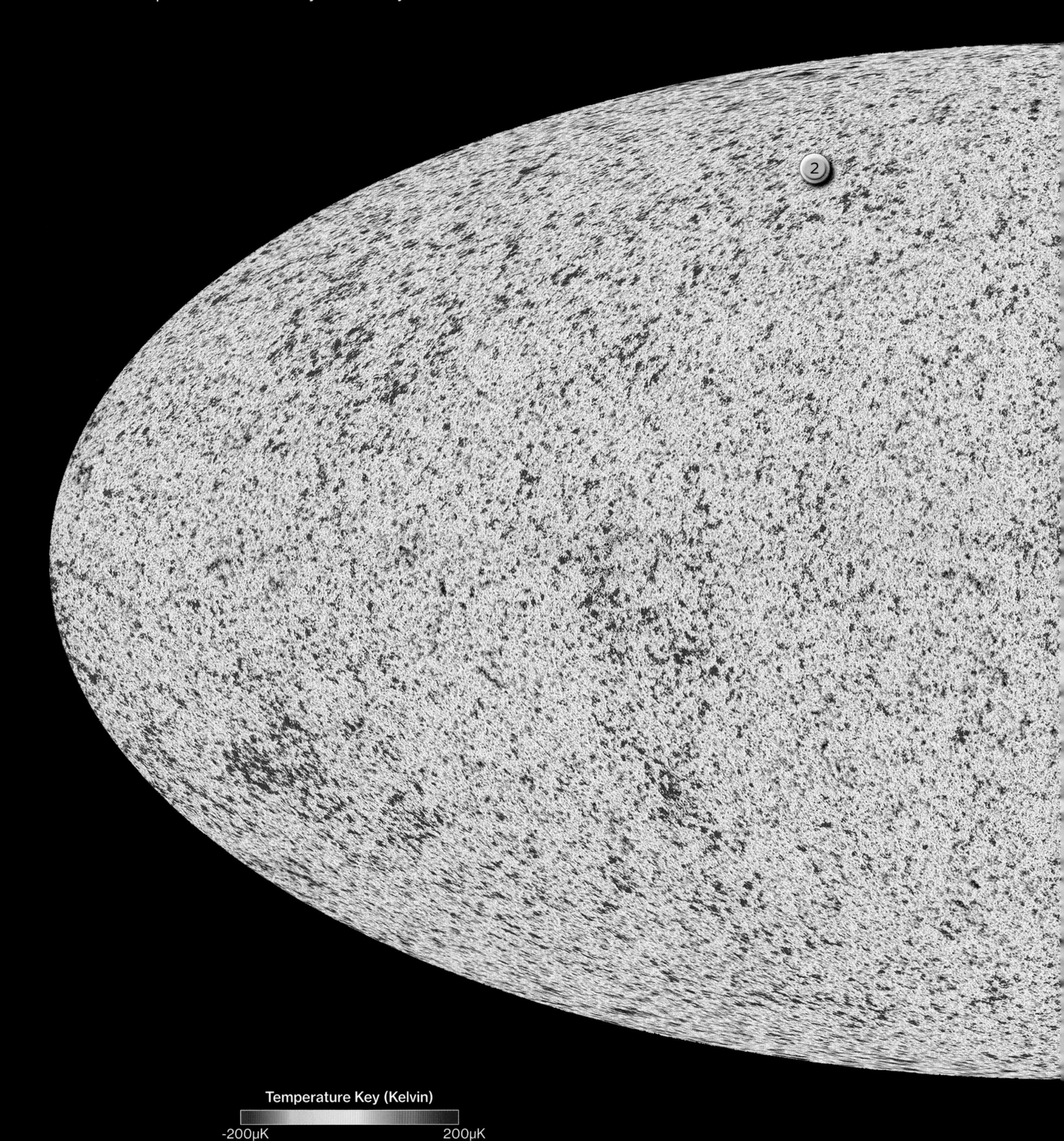

Temperature Key (Kelvin)

-200μK 200μK

TEMPERATURE REGIONS

1 WARM AREA

2 MID RANGE

3 COOLEST AREA

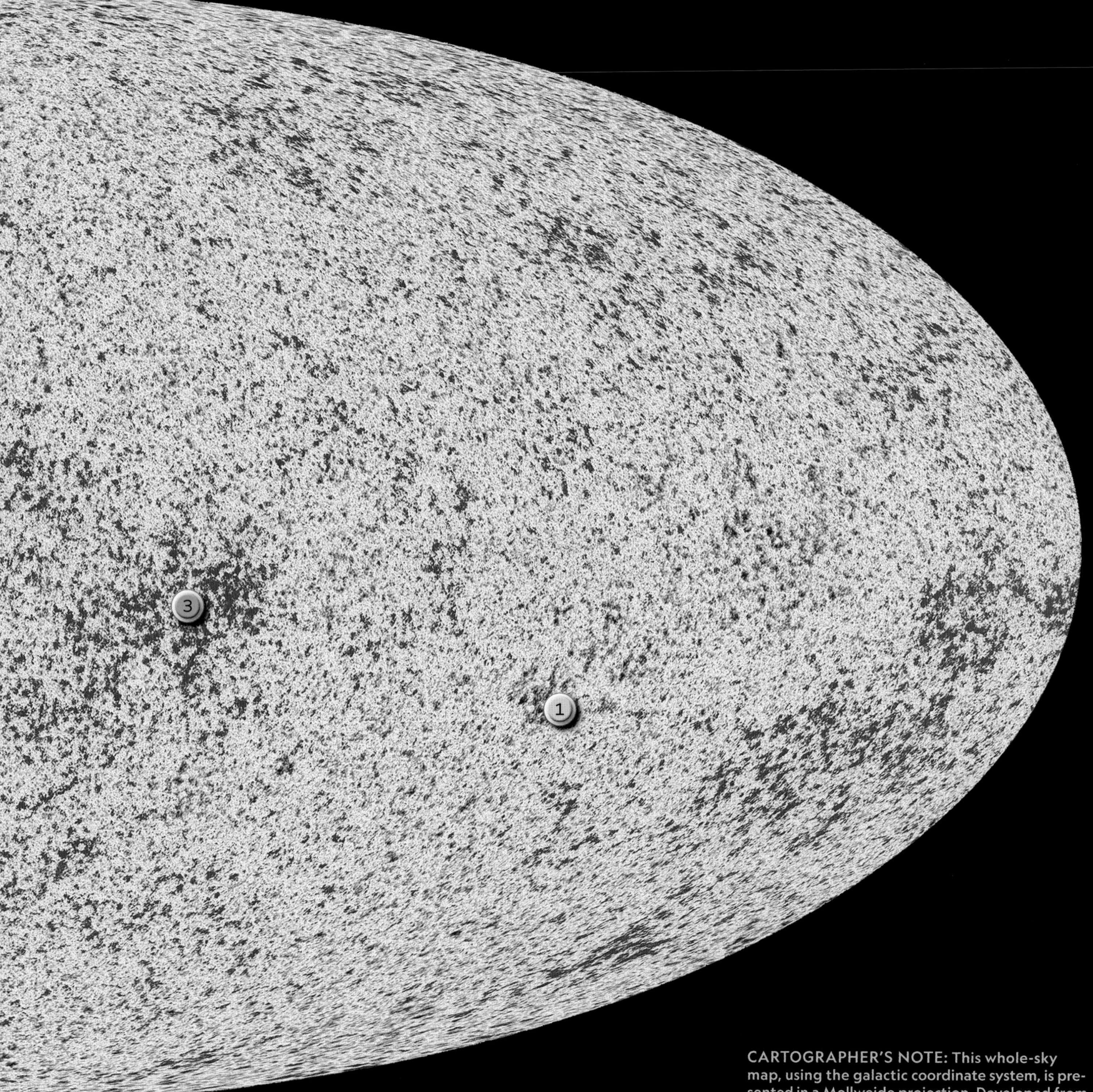

CARTOGRAPHER'S NOTE: This whole-sky map, using the galactic coordinate system, is presented in a Mollweide projection. Developed from data gathered by the European Space Agency's Planck probe, it presents differences in the temperature of background radiation using variable color. Warmer areas appear in red, cooler in blue.

As Penzias and Wilson began to scan the sky, a problem quickly arose. No matter which way they pointed their receiver, they detected a faint microwave signal, which showed up as a hiss in the earphones of their apparatus. In a situation like this, scientists always assume that there is something wrong with their electronics, and so Penzias and Wilson began the tedious job of finding the problem. They even evicted some pigeons that had nested in the receiver and had, as they put it delicately, coated part of the interior with a "white dielectric substance." Nothing helped—the hiss just wouldn't go away. Finally someone suggested that they go down the road to Princeton University, where cosmologists were working on the theory of something called the big bang. The theorists were suggesting that there ought to be a universal background of microwave radiation, an echo of the origin of the universe.

Let's take a moment to understand this prediction. If you watch the coals in a fire, you will notice that they change color as time goes by. They are white-hot when the fire is at its fiercest, but as the fire cools they turn red, then orange. The next day, you will feel warmth from the coals, even though they are no longer giving off visible light. From a physicist's point of view, the coals are giving off radiation whose wavelength increases as the temperature drops. In this case, we go from visible light (with a wavelength thousands of atoms across) to infrared radiation that we can feel but not see, but which has a longer wavelength than red light. The Princeton theorists were suggesting that the universe, like the coal in our example, started out very hot, and as it cooled the radiation associated with it went to longer and longer wavelengths. After billions of years, they argued, the radiation would be in the microwave range, with a wavelength of up to a meter. It was this radiation that Penzias and Wilson had discovered—that faint hiss was, in fact, nothing less than the birth cry of the cosmos! For their work, Penzias and Wilson received the Nobel Prize in physics in 1978.

Earth's atmosphere is transparent to some microwaves—which is why satellite TV works—but it absorbs others. To get a complete picture of the cosmic microwave background, we need to get above the atmosphere. A series of satellites has been launched specifically for this purpose.

EYES IN SPACE

The first of these satellite observatories was the Cosmic Background Explorer (COBE), which was launched in 1989 and operated for four years. It established beyond a doubt that the microwaves were characteristic of an object at a temperature about 3 degrees above absolute zero—2.725 kelvins, if you want to be precise. (This is about -270°C/-454°F.) It was also clear that to an accuracy of about one part in a hundred thousand, the microwave radiation is isotropic—that is, it is the same in all directions. The COBE results were the most precise measurement that had ever been made in cosmology. The two men most involved in COBE—John Mather and George Smoot—received the Nobel Prize in physics in 2006 for their work.

Another important result that came out of COBE was the first measurement of the deviations from uniformity in the background radiation. Although the radiation was *almost* the same in all directions, it had tiny variations. The technical term for these deviations is microwave

Seeing the microwave radiation from the early universe (opposite, upper left) is like seeing light scattered across clouds (upper right). Our instruments see the radiation as it existed when it was scattered by free electrons, before the formation of atoms. The electromagnetic spectrum (opposite, below) ranges from radio waves, whose wavelengths can be thousands of kilometers long, to gamma rays, whose wavelengths are smaller than the diameter of an atom. Microwaves are high-frequency radio waves whose wavelength can be up to a meter long.

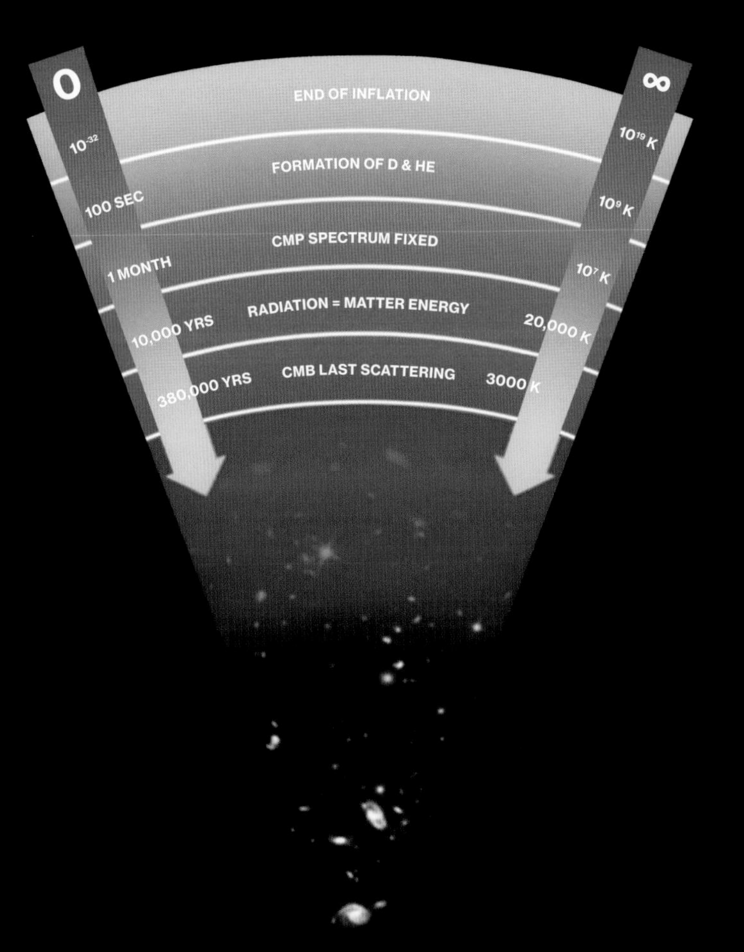

SURFACE OF LAST SCATTER

The microwaves we see come from what is called the surface of last scatter, and they have been traveling freely since then.

CLOUD SCATTER

Seeing microwaves from the surface of last scatter is like looking at light coming through a cloud—we see light that last scattered from the clouds' water molecules.

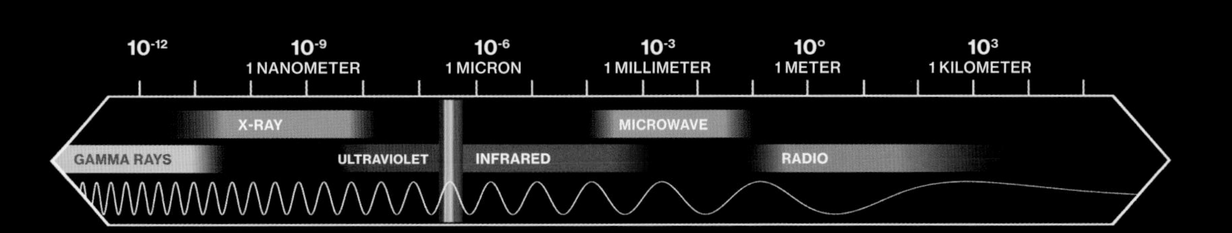

1965 PENZIAS AND WILSON

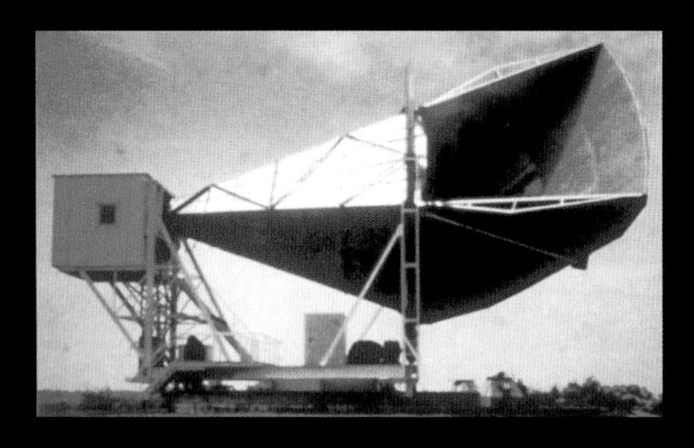

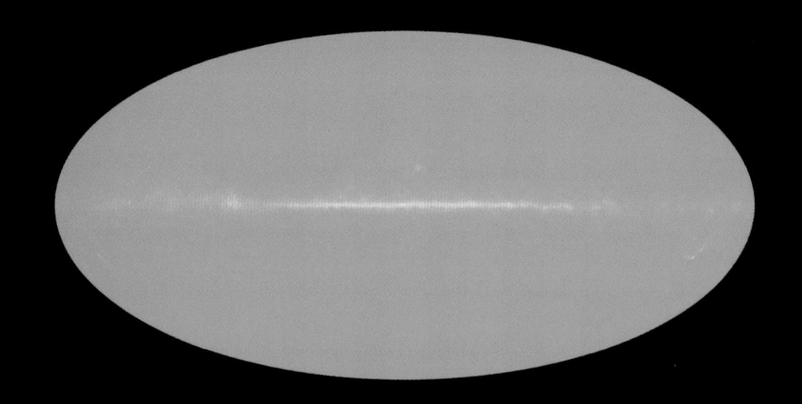

1992 COBE

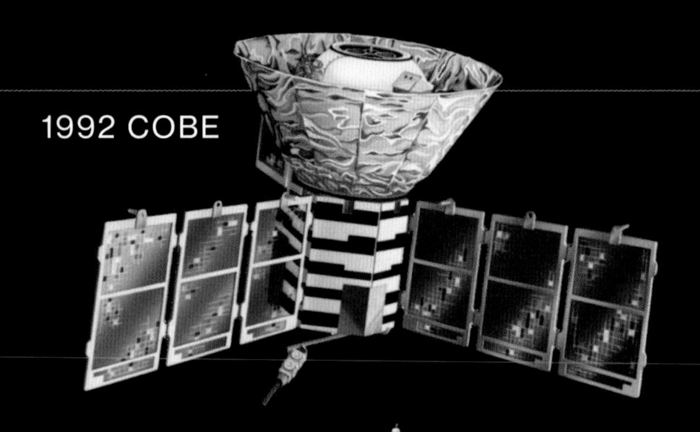

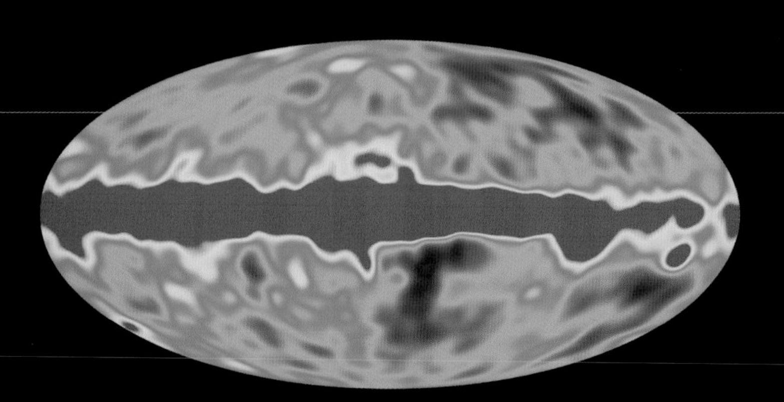

2003 WMAP

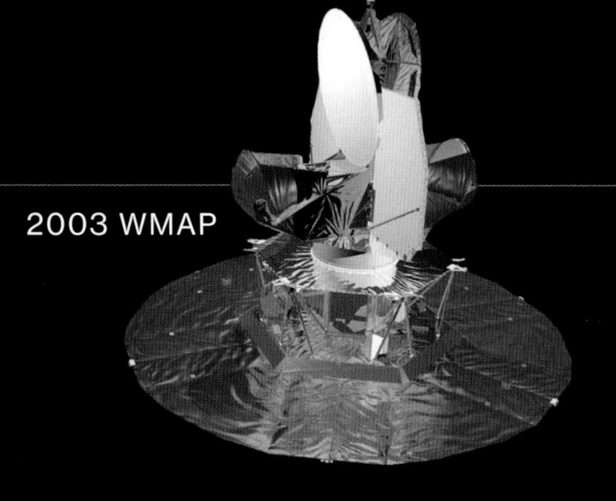

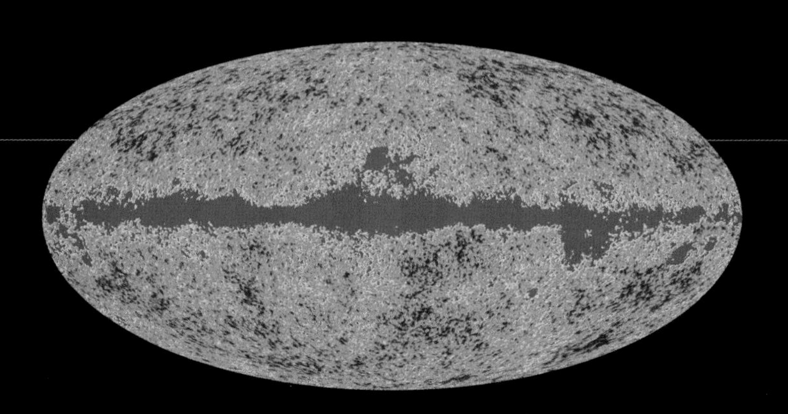

2009 PLANCK

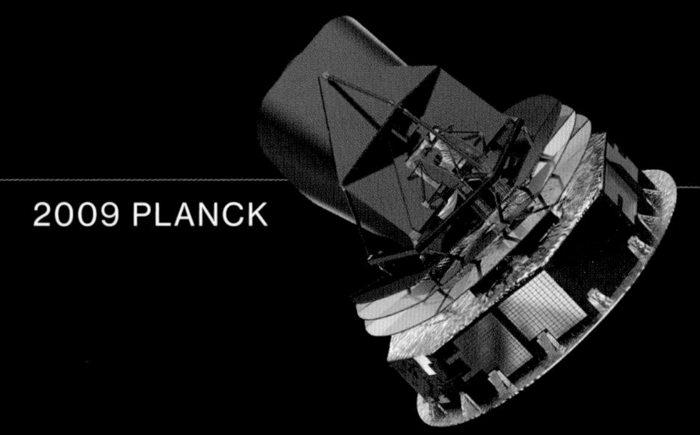

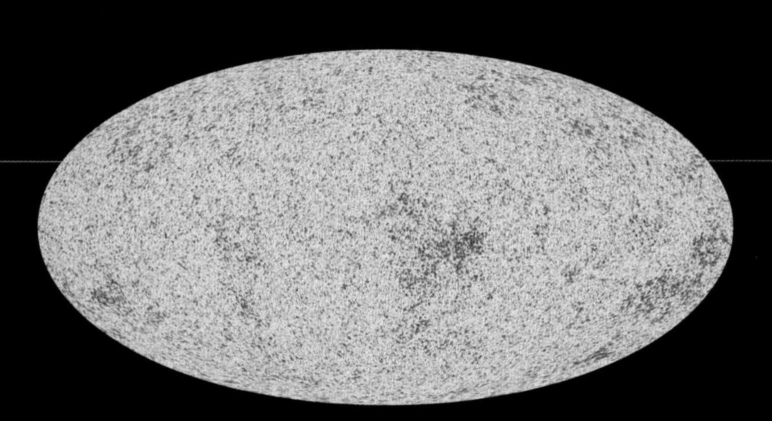

anisotropy. It turns out that these small differences are extremely interesting: They contain information about the state of the universe in the earliest stages of its existence. As we shall see on page 317, early in its life the universe was a plasma, something like the sun. Electrons and protons whizzed around independently, and if an electron happened to hook onto a proton to form an atom, the next collision would knock it free. Furthermore, there was a constant tug-of-war going on, with the particles trying to gather together under the influence of gravity and radiation breaking up the concentrations. When the universe was a few hundred thousand years old, the temperature dropped to the point where atoms could survive these collisions. At this point the universe became transparent, as will be explained on page 317. The radiation released at that time streamed outward and eventually became the cosmic microwave radiation we see today. Thus, when we see small differences in temperature, we are, in fact, seeing the matter concentrations in the universe as they were when it was a few hundred thousand years old—what one astronomer called the "ripples at the beginning of time." It is these tiny seeds that grew into the large-scale structures we see in the universe today.

21ST-CENTURY SCANS

The next satellite to study the microwave background was launched in 2001. It was named the Wilkinson Microwave Anisotropy Probe (WMAP) after prominent Princeton cosmologist David Wilkinson, who died a few years after launch. WMAP was not placed into orbit around Earth: It was sent on a three-month journey to a spot between Earth and the sun known as a Lagrange point. At these points, the gravitational forces of Earth and the sun balance in such a way that a satellite can remain in stable orbit. Lagrange points are becoming favorite places to park satellites because, in these locations, radiation from Earth does not interfere with the instruments.

WMAP did a much finer survey of the microwave sky. By comparing the precise data from WMAP with the predictions of their theories, cosmologists were able to provide solid evidence for the theories of the big bang that run through this part of the book. Many cosmologists, in fact, consider the ability of these theories to reproduce the uneven background radiation to be the best evidence we have for our picture of the structure and evolution of the universe. In addition, WMAP data allowed cosmologists to pinpoint the age of the universe at 13.7 billion years, with an accuracy of about 100,000 years. The satellite stopped taking data in August 2010, and by 2012 it had released most of the data it had collected.

In May 2009 the European Space Agency launched the next-generation microwave probe. It is named the Planck, after the early 20th-century German scientist Max Planck, one of the founders of quantum mechanics. Between its launch and decommissioning in 2013, the spacecraft made the most accurate map of the microwave background and anisotropies ever done.

Although there were no major changes in our understanding of the background, our knowledge became more precise. For example, the Planck measurement of the lifetime of the universe was 13.798, plus or minus .037 billion years—a far more accurate measurement than had been available up to that time.

As instruments have improved, so have maps of the cosmic microwave background. At top left is the large land-based microwave receiver Penzias and Wilson used to discover the radiation. Below are the satellites that have been launched since 1992 to map the radiation in increasing detail. On the right are maps of the microwave sky as seen by each of these instruments.

When we look out at the universe, we see many different kinds of galaxies. Most are like the Milky Way—calm, homey places where stars slowly turn the primordial hydrogen of the big bang into the other chemical elements. Only the occasional supernova (see pages 250–55) provides a bit of excitement. Astronomers, following Hubble, classify these normal galaxies by their shapes. With many gradations within each category, these shapes are spiral, elliptical, and irregular. • The Milky Way is one type of spiral galaxy. The best explanation for the spiral appearance of these galaxies states that pressure waves sweep around them, more or less like water sloshing in a bathtub. These waves trigger the formation of bright new stars in rotating arms.

GALACTIC ZOO

GALAXIES, CLUSTERS, AND SUPERCLUSTERS

Spiral

Barred spiral

Elliptical

Irregular

STARS IN GALAXIES: **10 MILLION TO 100 TRILLION**
MOST MASSIVE GALAXY: **M87, 6 TRILLION SOLAR MASSES**
LEAST MASSIVE GALAXY: **WILLMAN 1, ABOUT 500,000 SOLAR MASSES**
NEAREST GALAXY TO MILKY WAY: **CANIS MAJOR DWARF, 25,000 LIGHT-YEARS**
GALAXY CLUSTER MILKY WAY BELONGS TO: **LOCAL GROUP**
NUMBER OF GALAXIES IN LOCAL GROUP: **30–50**
LOCAL GROUP LARGEST MEMBERS: **MILKY WAY, ANDROMEDA GALAXY**
LOCAL GROUP IS PART OF: **VIRGO CLUSTER**
NUMBER OF GALAXIES IN THE VIRGO CLUSTER: **1,200–2,000**
DISTANCE TO THE VIRGO CLUSTER CENTER: **54 MILLION LIGHT-YEARS**
MASS OF THE VIRGO CLUSTER (SUN = 1): **1.2 QUADRILLION**

Nearby spiral galaxy M74. (Insets) Various galaxy shapes.

THE GALACTIC TUNING FORK

Seventy-five nearby galaxies are organized here according to Edwin Hubble's "tuning fork" classification.

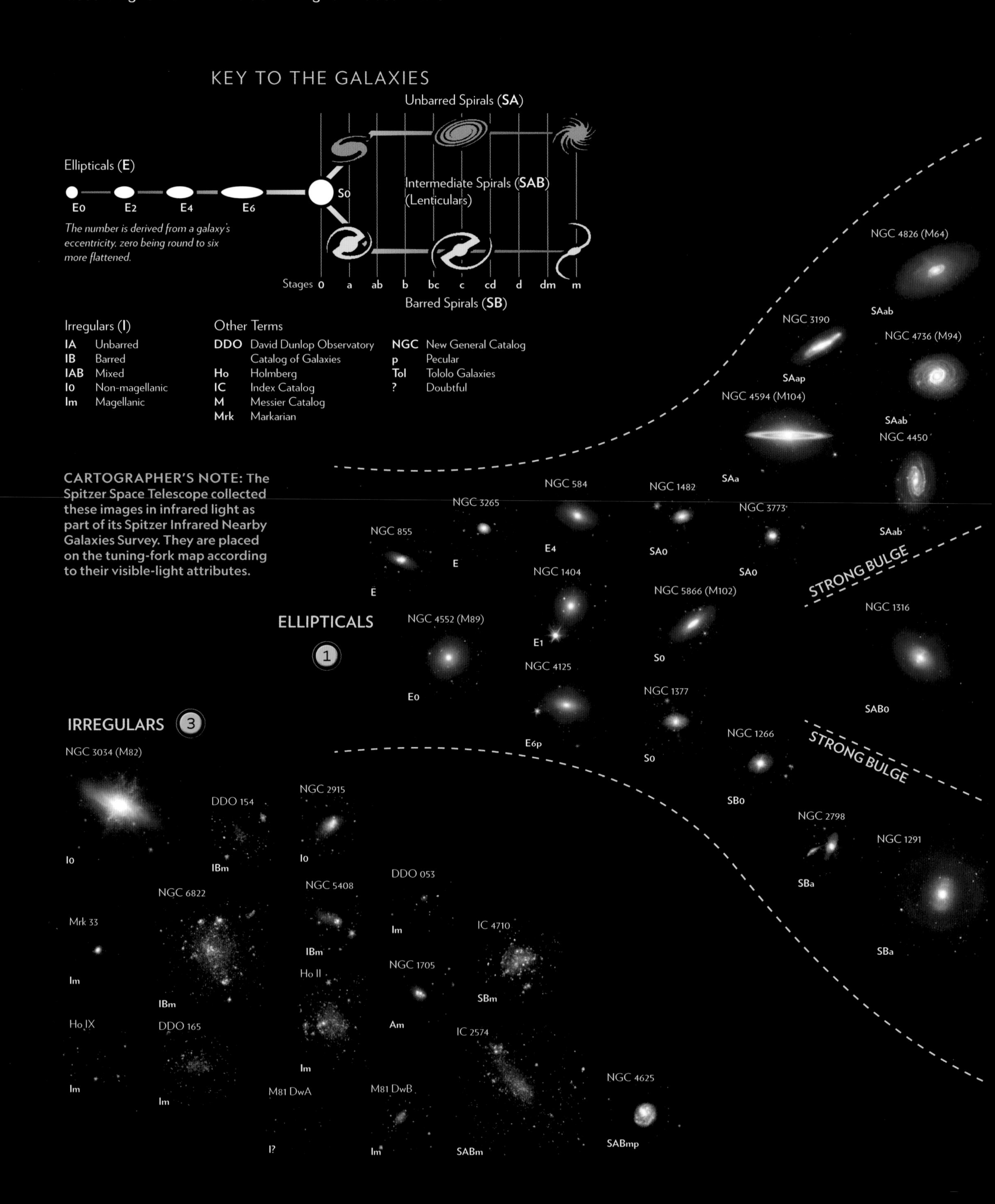

KEY FEATURES

1. **ELLIPTICAL GALAXIES:** These are made primarily of old stars (blue).
2. **UNBARRED SPIRAL GALAXIES:** Red and green colors mark areas of star birth.
3. **IRREGULAR GALAXIES:** These are often found in colliding clusters.

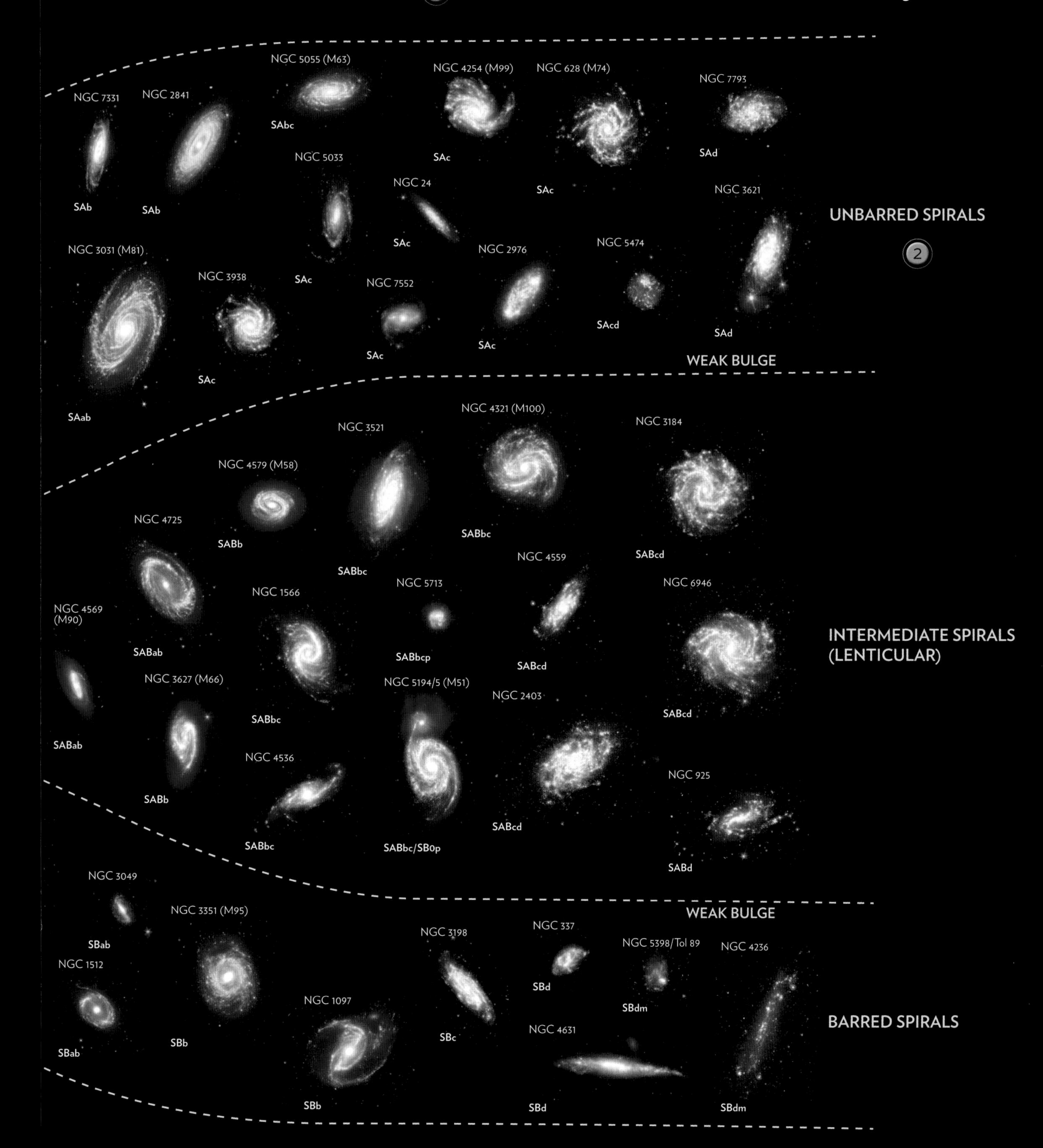

As well as spiral galaxies, the universe holds elliptical galaxies, which as the name implies are basically ovoid blobs of stars. They come in all sizes, from dwarfs to galaxies much larger than the Milky Way. Unlike what we see in spiral galaxies, there appears to be very little star formation in ellipticals.

Finally, irregular galaxies are the leftovers—I think of them as being like the pieces of dough that are left after the cookie cutters have done their work. Irregulars make up most of the galaxies in the universe.

ACTIVE GALAXIES

A small percentage of galaxies are not like our own Milky Way, though. These galaxies are wild and violent places, torn by massive explosions, sometimes ejecting huge jets of hot gases hundreds of light-years into intergalactic space. These are the so-called active galaxies and, like normal galaxies, they come in many shapes and forms. They all appear to have what astronomers call an active galactic nucleus, a small region at the core of the galaxy that is the source of its prodigious energy output. Our best current theory is that each of these galaxies has a large black hole at its center and that nearby matter falling into the black hole bunches up to form a very hot disk. It is this disk, presumably, that generates both the radiation and the jets in active galaxies.

The most important active galaxies are a class of very energetic objects known as quasars (short for "quasi-stellar radio source"). As their name implies, quasars usually emit a lot of their energy as radio waves and relatively little as visible light. In fact, after they were first discovered in the 1950s, it took more than a decade for astronomers to identify a visible object associated with a quasar. When they did, they were surprised to see that the light from the quasar was shifted far to the red, indicating that it was billions of light-years from Earth. Quasars can be seen at great distances. Scientists believe that, like all active galaxies, quasars are powered by material falling into a central black hole.

CLUSTERS AND SUPERCLUSTERS

Galaxies are not scattered randomly in space but tend to be bunched together in groups and clusters. The Milky Way, for example, is part of what is called the Local Group. The Local Group includes one other large spiral galaxy, the Andromeda Nebula, and some 30-plus irregular galaxies. It is about 10 million light-years across.

The Local Group, in turn, is part of a larger structure known as the Virgo supercluster. This supercluster contains at least a hundred groups and clusters and is about 110 million light-years across. It is one of literally millions of superclusters in the universe. The presence of clusters and superclusters provides evidence for the existence of dark matter (see pages 273–7). If you add up the gravitational force of all of the stars in all of the galaxies, it turns out that the force is too small to hold the clusters and superclusters together. Only if you add in dark matter can the structures keep from flying apart.

So it turns out that the region of the universe in our immediate vicinity is lumpy. The question is whether it remains that way when it's examined on a larger scale. Any attempt to answer this question would involve producing a large-scale, three-dimensional map of the galaxies in the universe—a difficult feat, because the Cepheid variable method Hubble used would be just too cumbersome for this big task. But all is not lost: Hubble's law (see page 289) actually provides a quick way of estimating these measurements. If we measure the redshift of the light from the galaxy, we can find the velocity with which the galaxy is moving away from us and then, from Hubble's law, its distance.

The first of these redshift surveys was completed in 1982 by astronomers Margaret Geller and John Huchra of the Harvard-Smithsonian Center for Astrophysics. Instead of discovering that galactic graininess merged into uniformity, they found a completely unexpected large-scale structure, a finding confirmed by many studies since. The easiest way to visualize the large-scale structure of the universe is to picture a huge mound of soap suds that you cut with a knife: The result would be thin films of soap surrounding empty bubbles. In the same way, the redshift surveys tell us that galactic superclusters are arranged along thin sheets surrounding empty spaces known as voids. The largest structure in the known universe, dubbed the Great Wall, is a filament of superclusters 500 million light-years long, 200 million light-years wide, but only 15 million light-years thick.

Thus, the structure of the universe remains interesting out to the largest scales we can see.

KEEN EYES IN SPACE

With the possible exception of Galileo's telescope, the Hubble Space Telescope is arguably the most important astronomical instrument ever built. Launched in 1990, it resides in a low Earth orbit, a bit more than 160 kilometers (100 mi) up. It is equipped with a telescope whose mirror is 2.4 meters (a little less than 8 feet) across.

The Hubble cannot see farther than other instruments – that distinction will always belong to the largest "light bucket" on Earth. But because the Hubble is above the planet's distorting atmosphere, it can see objects in much more detail. In fact, some writers have suggested that it be thought of as an "astronomical microscope" rather than as a telescope. Additionally, the Hubble can see radiation in the near ultraviolet and near infrared – frequencies that are absorbed in Earth's atmosphere.

The instrument encountered problems soon after launch, when it was discovered that some of its optics had been installed improperly. A repair mission by astronauts fixed the problem, and the Hubble has since then posted an impressive array of achievements. They include the following:

- A refined estimate of the distance to stars that allows us to calculate the age of the universe to an accuracy of 10 percent
- The discovery of the existence of dark energy
- The discovery of galactic black holes

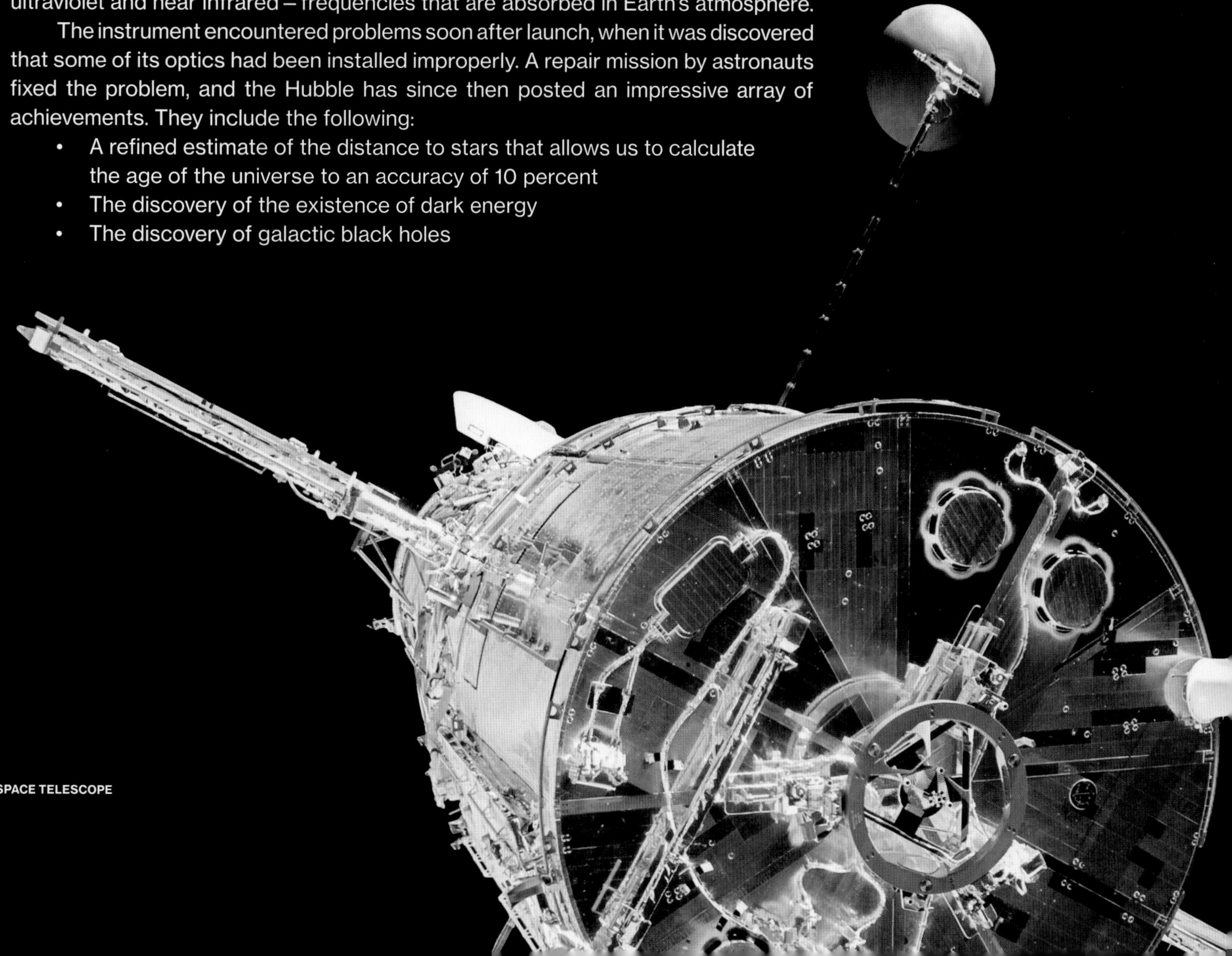

HUBBLE SPACE TELESCOPE

Once Hubble had established the existence of other galaxies and the universe's expansion, it became possible to talk with more confidence about how the universe began. The first point to remember is that the universe's beginning—the big bang—was not an explosion like a bomb blasting out into surrounding air. It was, instead, an expansion of space itself. • An analogy: Imagine that you are making raisin bread with a special transparent dough. If you were standing on any raisin as the dough was rising, you would see the other raisins moving away from you, because the dough between you and each of them is expanding. A raisin that started out twice as far away from you as another would be moving away twice as fast, because there is twice as much dough between you and the farther raisin.

BIG BANG

THE BEGINNING OF SPACE AND TIME

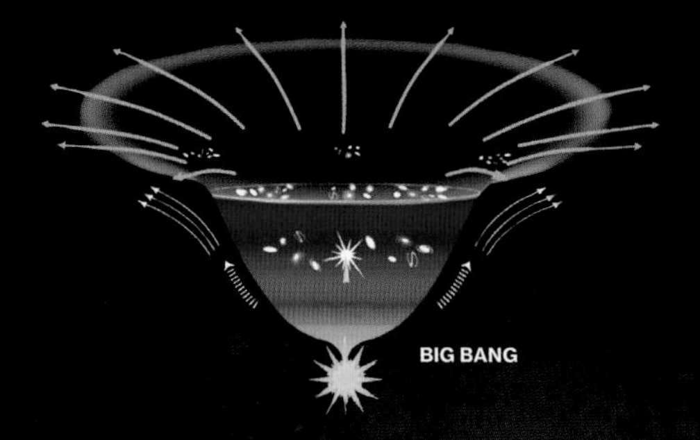

BIG BANG OCCURRED: **13.7 BILLION YEARS AGO**
MATTER DOMINATES: **FIRST 70,000 YEARS AFTER BIG BANG**
HYDROGEN AND HELIUM FORM: **380,000 YEARS AFTER BIG BANG**
FIRST STARS: **ABOUT 100 MILLION YEARS AFTER BIG BANG**
FIRST GALAXIES: **600 MILLION YEARS AFTER BIG BANG**
RECESSION RATE (CURRENT): **70.4 KM/SEC/MEGAPARSEC (43.7 MI/SEC/MEGAPARSEC)**
KEY BIG BANG PREDICTION: **MICROWAVE BACKGROUND SIGNAL**

1912: **FIRST EVIDENCE SEEN (BUT NOT UNDERSTOOD)**
1929: **EDWIN HUBBLE OBSERVES GALAXIES ARE RECEDING**
1949: **"BIG BANG" NAME GIVEN BY FRED HOYLE**
1964: **DETECTION OF MICROWAVE BACKGROUND SIGNAL**

Computer simulation of the big bang. (Inset) Diagram of expanding universe.

I BIG BANG

A representation of the evolution and expansion of the universe over 13.7 billion years

CARTOGRAPHER'S NOTE: A timeline map of the expansion of the universe includes data from the Planck probe. The vertical grid represents the size of the universe. From a period of rapid inflation after the big bang, the universe grew steadily until recently, when dark energy began to speed up expansion.

KEY FEATURES

1. Inflationary period
2. Dark energy speeds up expansion
3. Planck spacecraft

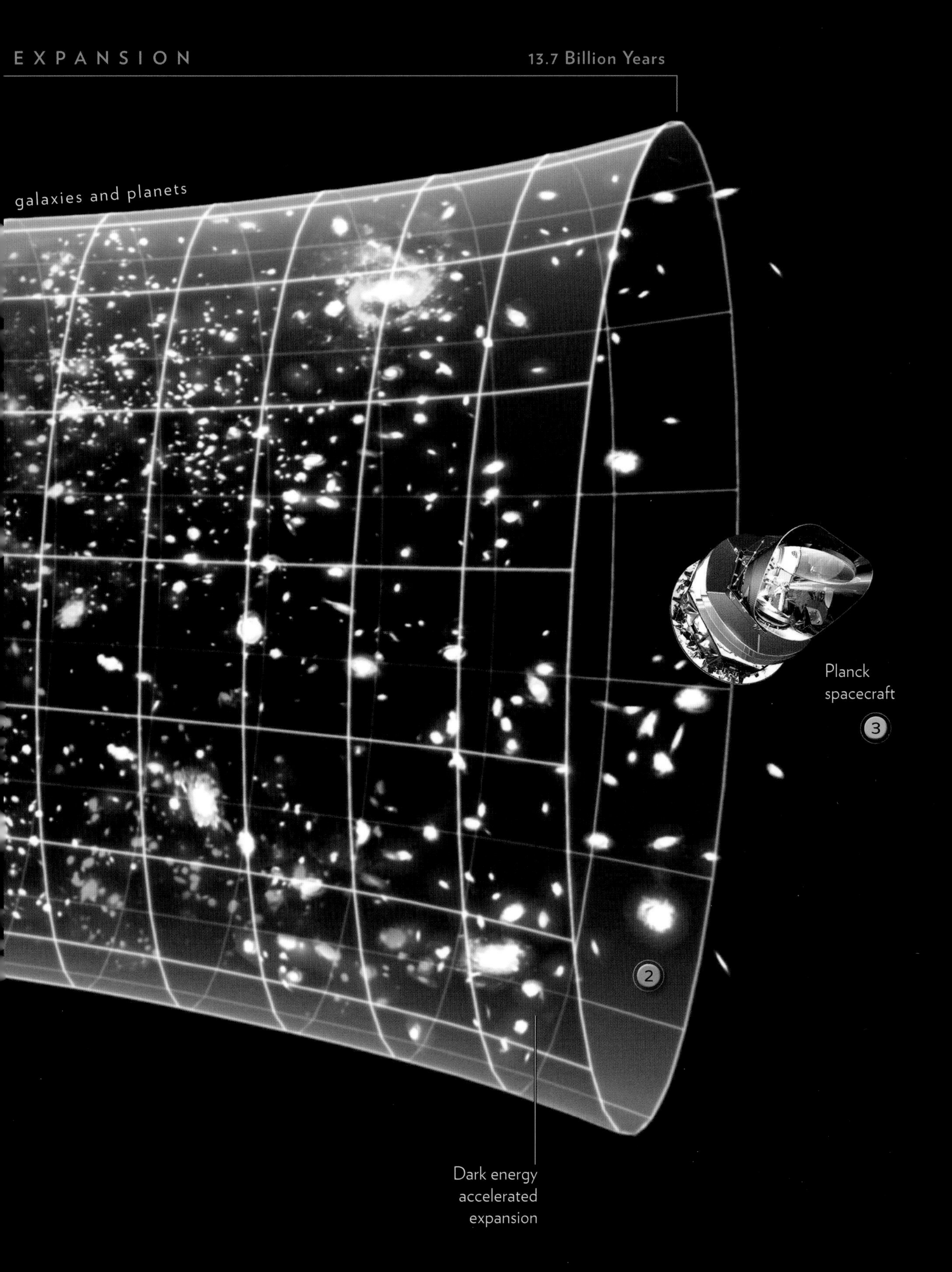

If you substitute galaxies for raisins in the big bang analogy, you have exactly what Hubble saw—a universal expansion. And just as no raisin is actually moving *through* the dough in our analogy, galaxies in the Hubble expansion are being carried along by the expansion of space itself, not moving through space.

It is easy to believe that the fact that everything is moving away from us means that we must be at the center of the universe, but thinking about our analogy can help clarify that as well. Stand on any raisin in the bread and you will see yourself as being stationary while all of the other raisins move away from you. Every raisin, in other words, sees itself as the center of a universal expansion, which means that while we on Earth may indeed see ourselves as the center of the Hubble expansion, so does everyone else, anywhere in the universe. In the words of the 15th-century theologian Nicolas of Cusa, "The universe has its center everywhere and its edge nowhere."

RUNNING THE FILM BACKWARD

We can learn more about our universe by thinking about the Hubble expansion as a film and then imagining that we run the film backward. In our reversed movie, the universe will keep shrinking down—eventually, in principle, reaching a single point. In other words, our universe had a definite beginning: a moment about 13.7 billion years ago, to be exact.

The fact that the universe began at a specific time in the past has important philosophical consequences. Before we learned about the Hubble expansion, we could imagine that the universe was simply eternal, with no beginning, no end, and no change. Or it could have been cyclic, which would be another kind of eternal universe. Or it might have been linear, with a beginning and an end. The only way to decide which of these options actually describes the universe is to make observations, which is, of course, exactly what Hubble did. We live in a universe that definitely had a beginning. Deciding what its end will be is a little more complicated, and we'll discuss the current thinking on this subject in the section "End of the Universe" (see pages 318–21).

It is a general property of materials that they tend to heat up when they are compressed and cool when they expand. Thus, we would expect that in its earlier stages the universe was hotter than it is now, simply because it was smaller and more compressed. In other words, we would expect that the universe had a hot beginning and has been cooling ever since. And in fact, we have found evidence for this in the cosmic microwave background (see pages 291–7).

CONDENSATIONS

Before we discuss the beginnings of the universe in the next sections, it's important to understand an important concept: that of transitions, or "condensations." The fact that the universe started out at a high temperature can give us a sense of how it developed in its earliest stages. Here's another analogy that may help you picture this: Imagine that you keep steam at a very high temperature and pressure and then release it suddenly. The steam expands, and cools as it does so, but at 100°C (212°F) something important happens. At this temperature, the steam condenses into water droplets. This pattern—long periods of expansion and cooling with sudden changes in the basic structure of the system—is what we see in the early development of the universe. Let's look at a key condensation—the formation of nuclei at about three minutes—to get a sense of how these transitions worked.

Before the universe was three minutes old, matter existed in the form of free protons and neutrons (the particles that make up the nuclei of atoms) and free

An artist's two-dimensional conception of the big bang over time ranges from the extreme energies of the beginning, at the white-hot center, to the cooler realms of later millennia, when matter began to condense into stars and galaxies.

electrons. If a proton and a neutron came together to form a simple nucleus, the next collision that nucleus experienced would be so violent that it would be torn apart. At three minutes, however, the temperature had fallen to the point that the free protons and neutrons could start to build nuclei. The entire constitution of the universe changed suddenly as a result of this condensation.

On page 254 we pointed out that all of the universe's heavy elements were made in reactions in supernovae. Now that we understand how nuclei were made in the big bang, we can see why this is necessary. Before three minutes, no nuclei can survive. After three minutes, you can start the process of putting protons and neutrons together to make nuclei, but you are doing it in an environment in which the universe is expanding, carrying particles away from each other and making interactions less frequent. This means that there is a narrow window of opportunity, probably less than a minute long, between the time when nuclei can stay together and the time when the density of the universe drops and nuclei no longer form because interactions are too rare. In this window various forms of hydrogen, helium, and lithium appear—everything else, as we saw, is created later in supernovae. The elements that were taken into stars and later forged into the entire periodic table, in other words, were made in a short burst when the universe was only a few minutes old.

The abundance of these light nuclei in the universe today is one of the strongest pieces of evidence we have

ATOMS FORMED WHEN TEMPERATURES REACHED

3,000

DEGREES KELVIN

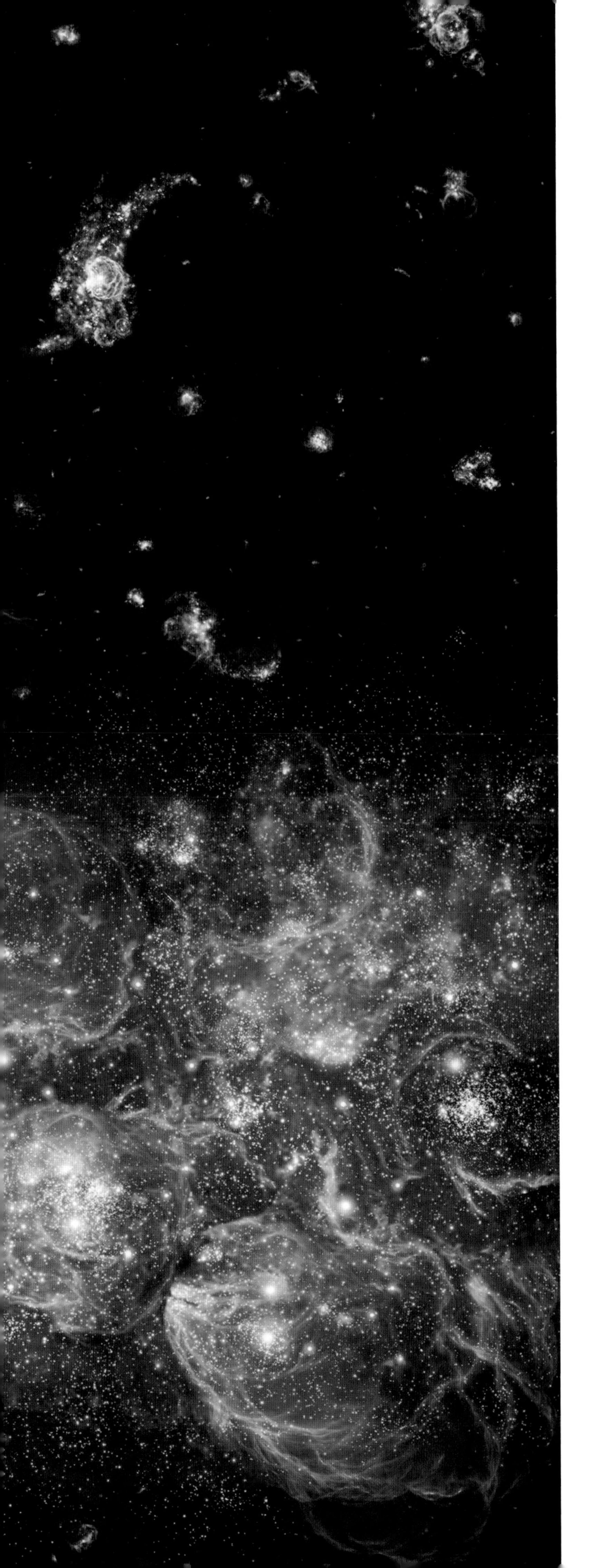

An illustration portrays the early universe, less than one billion years old, when the first stars and galaxies took shape from primordial hydrogen in a burst of star formation, and supernovae exploded across the sky.

for the big bang model. We can reproduce in our laboratories the kinds of energies that particles had when the universe was three minutes old, and our model of the Hubble expansion tells us how often collisions between these particles occurred. Thus, we can make extremely precise and unforgiving predictions of how much of each kind of light element was made in the big bang. The fact that these predictions are borne out by observations provides yet more evidence of the basic correctness of the big bang picture.

THE OBSERVABLE UNIVERSE

One final point: There is a distinction to be made between the universe—by definition, everything there is—and the observable universe—what we can actually see. We can get a rough notion of the size of the observable universe by noting that since the universe is 13.7 billion years old, the farthest objects we can possibly see are 13.7 billion light-years away. In this oversimplified picture, then, the observable universe would be a sphere of radius 13.7 billion light-years centered on Earth, a sphere that grows by 1 light-year every year. (A more detailed calculation would take account of the Hubble expansion and the fact that objects were closer to us when their light was emitted than they are now.)

Many cosmological models situate this sphere inside a much larger universe, like a weak candle in a huge cavern. By definition, of course, we can have no direct knowledge of anything outside the observable universe—a point to which we return when we discuss the multiverse.

One of the consequences of Edwin Hubble's discovery of the expanding universe was the realization that at earlier times the universe was smaller and hotter than it is today. One way to think about "hotter" is to note that when an ordinary substance is heated up, its constituent atoms and molecules move faster. This, in turn, means that when those constituents collide, they do so at higher velocities, and the collisions are more violent. We've already seen some examples of how this fact played into the history of the universe. Before the universe was three minutes old, it was too hot for nuclei to exist. Until the universe was several hundred thousand years old, it was too hot, and collisions were too violent, for atoms to exist.

BEGINNING OF THE UNIVERSE

FROM ENERGY TO MATTER

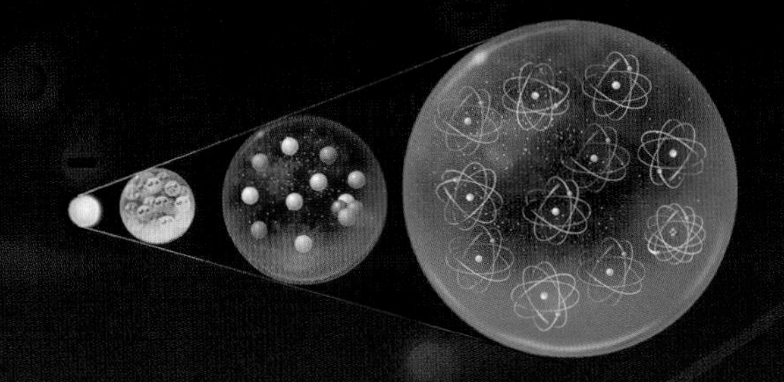

PLANCK ERA, QUANTUM GRAVITY: **0 – 10^{-43} SECONDS**

ELECTROMAGNETISM, STRONG, AND WEAK FORCES UNIFIED: **10^{-43} – 10^{-36} SECONDS**

UNIVERSE INFLATES EXPONENTIALLY: **UNKNOWN – 10^{-32} SECONDS**

STRONG FORCE SEPARATES: **10^{-36} – 10^{-12} SECONDS**

WEAK FORCE SEPARATES: **10^{-12} – 10^{-6} SECONDS**

HYDROGEN NUCLEI FORM: **10^{-6} – 1 SECONDS**

LEPTONS AND ANTI-LEPTONS ANNIHILATE: **1 – 10 SECONDS**

PLASMA OF NUCLEI, ELECTRONS, AND PHOTONS FORMS: **10 SECONDS – 380,000 YEARS**

NUCLEOSYNTHESIS, HELIUM NUCLEI FORM: **3 – 20 MINUTES**

FIRST STARS, QUASARS, AND GALAXIES: **150 MILLION – 1 BILLION YEARS**

SUN AND SOLAR SYSTEM FORM: **9 BILLION YEARS**

Artwork of elementary particles in the first three minutes after the big bang. (Inset) Formation of the first atoms.

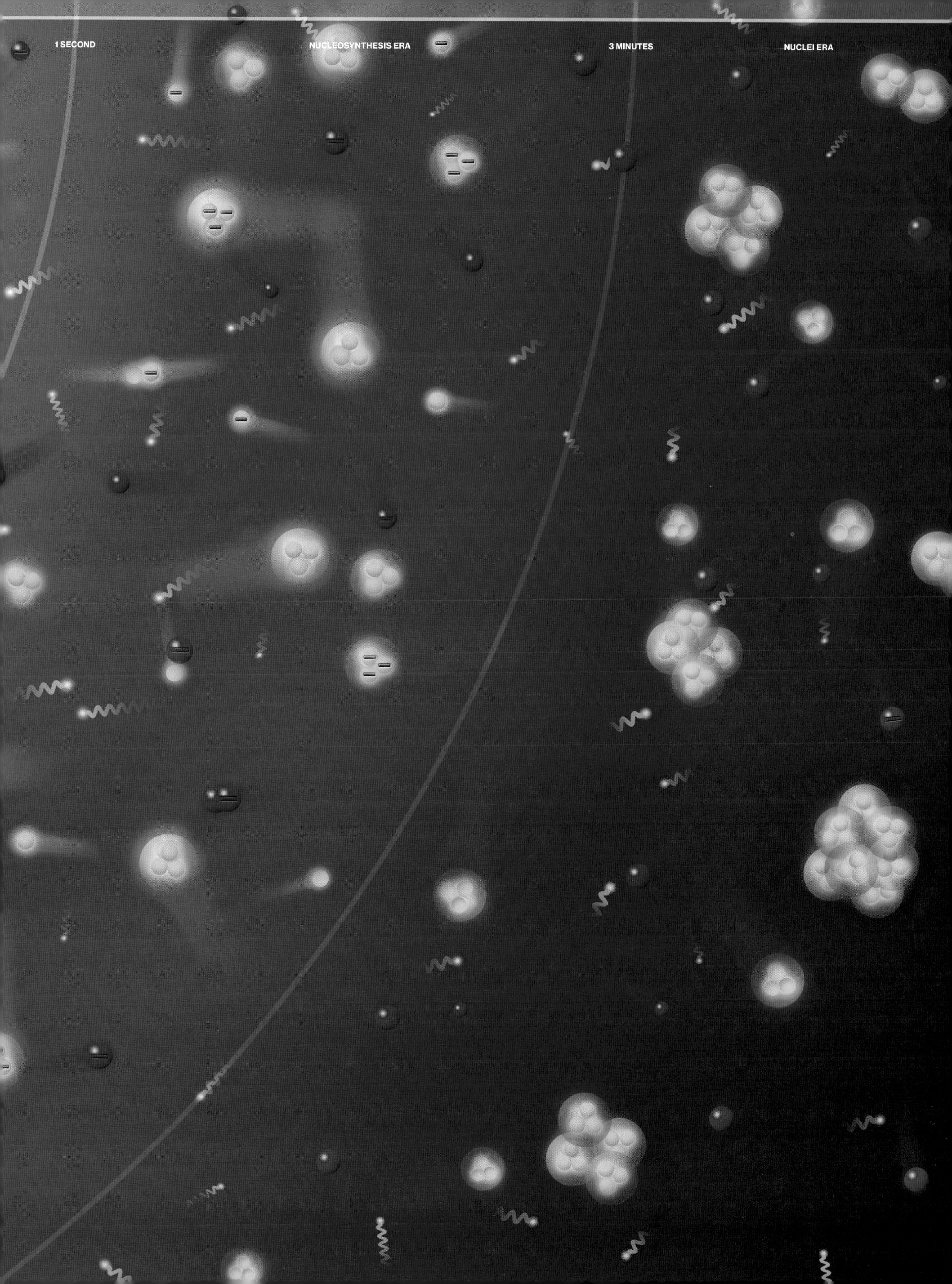
1 SECOND
NUCLEOSYNTHESIS ERA
3 MINUTES
NUCLEI ERA

Each milestone in the early minutes and years of the universe marked a kind of "condensation," in which the basic fabric of the universe changed. First, particles came together to form nuclei, and then, much later, nuclei and electrons condensed into atoms. In fact, we will see that the early history of the universe after the big bang consists of three separations followed by three such condensations.

The first three separations do not involve the basic constituents of matter, but the forces that act between them. Physicists recognize four distinct forces that act in our universe:

1. *Strong force*—the force that holds the nucleus of the atom together
2. *Electromagnetism*—the force that turns on the lights and keeps those notes stuck to your refrigerator door
3. *Weak force*—the force that governs some radioactive decays
4. *Gravity*—the most familiar

In our universe, these forces are distinct and quite different—the forces that hold together the nuclei in the atoms in our bodies are obviously different from the forces that hold the notes on our refrigerators. From the point of view of the theoretical physicist, however, when the temperature gets high enough, they become indistinguishable—the word they use is that the forces become "unified" when the energy is high enough.

THREE SEPARATIONS

Let's start at the very beginning, a point at which we run out of both experiment and theory. We believe that in the instant the universe was born, before it was 10^{-43} seconds old, all the forces—gravity, electromagnetism, strong, and weak—were unified into a single force. We do not yet have a well-tested theory to back up this belief. But we believe that at 10^{-43} seconds after the big bang, gravity separated, or "froze out" from the other unified forces.

The second separation came at the unimaginably small time of 10^{-35} seconds after the big bang—that's a decimal point followed by 35 zeroes. A well-tested theory, called the standard model, describes what happened then: The strong force separated from the combined electromagnetic and weak, or electroweak. Before 10^{-35} seconds, in other words, there were only two forces

Recent studies have shown that the universe is much stranger, and more mysterious, than we used to believe. The upper pie chart shows how matter is distributed in its various forms. Note that our familiar world constitutes only about 5 percent of the total. The lower chart shows the distribution when the universe was 380,000 years old.

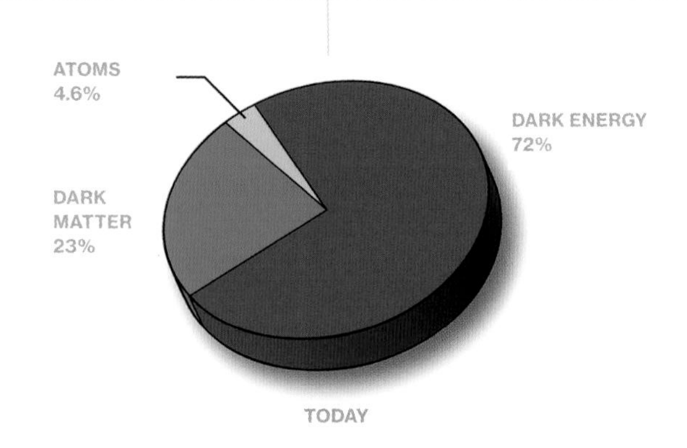

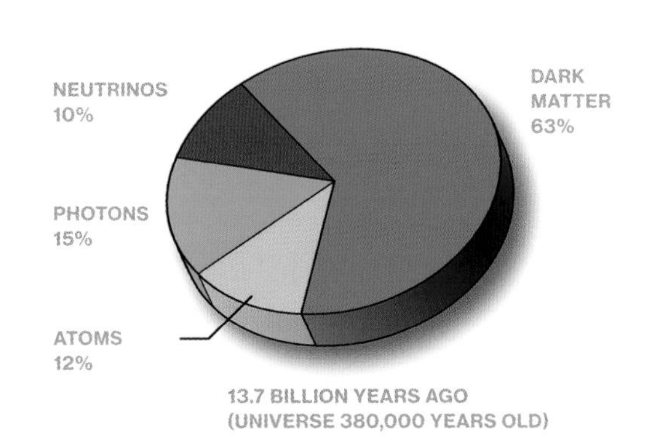

An artist's conception tracks the evolution of the universe from the big bang (lower left) to the formation of matter. The yellow indicates the earliest period, the Planck Era when the four forces were unified; orange, the period of rapid inflation; and red, the period when atoms formed.

10^{-43} SEC
10^{-32} SEC
3 MIN
300,000 YEARS
10^{27} °C
10^{13} °C
10000 °C

An artist's impression of what a quasar might have looked like early in its existence. The supermassive black hole in the center is surrounded by infalling clouds of gas, made luminous by their high level of compression.

operating in the universe (gravity and the unified strong-electroweak); after this time, there were three.

Several other important things happened at this time. The most important of these is that the universe went through a brief but intense expansion called inflation. In the space of about 10^{-35} seconds, the universe went from something smaller than a proton to something about the size of a grapefruit. (Remember that that this was an expansion *of* space, and not motion *in* space, and this means that it didn't violate Einstein's prohibition against faster-than-light travel.) The brainchild of physicist Alan Guth (now at MIT), the so-called inflationary universe is now a standard part of the story of our universe. And, as you might suspect, the introduction of inflation solved another essential problem in cosmology.

It has to do with the cosmic microwave background. As we pointed out, the radiation is uniform to four decimal places, and this, in turn, means that the different parts of the universe from which that radiation comes are all at the same temperature to the same level of accuracy. The problem was that if you simply extrapolated the Hubble expansion back in time, there was never a time when two parts of the sky 180 degrees apart would have been in contact long enough to establish that kind of uniformity. It would be as if you turned on the hot water tap in your bathtub and all the water in the tub warmed up instantly. What inflation does, in essence, is to say that the observed uniformity in temperature was established when the bathtub was a lot smaller and that it was simply maintained through the inflationary event.

After inflation, when the universe was a mere 10th of a nanosecond old, the third separation occurred. The unified electroweak force froze out into the electromagnetic force and the weak force. From that point on, the universe featured the same four forces we see today.

THREE CONDENSATIONS

We reach the next point when the universe is a mere 10 microseconds old. To understand what happened then, we have to talk a little about elementary particles.

Starting in the 1930s, physicists looking at the debris of collisions between cosmic rays and atomic nuclei discovered that there were particles besides the proton and neutron inside the nucleus of the atom. By the 1960s scientists had found that the so-called elementary particles, such as protons, weren't elementary at all: They were combinations of things more elementary still—things that were given the name "quark." (The name has an unusual provenance: There were three kinds of quarks in the original theory, and there was a line in James Joyce's *Finnegan's Wake* that read "Three quarks for Muster Mark.")

Before 10 microseconds, the universe consisted of a sea of quarks, these most fundamental constituents of matter. At 10 microseconds, the quarks condensed into elementary particles (protons and neutrons are the most familiar). The universe became a hot plasma of atomic nuclei.

At a few hundred thousand years, as the universe cooled, these nuclei captured electrons and became atoms. The formation of atoms was a particularly important event because it both released the radiation that became the cosmic microwave background and marked the point at which ordinary matter could start to collapse into galaxies.

Physicists say that the universe became "transparent" at this point. Until then, light in the form of photons could not move freely through space, blocked as the photons were by free-floating, energetic electrons. In fact, before the discovery of dark matter (see pages 273–7), this pretransparent state posed a problem. Photons—electromagnetic radiation—interact and exert a pressure in a plasma, and if ordinary matter tried to condense into galaxies before atoms formed, intense radiation would blow the concentration apart. Furthermore, calculations showed that by the time atoms formed, ordinary matter was spread too thin to be able to collect itself into galaxies. Dark matter, however, doesn't interact with radiation, so it was able to start clumping together before the formation of atoms. When the universe became transparent with the formation of atoms, then, ordinary matter was simply pulled into the concentrations of dark matter that had already formed, creating the galaxies we see today. Far from being a problem, dark matter actually resolved an old problem in cosmology.

The six transitions we've discussed so far can be summarized as follows: Before 10^{-43} seconds, only one unified force operated in the universe. In the next tiny fractions of a second, first gravity, then the strong force, then the weak and electromagnetic forces split apart. At 10 microseconds, quarks condensed into elementary particles, which at three minutes condensed into nuclei and after a few hundred thousand years picked up electrons to become atoms.

Before all of that, of course, there is the ultimate mystery—the creation of the universe itself. And, although scientists are starting to engage in serious speculation about this event, my own sense is that, for the moment at least, we ought to leave this discussion with words from the *Rubaiyat,* written by the poet and astronomer Omar Khayyám:

There was a door to which I found no key
There was a veil past which I could not see

At first glance, predicting the end of the universe seems simpler than tracing its beginnings. The only force acting between far-flung galaxies is gravity, and the only question is whether it's strong enough to reverse the Hubble expansion. The answer to that question depends on a single number—the amount of mass in the universe. • Traditionally, astronomers have distinguished between a situation in which there is not enough mass to halt the expansion—the so-called open universe—and a situation in which the expansion is eventually halted and reversed—a closed universe. The boundary between these two, in which the expansion slows to a halt in an infinite amount of time, is called a flat universe.

END OF THE UNIVERSE

IT ALL DEPENDS ON MASS

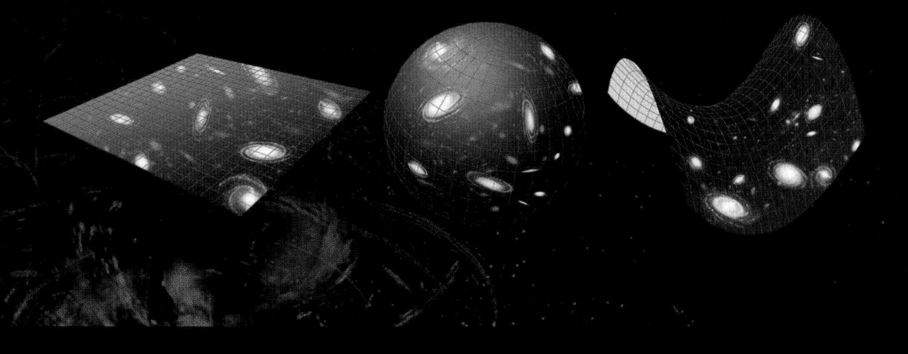

AGE OF THE UNIVERSE TODAY: **13.7 BILLION YEARS**

MILKY WAY AND ANDROMEDA COLLIDE: **3 BILLION YEARS FROM NOW**

SUN BECOMES RED GIANT, THEN WHITE DWARF: **5 BILLION YEARS FROM NOW**

GALAXIES BEYOND LOCAL SUPERCLUSTER REDSHIFT OUT OF SIGHT: **2 TRILLION (2×10^{12}) YEARS**

STAR FORMATION CEASES: **100 TRILLION (10^{14}) YEARS**

PROTONS PROBABLY DECAY AND DISAPPEAR: **10^{34} YEARS–10^{40} YEARS**

BLACK HOLES DOMINATE UNIVERSE: **10^{40} YEARS–10^{100} YEARS**

BLACK HOLES EVAPORATE: **10 YEARS**

POSSIBLE FATES: **BIG CHILL, BIG RIP, BIG CRUNCH**

Artwork of the "big rip."
(Inset) Flat, closed, and open universes.

For theoretical as well as observational reasons, astronomers have always assumed that the universe is flat and have tried to find enough mass to bring about this end. That critical amount of mass is said to "close the universe."

Unfortunately, if you count up all visible matter—all the stars and galaxies and nebulae—you get only about 5 percent of the mass you need to close the universe. If you add in dark matter (see pages 273–7), the number goes up to about 28 percent. This was the way things stood in 1998, when some rather astonishing observational results were announced.

For years astronomers had been trying to find a way of determining the end of the universe that doesn't involve adding up masses. When you look at very distant galaxies, you see light that was emitted billions of years ago. By measuring the redshift of that light, you can tell how fast the universe was expanding back then. We would expect that gravity should be slowing that expansion over time. Therefore, this method of tackling the fate of the universe is known as measuring the deceleration parameter.

To measure the deceleration parameter, you need a way of measuring the distance to very distant galaxies, billions of light-years away. Such galaxies are so remote that astronomers cannot make out individual stars, as Edwin Hubble did with his Cepheid variables, so a different standard candle is needed. Enter the type Ia supernova.

A type Ia supernova occurs when you have a double-star system in which one partner has gone through its life cycle (see page 254) and become a white dwarf. If the white dwarf pulls material off its normal-size partner, it can

MICHAEL TURNER

"It was probably the most anticipated surprise of all time."

So says University of Chicago cosmologist Michael Turner, speaking of the discovery of the accelerated Hubble expansion in the 1990s. Now director of the Kavli Institute for Cosmological Physics, Turner was one of the organizers of the first cosmology group set up at a particle physics facility – in this case, the Fermi National Accelerator Laboratory outside of Chicago.

Those were exciting times in cosmology. Dark matter had been discovered; the inflationary scenario was explaining the basic geometry of the universe. In spite of all of this, however, a piece of the puzzle seemed to be missing. Turner was among a small group of theoreticians who suggested that that missing piece might be the energy of empty space – the energy of the vacuum itself, often referred to as the cosmological constant. When the accelerated expansion was announced in 1998, everyone recognized that whatever was causing the acceleration was that missing piece.

And what does Michael Turner think dark energy is?

"Half the days of the week I think it's the cosmological constant. The rest of the week I think it's something much more fundamental."

become so massive that nuclear reactions begin and the entire star explodes. The energy released in this sort of event is enormous, and for brief periods the supernova can outshine an entire galaxy. Because all white dwarfs are basically the same size, and because the explosion can be seen from so far away, Type Ia supernovae make excellent standard candles for determining the distance to the farthest galaxies.

By combining the knowledge of the distance to a galaxy (from type Ia supernovae) with the knowledge of how fast it is receding from us (from the measured redshift), we can deduce the universe's rate of expansion in the distant past. Studying these supernovae in the 1990s, astronomers expected that they would find that the expansion of the universe is slowing down. Instead, to everyone's amazement, just the opposite is true—the universe is expanding faster now than it did billions of years ago. The Hubble expansion is accelerating!

DARK ENERGY

There is only one way to explain this astonishing fact: There must be a force in the universe capable of overcoming the inward pull of gravity. Just as gravity always pulls things together, this new force must push them apart. This new phenomenon was given the name dark energy, by University of Chicago cosmologist Michael Turner, but it should not be confused with the dark matter we discussed on pages 272–7. However, since energy and mass are equivalent (remember $E = mc^2$), dark energy can be thought of as contributing to the mass of the universe.

The final tally on the mass of the universe, then, is as follows:

Ordinary matter: about 5 percent

Dark matter: about 23 percent

Dark energy: about 73 percent

Altogether, this mass is enough to close the universe.

It is a sobering realization that the stuff we're made of, the stuff we're used to thinking of as the basic fabric of the universe, is actually just a small part of what's out there.

Once type Ia supernovae were developed as a standard candle, it was possible to trace the history of the Hubble expansion. It appears that for the first five billion years or so, the expansion did indeed slow down. In this era, matter was much more densely packed than it is now, and the inward pull of gravity dominated. As matter became more widely scattered, however, gravity weakened and the outward force of dark energy took over. The expansion started to accelerate, as it has been doing ever since.

Given this new understanding of the universe, what can we say about how it will end? The answer depends on the properties of dark energy, which we do not understand. Nevertheless, we can lay out some possibilities.

BIG CHILL, BIG RIP, OR BIG CRUNCH

In the "big chill" scenario, the Hubble expansion will continue forever. Matter will be spread out farther and farther, and the universe will end as a cold, empty place with occasional bits of matter floating around.

A popular idea about dark energy is that it represents a cost of creating spacetime. If so, as the universe expands and the amount of space increases, the amount of dark energy will increase as well. If the amount of dark energy increases with time, the rate of acceleration will also increase until everything—planets, atoms, nuclei—is torn apart in the "big rip." This would be a spectacular (but probably unlikely) ending for the universe.

If the amount of dark energy is fixed, then the expansion will eventually dilute its effects until gravity takes over again at some point in a "big crunch." This would leave us where we started, with a choice between open, closed, and flat, with most evidence pointing toward the last of these.

MYSTERIES

| EPILOGUE |

One of the most important developments in our understanding of the heavens was the realization that to study the largest thing we know of—the universe—we have to study the smallest things we know of—the particles that make up the fundamental building blocks of matter. In recent years, the strange and complex world of string theory has led to the equally strange idea of parallel universes. If experiments validate multidimensional string theories, we will end up with a very different view of reality. Just as Copernicus taught us that Earth is not the center of the universe and Hubble taught us that our Milky Way is just one among billions of galaxies, string theorists are telling us that our entire universe may be just one of a huge number of possible universes. The entire collection of universes is termed the multiverse.

STRING THEORY & THE UNIVERSE

In the last couple of centuries, physicists have delved deeper and deeper into the building blocks of matter. Here is a quick summary of our plunge into the very small:

19th century—we find that matter is made from atoms.

Early 20th century—we find that the atom has a nucleus.

Mid-20th century—we find that the nucleus is made up of elementary particles.

Late 20th century—we find that the elementary particles are made from quarks.

Today—we speculate that the quarks might be made from strange things called strings.

Physicist Steven Gubser opens his excellent *The Little Book of String Theory* with the statement "String theory is a mystery," and truer words were never spoken. The basic idea of these theories is that quarks are made from tiny entities called strings. As the name implies, you can think of them as analogous to the strings of a violin or guitar, with different quarks corresponding to different patterns of the strings' vibration. There are many versions of string theories, but for our purposes we can concentrate on a couple features that make strings particularly interesting:

- They unify gravity with the other fundamental forces (see opposite) and thus can describe the earliest stages of the big bang.
- They are difficult mathematically and typically involve vibrations of strings in 10 or 23 dimensions.

MANY DIMENSIONS

This second point is so unfamiliar to most of us that we had better deal with it first. Our everyday world has four dimensions—three in space and one in time. The space

String theory imagines a multidimensional universe composed of string and branes. An artist's conception (opposite) of how branes might fold together in such a universe

dimensions are front and back, up and down, and left and right. We're less used to thinking of time this way, but "before and after" are familiar concepts.

When physicists started developing string theories, they discovered that the only way to keep their calculations from yielding infinities was to add more dimensions. (In physics jargon, we say that the theories are "renormalizeable" only in this multidimensional world.) So if the theories make sense only in 10 or 23 dimensions, but we live in a world of four dimensions, what is to be done?

A simple analogy illustrates the way string theorists get around this problem. Think about a garden hose lying on a lawn. If you look at the hose from far away, it is a line. If you wanted to move along the hose, you would have only two choices—forward and back. This means that, as seen from a distance, the hose is a one-dimensional object. If you get close to the hose, however, you see that it actually has three dimensions—it also has both left and right and up and down. In the same way, string theorists argue, the strings create a four-dimensional world when observed from far away, and the extra dimensions become apparent only when we get close in, something we cannot do with our current technologies. Physicists say the extra dimensions "compactify."

GRAVITY

It is, however, the first characteristic mentioned above—the unification of gravity with other forces—that is of most interest to scientists. It seems that string theories are able to resolve a century-long split in our view of nature. The early 20th century saw two major scientific revolutions. One was Albert Einstein's theory of relativity, which remains our best description of gravity (see pages 270–71), and the other was quantum mechanics, which is our best description of the subatomic world. The problem is that these two theories look at forces very differently. In relativity the force of gravity arises from the warping of space and time by the presence of mass. Gravity is, in other words, a result of altered geometry. In quantum mechanics, on the other hand, forces arise from the exchange of particles—an essentially dynamic approach. At the moment, the strong, electromagnetic, and weak forces are all described in this way, whereas gravity is explained by geometry. Reconciling these two competing points of view has been an outstanding problem in theoretical physics for decades.

String theories resolve this long-standing dilemma. In these theories, gravity is the result of the exchange of (as yet undiscovered) particles called gravitons. Thus, gravity is not fundamentally different from the other three forces. We can, in fact, imagine the way that different scientists through the ages would have answered the simple question "Why don't I just float up out of the chair I'm sitting in?"

Isaac Newton: Because Earth and you exert a gravitational force on each other.

Albert Einstein: Because the mass of Earth warps the spacetime grid at its surface.

String theorist: Because there is a flood of gravitons being exchanged between you and Earth.

These explanations are not contradictory, but complementary. Apply string theory to massive objects and you get the results of relativity; apply relativity to ordinary objects and you're back to Newton.

BRANES

Newer versions of string theories involve objects called branes (from "membranes"), which you can think of as sheets flopping around in multidimensional space (imagine moving a string perpendicular to its length so that its path traces out a sheet). Physicists also talk of an ultimate theoretical version of strings, dubbed M theory, that will mark the end of our quest to know the fundamental structure of matter. M theory hasn't been written down yet, although there are a lot of extremely bright people trying to accomplish the task.

Having written this quick introduction to the frontiers of modern theories, I have to say that there is a serious debate in the physics community about whether they really constitute science. The difficulty of the mathematical descriptions has prevented string theorists from making predictions that can be tested by experiment. Skeptics argue that without the traditional interplay between theory and experiment, string theory is just mathematics. Defenders counter that many of the general features of the theories, such as the prediction that there is a class of so-called supersymmetric particles that has not yet been discovered, are indeed testable. At the moment, the failure of experimental physicists to find these supersymmetric particles using the Large Hadron Collider in Geneva, Switzerland—the world's most powerful accelerator—has cast a pall over string theory. Whether this will change in the future is unknown.

In any case, string theories give us a way of bridging that final gap we discussed, the gap that separates the well-thought-out history of the big bang from the unknowns surrounding the unification of gravity. They allow us to confront the ultimate question: How did the whole thing start?

British theoretical physicist Michael Green is one of the pioneers of string theory.

THE MULTIVERSE

Just as there are many different versions of string theory, there are many different versions of the multiverse. Some seem truly bizarre, with multidimensional branes (see pages 325–6) colliding to generate each new universe. The most common picture, however, is most easily visualized as bubbles in boiling water. We can think of each bubble as representing a universe like our own, full of galaxies. Our own bubble is expanding, and in some versions of the multiverse theory, a bubble like ours continuously creates small bubbles at its surface, with each small bubble experiencing its own version of the big bang. Our own universe, in other words, might be shedding little baby universes even as you are reading these words.

STRING THEORY LANDSCAPE

String theories predict the existence of the multiverse because of something called the string theory landscape. Imagine a cosmic pinball game, where balls roll over a surface full of hills and valleys. We know that the ball will eventually settle in one of the valleys—not necessarily the deepest valley on the board, since the ball can get trapped in a shallow valley as well. A valley capable of trapping the ball, even though it is not the deepest valley, is called a false vacuum in the jargon of theoretical physics.

According to string theory, when we map out the surface corresponding to possible energy states in multiple dimensions, we find a huge number of false vacua (valleys)—some 10^{500} of them, in fact. Each of these represents a possible place where the pinball could wind up, or, in string theory, a possible universe. 10^{500} is a huge number—a 1 followed by 500 zeroes! For all intents and purposes,

An artist's conception of the multiverse.

Artwork depicts the simultaneous creation of multiple universes with differing physical laws. Our universe is in the center, and a universe without matter is at right.

we can say that there are an infinite number of possible universes in the string theory landscape. And, if universes like ours are really shedding baby universes, we can think of each shedding event as rolling another pinball over the hills and valleys. If we roll enough pinballs, we can expect that eventually most of the false vacua will be filled. In the string theory multiverse, then, any possible universe will eventually appear somewhere in the landscape.

As was the case with string theory, there is a philosophical debate about the multiverse. It centers on the fact that in most versions of the theory there is really no way for one universe to communicate with another, and hence no way to have direct experimental confirmation of the existence of a universe outside of our own. On the other hand, if it turns out that some version of string theory passes the experimental tests to which we can subject it, and if that theory also predicts the existence of the multiverse, then we would have to take that prediction seriously. And this is important because it may be that the existence of the multiverse could solve a very deep and long-standing problem in our understanding of the universe.

FINE-TUNING PROBLEM

It's called the fine-tuning problem. It can be stated in many ways, but you can get a sense of it by thinking about the gravitational force. If this force were much stronger than it actually is, the big bang would have collapsed soon after it began, simply because the stronger gravity would have ended the expansion before it really got started.

Similarly, if the gravitational force were weaker, it would not have been strong enough to gather matter together into stars or planets. In both cases, the universe would not have produced living creatures capable of asking questions about gravity. So the gravitational force has to be fine-tuned—restricted to certain values—in order for life to develop.

This sort of fine-tuning seems to be a feature of all of nature's constants. Theoretical calculations, for example, indicate that changes of only a few percent in the strong force (the one that holds the nucleus together) or the electromagnetic force (the one that keeps electrons locked in orbits in the atom) would prevent the creation of atoms like carbon and oxygen, eliminating the possibility of life as we know it. In the same way, a number called the cosmological constant, which some theorists believe is involved in the acceleration of the Hubble expansion (see page 321), is measured to be almost (but not quite) zero in our universe. And yet when physicists use quantum mechanics to calculate this number, they are off by 120 orders of magnitude—about as far off from the real value as it is possible to be. (An order of magnitude is one power of 10, so 120 orders of magnitude is a 1 followed by 120 zeroes.) Some as-yet-unknown effect must be almost canceling this large value, but all we can say now is that the cosmological constant seems to be fine-tuned with a vengeance!

This fine-tuning of the forces and constants of nature has always been a problem for scientists. Why should we be in a universe where all these numbers are exactly as they are? Some theologians have even advanced fine-tuning as proof of the existence of God.

ANTHROPIC PRINCIPLE

The string theory multiverse supports another view, however, related to an old concept called the anthropic principle. Supporters of this view point out that the question "Why are the constants of nature as they are?" is incorrectly posed. The question should really be "Why are the constants of nature as they are, *given that an intelligent living creature is asking the question?"* In a universe incapable of producing life, the question would never get asked, so the mere fact that it is being posed is already a statement about the type of universe we're in.

I should point out that there are actually two versions of the anthropic principle: weak and strong. The weak principle is the argument given above—that we must live in a universe capable of producing life because the question is being asked. The strong version asserts that there is some as-yet-undiscovered law that says the universe *must* be such that life can exist. Most scientists prefer the weaker version.

My old statistics professor used to talk about what he called the "golf ball on the fairway" problem. Before a golfer swings, the chance that the ball will land on a particular blade of grass is tiny—but the ball will eventually land on some blade of grass. If it hadn't landed there, it would have landed somewhere else equally improbable. In the same way, when we contemplate the string theory landscape, it's no use asking why we are in this particular improbable universe because if we weren't here, some version of us would be somewhere else equally improbable.

Looking at things this way leads to some interesting thoughts. The number of possible universes is so large, for example, that the subset capable of producing life is probably large as well. This leads to the standard science-fiction scenario where there is another you reading these words in another universe, except that maybe you have a tail and green scales. The multiverse represents the triumph of the Copernican worldview: the ultimate removal of mankind from the center of existence.

CARINA NEBULA

An infrared image from the European Southern Observatory's Very Large Telescope luminously outlines the structures of the Carina Nebula, a region of intense star formation 7,500 light-years from Earth. At lower left is the unstable star Eta Carinae.

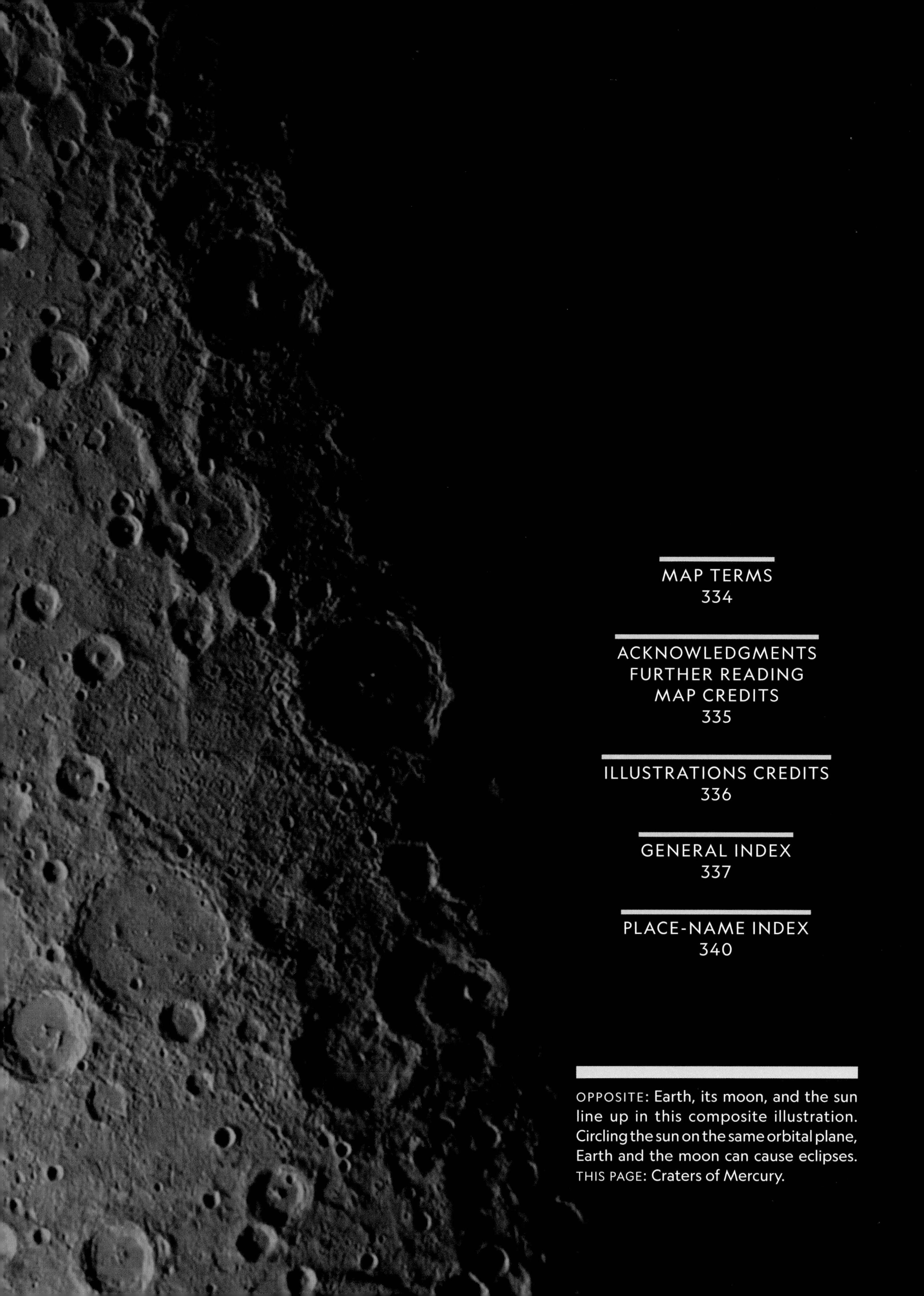

OPPOSITE: Earth, its moon, and the sun line up in this composite illustration. Circling the sun on the same orbital plane, Earth and the moon can cause eclipses. THIS PAGE: Craters of Mercury.

MAP TERMS

TERM	PLURAL	DEFINITION
Albedo feature		Geographic area distinguished by amount of reflected light
Arcus	**arcūs**	Arc-shaped feature
Astrum	**astra**	Radial-patterned features on Venus
Catena	**catenae**	Chain of craters
Cavus	**cavi**	Hollows, irregular steep-sided depressions usually in arrays or clusters
Chaos	**chaoses**	Distinctive area of broken terrain
Chasma	**chasmata**	A deep, elongated, steep-sided depression
Collis	**colles**	Small hills or knobs
Corona	**coronae**	Ovoid-shaped feature
Crater	**craters**	A circular depression
Dorsum	**dorsa**	Ridge
Eruptive center		Active volcanic centers on Io
Facula	**faculae**	Bright spot
Farrum	**farra**	Pancake-like structure, or a row of such structures
Flexus	**flexūs**	A very low ridge with a scalloped pattern
Fluctus	**fluctūs**	Flow terrain
Flumen	**flumina**	Channel on Titan that might carry liquid
Fossa	**fossae**	Long, narrow depression
Insula	**insulae**	Island, an isolated land area (or group) surrounded by, or nearly surrounded by, a liquid area
Labes	**labēs**	Landslide
Labyrinthus	**labyrinthi**	Complex of intersecting valleys or ridges.
Lacuna	**lacunae**	Irregularly shaped depression on Titan having the appearance of a dry lake bed
Lacus	**lacūs**	Lake or small plain; on Titan, a "lake" or small, dark plain with discrete, sharp boundaries
Lenticula	**lenticulae**	Small dark spots on Europa
Linea	**lineae**	A dark or bright elongate marking, may be curved or straight
Lingula	**lingulae**	Extension of plateau having rounded lobate or tongue-like boundaries
Macula	**maculae**	Dark spot, may be irregular
Mare	**maria**	Sea; large circular plain; on Titan large expanses of dark materials thought to be liquid hydrocarbons
Mensa	**mensae**	A flat-topped prominence with cliff-like edges
Mons	**montes**	Mountain
Oceanus	**oceani**	A very large dark area on the moon
Palus	**paludes**	Means "swamp"; small plain
Patera	**paterae**	An irregular crater or a complex one with scalloped edges
Planitia	**planitiae**	Low plain
Planum	**plana**	Plateau or high plain
Plume	**plumes**	Cryo-volcanic features on Triton
Promontorium	**promontoria**	Cape; headland
Regio	**regiones**	A large area marked by reflectivity or color distinctions from adjacent areas or a broad geographic region
Reticulum	**reticula**	Reticular (netlike) pattern on Venus
Rima	**rimae**	Fissure
Rupes	**rupēs**	Scarp
Scopulus	**scopuli**	Lobate or irregular scarp
Sinus	**sinūs**	Means "bay"; small plain
Sulcus	**sulci**	Subparallel furrows and ridges
Terra	**terrae**	Extensive landmass
Tessera	**tesserae**	Tile-like, polygonal terrain
Tholus	**tholi**	Small domical mountain or hill
Unda	**undae**	Dunes
Vallis	**valles**	Valley
Vastitas	**vastitates**	Extensive plain
Virga	**virgae**	A streak or stripe of color

ACKNOWLEDGMENTS

The author thanks Vera Rubin of the Carnegie Institution and Michael Turner of the University of Chicago for their participation in developing this manuscript, and editors Susan Tyler Hitchcock and Patricia Daniels, as well as cartographer Matthew Chwastyk, for their help in bringing the project to completion.

ABOUT THE AUTHOR

James Trefil is Clarence J. Robinson Professor of Physics at George Mason University in Fairfax, Virginia. His previous books include *Other Worlds* (National Geographic Books, 1999), and he served as Principal Science Consultant for *The Big Idea* (National Geographic Books, 2012). He has received numerous awards for his writing, most recently the Science Writing Award given by the American Institute of Physics, and in 2011 was awarded an honorary degree from the University of Sts. Kiril and Methodius in Skopje, Macedonia. His most recent book is *Science in World History.*

FURTHER READING

More Great Astronomy and Space Books From National Geographic

Teasel Muir-Harmony, *Apollo to the Moon: A History in 50 Objects* (2018)

James Trefil, *The Story of Innovation: How Yesterday's Discoveries Lead to to Tomorrow's Breakthroughs* (2017)

Buzz Aldrin, *No Dream Is Too High: Life Lessons From a Man Who Walked on the Moon* (2016)

Leonard David, *Mars: Our Future on the Red Planet* (2016)

Buzz Aldrin, *Mission to Mars: My Vision for Space Exploration* (2015)

Marc Kaufman, *Mars Up Close: Inside the Curiosity Mission* (2014)

Howard Schneider, *National Geographic Backyard Guide to the Night Sky* (2013)

The Big Idea: How Breakthroughs of the Past Shape the Future (2011)

J. Richard Gott and Robert J. Vanderbei, *Sizing Up the Universe: The Cosmos in Perspective* (2010)

MAP CREDITS

CARTOGRAPHER
Matthew W. Chwastyk, National Geographic

ALL MAPS
Place names: Gazetteer of Planetary Nomenclature, Planetary Geomatics Group of the USGS (United States Geological Survey) Astrogeology Science Center http://planetarynames.wr.usgs.gov

IAU (International Astronomical Union): http://iau.org

NASA (National Aeronautics and Space Administration): http://www.nasa.gov

SOLAR SYSTEM PP. 52–53, INNER PLANETS PP. 60–61, OUTER PLANETS, PP. 120–121
All images: NASA, JPL (Jet Propulsion Laboratory, California Institute of Technology), Johns Hopkins University Applied Physics Laboratory, Carnegie Institution of Washington

MERCURY PP. 64–67
Global Mosaic: MESSENGER (MErcury Surface, Space ENvironment, GEochemistry, and Ranging), NASA, Johns Hopkins University Applied Physics Laboratory, Carnegie Institution of Washington

VENUS PP. 74–77
Global Mosaic: Magellan Synthetic Aperature Radar Mosaics, NASA, JPL (Jet Propulsion Laboratory, California Institute of Technology)

EARTH PP. 83–87
Paleogeography: C.R. Scotese, Paleomap Project

Surface Satellite Mosaic: NASA Blue Marble, NASA's Earth Observatory

Bathymetry: ETOPO1/Amante and Eakins, 2009

EARTH'S MOON PP. 94–97
Global Mosaic: Lunar Reconnaisance Orbiter, NASA, Arizona State University

MARS PP. 104–107
Global Mosaic: NASA Mars Global Surveyor; National Geographic Society

Moon images: Phobos, Diemos, NASA, JPL (Jet Propulsion Laboratory, California Institute of Technology), University of Arizona

CERES PP. 118–119
Global Mosaic: NASA, JPL-Caltech (Jet Propulsion Laboratory, California Institute of Technology), UCLA/MPS/DLR/IDA

JUPITER PP. 124–125
Global Mosaic: NASA Cassini Spacecraft, NASA, JPL (Jet Propulsion Laboratory, California Institute of Technology), Space Science Institute

MOONS OF JUPITER PP. 132–139
All Global Mosaics: NASA Galileo Orbiter NASA, JPL (Jet Propulsion Laboratory, California Institute of Technology), University of Arizona

SATURN PP. 130–131
Global Mosaic: NASA Cassini Spacecraft NASA, JPL (Jet Propulsion Laboratory, California Institute of Technology)

MOONS OF SATURN PP. 148–161
All Gloabl Mosaics: NASA Cassini Spacecraft NASA, JPL (Jet Propulsion Laboratory, California Institute of Technology) Space Science Institute

SATURN'S RINGS PP. 168–169
NASA Cassini Spacecraft NASA, JPL (Jet Propulsion Laboratory, California Institute of Technology) Space Science Institute

URANUS PP. 174–175, URANUS'S MOONS PP. 176–179, NEPTUNE PP. 180–181, TRITON P. 182
Global imagery: NASA Voyager II, NASA, JPL (Jet Propulsion Laboratory, California Institute of Technology)

PLUTO P. 193
Global imagery: NASA, Johns Hopkins University Applied Physics Laboratory, Southwest Research Institute, Lunar and Planetary Institute

THE MILKY WAY PP. 210–211
Artwork: Ken Eward, National Geographic Society

THE SUN PP. 220–221
Artwork: Moonrunner Design, National Geographic Society

THE UNIVERSE PP. 282–283
Artwork: Ken Eward, National Geographic Society

COSMIC MICROWAVE BACKGROUND PP. 292–293
Mosaic: Planck Mission, ESA and Planck Collaboration

HUBBLE TUNING FORK PP. 300–301
Images: SINGS (Spitzer Infared Nearby Galaxies Survey, NASA, JPL-Caltech

THE BIG BANG PP. 306–307
Diagram: NASA, JPL-Caltech (Jet Propulsion Laboratory, California Institute of Technology)

ILLUSTRATIONS CREDITS

1, NASA/NOAA/GSFC/Suomi NPP/VIIRS/Norman Kuring; 2–3, NASA, ESA, and The Hubble Heritage Team (STScI/AURA); 4–5, X-ray: NASA/CXC/SAO/J.Hughes et al, Optical: NASA/ESA/Hubble Heritage Team (STScI/AURA); 6–7, NASA/JPL/Space Science Institute; 8–9, NASA, ESA, and J. Maíz Apellániz (Instituto de Astrofísica de Andalucía, Spain); 10–11, E.J. Schreier (STScI), and NASA; 12–13, NASA; 15, NASA; 16, NASA; 19, NASA; 20, ESA; 23, Cycler graphic created by 8i in collaboration with the Buzz Aldrin Space Institute. Earth, Mars and Moon elements by NASA.; 24, NASA; 26, NASA; 27, Rebecca Hale, NG Staff; 28–9, NASA; 30, © Visual Language 1996; 31, Jean-Leon Huens/National Geographic Creative; 32, NASA, ESA, and the Hubble SM4 ERO Team; 34, NASA, ESA, J. Merten (Institute for Theoretical Astrophysics, Heidelberg/Astronomical Observatory of Bologna), and D. Coe (STScI); 39 (UP), © Visual Language 1996; 48–9, NASA/JPL-Caltech; 50–51, David Aguilar; 54–5, NASA/JPL-Caltech; 55 (INSET), NASA/JPL-Caltech; 57, NASA/JPL-Caltech/T. Pyle (SSC-Caltech); 58, Ron Miller/Stocktrek Images/Corbis; 62–3, NASA/Johns Hopkins University Applied Physics Laboratory/Arizona State University/Carnegie Institution of Washington. Image reproduced courtesy of Science/AAAS; 63 (INSET), NASA/Johns Hopkins University Applied Physics Laboratory/Carnegie Institution of Washington; 68, Science Source; 69, NASA/Johns Hopkins University Applied Physics Laboratory/Carnegie Institution of Washington; 70, Pierre Mion/National Geographic Creative; 71, Rick Sternbach; 72, NASA/JPL/USGS; 73, NASA/JPL; 78, Science Source; 79 (UP), ESA/VIRTIS/INAF-IASF/Obs. de Paris-LESIA/Univ. of Oxford; 79 (LO), ESA/VIRTIS/INAF-IASF/Obs. de Paris-LESIA/Univ. of Oxford; 80, J. Whatmore/ESA; 82–3, NOAA/NASA GOES Project; 88, Planetary Visions Ltd./Science Source; 89, Alain Barbezat/National Geographic My Shot; 91 (UP), NG Maps; 91 (LO), Theophilus Britt Griswold; 92–3, Carsten Peter/National Geographic Creative; 92 (INSET), NG Maps; 98, Christian Darkin/Science Source; 99, NASA; 100, Designua/Shutterstock; 101, Michael Melford/National Geographic Creative; 102–3, NASA/JPL-Caltech/MSSS; 102 (INSET), NASA/JPL/Malin Space Science Systems; 108, Science Source; 109, NASA/JPL/University of Arizona; 110, Science Source; 111, NASA/JPL-Caltech; 111 (LO), NASA/JPL/Malin Space Science Systems; 112, Ian Dagnall/Alamy Stock Photo; 113, NASA/JPL-Caltech/ESA/DLR/FU Berlin/MSSS; 114–15, Mark Garlick/Science Source; 115 (INSET), NASA/JPL/USGS; 117, Sanford/Agliolo/Corbis; 122–3, NASA/JPL-Caltech/SwRI/MSSS/Gerald Eichstadt/Sean Doran; 123 (INSET), NASA/JPL/University of Arizona; 126–7, NASA/JPL-Caltech/SwRI/MSSS/Gerald Eichstadt; 126, Science Source; 128, H. Hammel (SSI), WFPC2, HST, NASA; 129, H. Hammel (SSI), WFPC2, HST, NASA; 130–31, NASA/JPL/USGS; 130 (INSET), NASA/JPL/DLR; 144–5, NASA/JPL-Caltech/Space Science Institute; 140 (INSET), NASA/JPL; 141, NASA/JPL/University of Arizona; 142, NASA/JPL/University of Arizona; 143, NASA/JPL; 144 (INSET), GeorgeManga/iStock; 162, Science Source; 163, NASA/JPL/Space Science Institute; 164, Davis Meltzer/National Geographic Creative; 165, NASA/JPL; 166–7, NASA/JPL; 166 (INSET), NASA/JPL/Space Science Institute; 168–9, NASA/JPL/Space Science Institute; 170, Ron Miller; 171, Ludek Pesek/Science Source; 172–3, John R. Foster/Science Source; 172 (INSET), Science Source; 183, NASA/JPL/Universities Space Research Association/Lunar & Planetary Institute; 184, Science Source; 185, Mark Garlick/Science Source; 186–7, NASA/JPL; 187, Science Source; 188–9, NASA/Johns Hopkins University Applied Physics Laboratory/Southwest Research Institute; 188 (INSET), NASA, ESA, H. Weaver (JHUAPL), A. Stern (SwRI), and the HST Pluto Companion Search Team; 190, Bettmann/Corbis; 191, Science Source; 192, NASA/Johns Hopkins University Applied Physics Laboratory/Southwest Research Institute; 194–5, NASA/Johns Hopkins University Applied Physics Laboratory/Southwest Research Institute; 196–7, Dan Schechter/Getty Images; 197 (INSET), NASA/JPL/UMD; 198, ESA/Rosetta/NAVCAM–CC BY-SA IGO 3.0; 200–201, Mark Garlick/Science Source; 200 (INSET), David Aguilar/National Geographic Creative; 202–3, NASA/Johns Hopkins University Applied Physics Laboratory/Southwest Research Institute; 204, Wampa76/iStock; 205, NASA/JPL-Caltech/R. Hurt (SSC); 206–7, Viktar Malyshchyts/Shutterstock; 208–9, Credit for Hubble image: NASA, ESA, and Q.D. Wang (University of Massachusetts, Amherst). Credit for Spitzer image: NASA, Jet Propulsion Laboratory, and S. Stolovy (Spitzer Science Center/Caltech); 212–13, Dr. Fred Espenak/Science Source; 213 (INSET), Jon Lomberg/Photo Researchers, Inc; 214, Harvard College Observatory/Science Source; 215, NASA, ESA, and the Hubble Heritage Team (STScI/AURA); 217, NASA/JPL-Caltech; 218–19, Sepdes Sinaga/National Geographic My Shot; 219 (INSET), SOHO (ESA & NASA); 220–21, SOHO/EIT; 223, James P. Blair/National Geographic Creative; 224–5, NASA/SDO/GOES-15; 226–7, Robert Stocki/National Geographic My Shot; 227 (INSET), Sebastian Kaulitzki/Shutterstock; 229, Ralph White/Corbis/Getty; 230–1, Dana Berry/National Geographic Creative; 230 (INSET), NASA/Kepler; 232–3, NASA/Ames/JPL-Caltech; 234–5, NASA/JPL-Caltech; 235, David Aguilar/National Geographic Creative; 236–7, ESO/M. Kornmesser; 238–9, Dr Verena Tunnicliffe, University of Victoria (UVic); 239, NASA/JPL-Caltech; 240–41, NASA/JPL-Caltech; 241, NASA/JPL-Caltech; 242–3, SETI; 242 (INSET), NASA; 244, SHNS/SETI Institute/Newscom/File; 246–7, NASA, ESA, and the Hubble SM4 ERO Team; 247 (INSET), Mark Garlick/Science Source; 249, Mark Garlick/Science Source; 250–51, NASA/JPL-Caltech/O. Krause (Steward Observatory); 250 (INSET), NASA, ESA, P. Challis and R. Kirshner (Harvard-Smithsonian Center for Astrophysics); 252–3, David A. Hardy/www.astroart.org; 252, NASA, ESA, and the Hubble Heritage STScI/AURA)-ESA/Hubble Collaboration. Acknowledgement: Robert A. Fesen (Dartmouth College, USA) and James Long (ESA/Hubble); 254 (UP LE), NASA, ESA and H.E. Bond (STScI); 254 (UP CTR), NASA, ESA and H.E. Bond (STScI); 254 (UP RT), NASA and The Hubble Heritage Team (AURA/STScI); 255, NASA, ESA, and the Hubble Heritage Team (STScI/AURA); 256–7, Mark Garlick/Science Source; 256 (INSET), X-ray Image: NASA/CXC/ASU/J. Hester et al. Optical Image : NASA/HST/ASU/J. Hester et al.; 259, Jonathan Blair/National Geographic Creative; 260–61, NASA, ESA, J. Hester and A. Loll (Arizona State University); 262–3, X-ray: NASA/CXC/CfA/R.Kraft et al.; Submillimeter: MPIfR/ESO/APEX/A.Weiss et al.; Optical: ESO/WFI; 262 (INSET), Mark Garlick/Science Source; 265, Detlev Van Ravenswaay/Science Source; 266, Digitized Sky Survey; 266–7, NASA/CXC/M.Weiss; 268–9, ESA/C.Carreau; 268 (INSET), ESA/D. Ducros 2010; 271, Caltech/MIT/LIGO Lab; 272–3, X-ray: NASA/CXC/M.Markevitch et al. Optical: NASA/STScI; Magellan/U.Arizona/D.Clowe et al. Lensing Map: NASA/STScI; ESO WFI; Magellan/U. Arizona/D.Clowe et al.; 273 (INSET), Volker Springel/Max Planck Institute for Astrophysics/Science Source; 274–5, Richard Nowitz/National Geographic Creative; 276, NASA, ESA, and R. Massey (California Institute of Technology); 278–9, NASA/JPL-Caltech/K. Su (Univ. of Arizona); 280–81, NASA/JPL-Caltech/UCLA; 284–5, NASA, ESA, M. Livio and the Hubble Heritage Team (STScI/AURA); 284 (INSET LE), Royal Astronomical Society/Science Source; 284 (INSET CTR), Royal Astronomical Society/Science Source; 284 (INSET RT), Royal Astronomical Society/Science Source; 287 (LO), Science Source; 288–9, Chris Butler/Science Source; 289, David Parker/Science Source; 290–91, ESA; 291 (INSET), NASA/NSF; 295 (UP), NASA/WMAP Science Team; 295 (LO), Equinox Graphics/Science Source; 296 (UP LE), NASA; 296 (CTR LE UP), NASA; 296 (CTR LE LO), NASA; 296 (LO LE), ESA and the Planck Collaboration; 298–9, NASA, ESA, and the Hubble Heritage (STScI/AURA)-ESA/Hubble Collaboration; 299 (INSET LE), NASA, ESA, K. Kuntz (JHU), F. Bresolin (University of Hawaii), J. Trauger (Jet Propulsion Lab), J. Mould (NOAO), Y.-H. Chu (University of Illinois, Urbana), and STScI; 299 (INSET CTR LE), NASA, ESA, and The Hubble Heritage Team (STScI/AURA); 299 (INSET CTR RT), NASA, ESA, and the Hubble Heritage (STScI/AURA)-ESA/Hubble Collaboration; 299 (INSET RT), NASA, ESA, and The Hubble Heritage Team (STScI/AURA); 300–301, NASA/JPL-Caltech/K. Gordon (STScI) and SINGS Team; 303, NASA; 304–5, Mehau Kulyk/Science Source; 304 (INSET), Ann Feild (STScI); 309, Don Dixon/cosmographica.com; 310–11, Adolf Schaller for STScI; 312–13, Mark Garlick/Science Source; 312 (INSET), BSIP/Science Source; 314, NASA/WMAP Science Team; 315, Jose Antonio Peñas/Science Source; 316, NASA/ESA/ESO/Wolfram Freudling et al. (STECF); 318–19, Moonrunner Design Ltd./National Geographic Creative; 319 (INSET), Mark Garlick/Science Source; 320, Fermilab; 322–3, NASA, ESA, S. Beckwith (STScI) and the HUDF Team; 324, Wikipedia; 326, Corbin O'Grady Studio/Science Source; 327, Detlev Van Ravenswaay/Science Source; 328, Mark Garlick/Science Source; 330–31, ESO/T. Preibisch; 332, NASA/JPL/Space Science Institute; 333, NASA/Johns Hopkins University Applied Physics Laboratory/Carnegie Institution of Washington

GENERAL INDEX

PLACE-NAME INDEX

Using this index

The International Astronomical Union regulates the naming of extraterrestrial features. The convention adopted is to use Latin for generic terms (e.g., Utopia Planitia on Mars, planitia meaning "low plain"). A listing of those terms appears on page 334. The grid system used to demarcate the location of mapped features is based upon the graticule. Three similar systems are used in this atlas depending on how the maps are presented. Terrestrial planets, Mercury, Venus, Earth, Mars, along with Earth's moon use the first as illustrated below (figure 1). Letter coordinates appear around the perimeter with numbers along the equator. Where both hemispheres of a moon or dwarf planet are presented together, as in the satellites of Jupiter and Saturn, a second method of finding features is provided (figure 2). The central meridian of each hemisphere is marked with letters, the equator with numbers that are continuous between the two hemispheres. A final method is used to accommodate bodies in the outer solar system where data is only available in the southern hemisphere. Letters encircle, designating longitude, while numbers represent latitude in rings around the moon (figure 3). The features collected in this index are referenced to these systems. Names without a generic are craters with the exception of those from Earth. Entries with two grid references are features that cross between two mapped hemispheres.

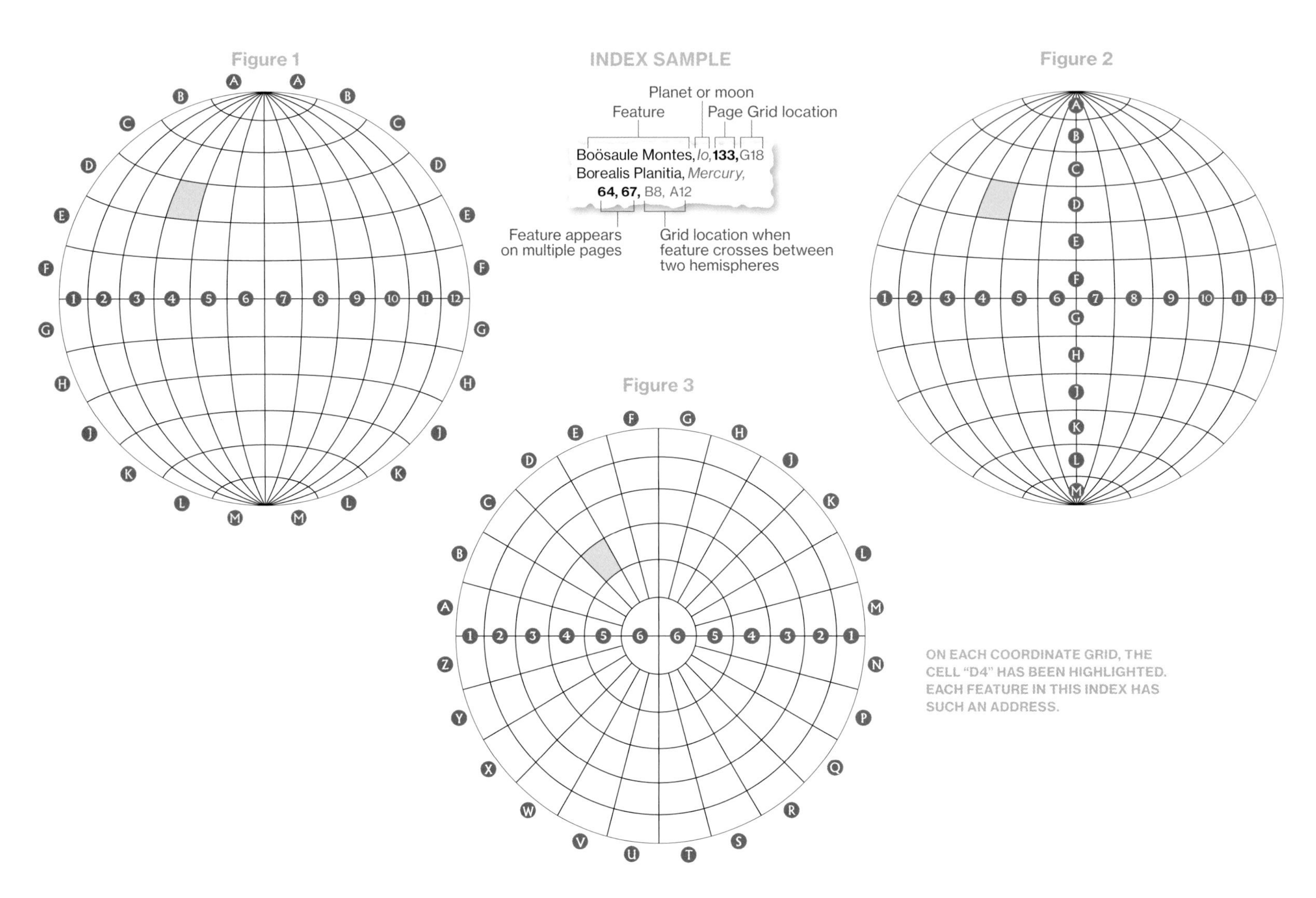

Since 1888, the National Geographic Society has funded more than 13,000 research, exploration, and preservation projects around the world. National Geographic Partners distributes a portion of the funds it receives from your purchase to National Geographic Society to support programs including the conservation of animals and their habitats.

National Geographic Partners
1145 17th Street NW
Washington, DC 20036-4688 USA

Get closer to National Geographic explorers and photographers, and connect with our global community. Join us today at nationalgeographic.com/join

For information about special discounts for bulk purchases, please contact National Geographic Books Special Sales: specialsales@natgeo.com

For rights or permissions inquiries, please contact National Geographic Books Subsidiary Rights: bookrights@natgeo.com

Produced by Print Matters Productions, Inc.

ISBN: 978-1-4262-1969-6

Printed in China

18/PPS/1